Cognitive, Conative and Behavioral Neurology

Michael Hoffmann

Cognitive, Conative and Behavioral Neurology

An Evolutionary Perspective

 Springer

Michael Hoffmann MD, PhD
Professor of Neurology
Orlando VA Medical Center
University of Central Florida
Orlando, FL, USA

ISBN 978-3-319-81446-9   ISBN 978-3-319-33181-2   (eBook)
DOI 10.1007/978-3-319-33181-2

Printed on acid-free paper

This Springer imprint is published by Springer Nature
The registered company is Springer International Publishing AG Switzerland

For my wife Bronwyn, daughter Jenna Leigh and son Michael

Preface

Human mind disorders are both fascinating and manifold in their presentations. The vast panoply of cognitive and behavioral syndromes spans several clinical disciplines, principally neurology, psychiatry, and psychology, each with their distinctive cultural and methodological approaches. Each discipline has made pivotal contributions in understanding the science of the mind. Yet their areas of concentrations across syndromes, disease states, and brain circuitry differ. What is becoming more pertinent today is the cross-disciplinary approach to patient diagnosis, treatment, and care. That premise is a primary focus of the book with the hope of fostering closer ties and collaborations amongst the primary clinical brain disciplines.

A unifying concept and approach is welding of the known syndromes into evolutionary origins. Evolutionary medicine has been touted as the most important basic science. By deciphering the origins of our anatomy, physiology, and brain circuitry and how they were exapted from prior structures goes a long way to understanding some of the more perplexing higher cortical function disorders. Einstein once retorted to the question; "How did you come up with the theory of relativity? By ignoring an axiom"! So it is today, with the study of the human mind. Until very recently everyone "knew" that we acquired bigger brains than apes, from which greater intelligence flowed and we then started making tools. Because our hands had to be freed up, this inaugurated bipedality. However, the reverse is true. We walked before we thought. Current evidence even suggests we spent over a million years in water, wading, fishing, swimming and in the process acquired unique attributes such as substantial fat layers, a tenfold adipocyte increase, nakedness, a descended larynx, and large sinuses in addition to bipedality. Such evolvement is critical to understanding how the brain circuitry evolved and our minds. Hence the evolutionary perspective on human cognition, conation, and behavior.

Orlando, FL, USA   Michael Hoffmann

Acknowledgments

The authors publications and images included in the figures were derived from four separate IRB approved cognitive stroke registries using cognitive vascular disorders as the brain lesion model.

1. **The NIH-NINDS Stroke Data Bank (New York)**
   Under the following contracts;
   N01-NS 2-2302, N01-NS-2-2384, N01-NS-2-2398, N01-NS-2-2399, N01-NS-6-2305N01-NS 2-2302, N01-NS-2-2384, N01-NS-2-2398, N01-NS-2-2399, N01-NS-6-2305
   Status of stroke research fellow (1990-1991)

2. **The Durban Stroke Data Bank. IRB approval University of Natal, Durban, South Africa (memorandum dated signed)**
   Status: Principal Investigator (1992–1998)

3. **The USF-TGH Stroke Registry. IRB # 102354 (University of South Florida)**
   Status: Principal Investigator (2002–2006)

4. **The USF-Cognitive Stroke Registry. IRB # 106113 (University of South Florida)**
   Status: Principal Investigator (2007–2010)

Additional notice regarding references

1. Source article for frontal chapter: Open source article updated and adapted from:
   Hoffmann M. The human frontal lobes and frontal network systems: an evolutionary, clinical, and treatment perspective. ISRN Neurol. 2013;2013:892459. doi: 10.1155/2013/892459. Epub 2013 Mar 14.PMID: 23577266. Free PMC Article
2. Source article for neuroradiology and treatment: Open source article updated and adapted from:
   Hoffmann M. The human frontal lobes and frontal network systems: an evolutionary, clinical, and treatment perspective. ISRN Neurol. 2013;2013:892459. doi: 10.1155/2013/892459. Epub 2013 Mar 14.PMID: 23577266. Free PMC Article
3. Source article for part of the introductory remarks:
   From PhD thesis, Department of Behavioral Medicine, University of KwaZulu Natal, April 2014

Contents

1   Introduction 1

2   Cognitive Archeology, Cognitive Neurology,
and the Unraveling of the Connectome: Divulging Brain
Function from Fractured Skulls and Fractured Minds 11

3   Neurochemistry 35

4   Vision: Elementary and Complex Visual Processing 51

5   Temporal Lobe Syndromes 83

6   Memory Syndromes 99

7   Left Hemisphere Syndromes: Apraxias 131

8   Parietal Lobe Syndromes 145

9   Right Hemisphere Syndromes 157

10   Language, Aphasias, and Related Disorders 187

11   Acquired Cultural Circuits 221

12   Frontal System Syndromes 247

13   Neuroimaging and Treatments Perspectives 297

Index 313

Introduction

Cognition and the brain has a long history dating back to the Greeks, notably Hippocrates and more recently the emergence of European neurology, psychiatry, and psychology during the nineteenth century. Several disciplines have been concerned with the study of cognition or higher cortical functions of the brain and each has its unique area of expertise and important contributions, along with its own culture and language. Today the principal clinical disciplines concerned include behavioral neurology, neuropsychiatry, and neuropsychology. In addition, the disciplines of speech and language, experimental psychology, cognitive neuroscience, and physical medicine and rehabilitation are also intimately involved. Each brain discipline has its own language of describing deficits and syndromes that may be confusing at times to the general student or clinician. However each discipline also has accumulated unique insights and experience that further the insights into our understanding of cognition in a generic sense.

Why the neuroarcheological perspective? With hindsight comes insight and foresight. If we know how something is constructed we often have better insights into how it may be "fixed." In behavioral neurology and cognition, there are a bewildering number of fascinating, obscure, and perplexing syndromes. Evolutionary neurology and the allied study of neuroarcheology help us understand a number of these presentations. Other books on evolutionary aspects of clinical brain sciences have already been published,

including evolutionary psychiatry and evolutionary psychology texts. This book concerns the paleoneurological perspective with a special emphasis on the mind, cognition, conation, and behavior. There has been a recent surge in evolutionary medicine interest and proposals for the inclusion of an evolutionary biology course have been made for the study of medicine and already in place at several university centers in the USA and Europe. In addition, explaining cognitive neurological syndromes to patients afflicted with these disorders becomes more understandable and meaningful if the explanation of how something evolved is shared.

The Clinical Method in Neurology with Specific Relevance to Cognitive Disorders

Localize as Far as Possible: Anatomically or Chemically

The brain is an electrochemical organ housed in a complexity of neuroanatomical networks. Hence, syndromes may be regarded as anatomically addressed, network related, electrically related, or chemically related. As physicians we assemble information during patient interviews in a sequential manner, called successive estimations, a computational mechanism used by the brain itself. The initial intention is to arrive at an anatomical diagnosis, as this dictates the overall likelihood of

© Springer International Publishing Switzerland 2016
M. Hoffmann, *Cognitive, Conative and Behavioral Neurology*,
DOI 10.1007/978-3-319-33181-2_1

a number of illnesses and likely pathophysiological entities, such as Wallenberg's syndrome and its lateral medullary brainstem localization. Sometimes the diagnosis is primarily "chemically" based, as for example with Parkinson's disease. Syndromes are suggested when a number of disparate units are involved. When no diagnosis is easily forthcoming, the best strategy is to generate a differential based on the general classification of diseases (infectious, vascular, neoplastic, inflammatory, toxic, metabolic).

The Clinical Method Is the Cornerstone of Medicine, but Remains Fallible

The interpretation of initial signs and symptoms is crucial. Experience helps, but none is as important as a peer-reviewed oversight, such as promulgated in the original philosophy of the "Grand Rounds." This type of clinical meeting format had a very sound basis as the collective review of a "problem patient" is often humbling to say the least in how many different opinions of a symptom or sign elicitation might have.

Note Taking Is Critical and Databases May Be Instrumental

A vast amount of information is gathered from the patient and family in a short period of time. Digitizing information directly into a database, for future review with subsequent neuroradiological, laboratory, and paraclinical testing, is invaluable as some obscure diagnoses reveal themselves only during protracted follow-up.

Establish the Reliability and the Source of Information

In neurology and in cognitive neurology in particular, one has to question the very source of the information—the patient. Most neurological disease entities involve higher cortical function deficits to a greater or lesser extent, including

peripheral nervous system entities such as muscular dystrophy and motor neuron disease. People with certain higher cortical function deficits may even deny the presence of a hemiplegia such as occurs with anosognosia. As another example, the typical, moderately impaired Alzheimer's patient usually offers no main complaint or none at all, when first seen in consultation.

The Unique Challenges with Diagnosing and Measuring Cognitive Neurological Deficits

1. The brain, the patient, and the doctor may get it "all wrong" (Fig. 1.1).
2. How can the brain get it wrong? It is well known that cognition fluctuates during the course of the day even in normal people. In the

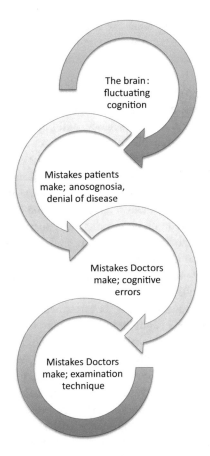

Fig. 1.1 What can go wrong with cognitive assessment?

context of neurological disease this may be even more profound, as occurs with progressive Lewy body disease for example. Neurological syndromes may fluctuate from hour to hour or day to day and confuse medical students, residents, and attending physicians alike. Furthermore, because of the burden of immense sensory input, the brain is prone to taking shortcuts in processing visual information for example.

3. We are told that we must listen to the patient, for they are telling us the diagnosis. Yet, the patient may deny deficits (anosognosia) or diminish their importance (anosodiaphoria).

4. Neurological lesions may be silent both to the patient and the doctor. Examples include silent strokes depicted only by MRI imaging or only by PET brain scanning in individuals with high cognitive reserve and incipient Alzheimer's dementia.

5. Neurological syndromes may be present only when concomitant metabolic upset, infection, or medication leads to the unmasking of deficits such as a mild aphasia after a stroke syndrome that has either completely or almost completely recovered.

6. Syndromes we normally attribute to one part of the brain, for example frontal lobe syndromes, may arise from the very posterior part of the brain or even brainstem and cerebellum. This observation underlies the complexity of the networks and the various diaschisis effects.

7. The primary syndrome may be due to a lesion that may allow the emergence of others. These have been referred to as "suppressed networks" leading to a phylogenetic syndrome. An example would be the artistic ability after left hemisphere lesions due to stroke or dementia and imitation behavior due to frontal lobe lesions.

8. The pathophysiology may be due to neuronal destruction, network destruction, and electrical, chemical, or autoimmune perturbation. This impacts the type of imaging tools that need to be deployed in aiding a precise diagnosis, whether EEG, PET brain imaging, or MRI diffusion tensor imaging.

9. The presenting syndrome may be due to an electrical phenomenon such as with frontal lobe or temporal lobe epilepsy which may be intermittent itself and associated with a differing interictal presentation, for example the "forced normalization" syndrome.

10. Chemical aberrations may present as a more obvious syndrome such as Parkinson's but also more covert presentations with diagnostic difficulty as with the serotonin syndrome.

11. The clinician may make errors in the recognition and evaluation of the patient's symptoms, signs, and investigative information. Groopman assembled more than a dozen types of errors doctors make in patient evaluation including commission errors, confirmation bias and anchoring, misattributing and misinterpreting abnormal investigative data, and the "Zebra Retreat"—avoiding contemplation of rare diagnoses [1].

12. The clinical method has moved beyond mere anatomical localization with respect to lesions. The complexity of brain networks include approximately 100,000 miles of fiber tracts and the hub failure hypothesis which posits that certain heteromodal association cortices are most prone to failure no matter what the pathophysiological process demands. Hence it may be more appropriate to think more in terms of networktopathies and the pervasiveness of the frontal network systems that influence all other brain regions (Fig. 1.2) [2].

The Uniquely Important Frontal Network Systems: The Frontal Lobes Connect to All Other Regions of the Brain and Cognitive Syndromes Almost Always Involve Frontal Network System Impairments.

Frontal lobe lesions and the consequent brain behavior relationship is a science that is at best only a few decades old, gaining momentum only by the 1980s. Two pivotal frontal lobe brain behavior studies were reported in the nineteenth century within a few years of each other, one in the USA, and the other one in France. Dr. John Harlow described the story of Phineas Gage's

Fig. 1.2 The traditional
"cognitive compass" (*left*)
and the concept of
metacognition (*right*): The
frontal lobes are
reciprocally connected to
and supervisory to all other
networks. *FNS* frontal
network systems, *RHS* right
hemisphere syndromes,
LHS left hemisphere
syndromes, *PNS* posterior
network syndromes

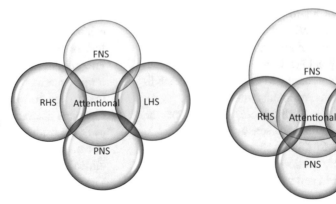

survival and profound behavioral impairment after a 3½ foot, 1 in. diameter iron tamping rod inadvertently passed through his frontal lobes in 1848 and reported this in the Publications of the Massachusetts Medical Society in 1868 [3, 4]. The second was Paul Broca's expressive aphasia, pathological study implicating the posterior, inferior frontal lobe in expressive speech [5]. Soon thereafter, other important contributions in the field such as Brodmann's cyto-architectonic brain studies and the emerging field of experimental psychology and Freud's writings prompted the American neurologist, Tilney, in 1928 to suggest that the human evolutionary period should be called the age of the frontal lobes [6]. These predictions did not materialize for most of the twentieth century and psychiatry as a discipline did not emphasize brain behavior relationships and neurology became relatively restricted to so-called elementary neurological syndromes of sensorimotor, visual, and coordination systems. Penfield's contribution in the 1950s (intraoperative stimulation) was significant in that although he had a motor response to stimulation of the motor cortex, frontal lobe stimulation revealed no response at all [7]. Pick's disease was described in 1892 and although it could have served as a very revealing pathology of differing frontal functions (as frontotemporal lobe dementia does today) it was largely ignored because of the fact that several different types of pathology in addition to Pick bodies can cause the so-called frontotemporal lobe dementia (FTLD), Pick's being a very infrequent pathology while FTLD was a common dementia. Hence, because of this

pathology-clinical mismatch that occurred over the next century, the FTLD clinical syndrome received little attention [8]. Luria's unique and seminal contributions to frontal lobe function, gleaned largely from traumatic brain injury patients, still profoundly influence the neuropsychology today [9]. Similarly, Lhermitte's innovative style of frontal testing outside the usual office or hospital setting illuminated how simple techniques may be very informative in discerning frontal brain behavior relationships. He described how field-dependent behavior syndromes frequently emerge, consequent to disruptions of the mirror neuron network in the brain [10, 11].

Other reasons why this has been so difficult include the fact that patients with frontal lobe damage rarely initiate clinical evaluation and we test what we have tests for. Tests for many symptoms, syndromes, and behaviors that cause someone to be irascible facetious, puerile, profane, lacking curiosity, and have aspontaneity and lack of foresight do not exist. A number of researchers have devised metric tests that sample various components of frontal function. These include the Wisconsin Card Sorting test [12], Stroop test [13], Iowa Gambling Test [14], BRIEF [15], FRSBe [16], DKEFS [17], FAB [18], FBI [19], various trail making tests such as the CTMT [20], Tower of London Test [21], and the EXIT [22]. Mesulam emphasized the frequent and surprising paucity of formal neuropsychological deficits associated with lesions of the frontal lobe lesions, some of whom may have normal scores in all tests. Behavioral tests however are more likely to elicit abnormalities [23]. He also proposed the term frontal network syndrome (as opposed to frontal lobe syndrome), in view of the most frequent causes being multifocal, subcortical processes such as cerebrovascular disease, multiple sclerosis, traumatic brain injury, and toxic metabolic encephalopathies rather than lesions of the frontal lobes themselves [24–26].

With a single, landmark, case report launching clinical interest in frontal syndromes just over 100 years ago, the foregoing discussion has attempted to elucidate the problems encountered and their probable reasons why further testing, interest, and the delineation of frontal syndromes have been lagging. Clinical experience with acute neurological patients such as stroke encephalopathy, multiple sclerosis, seizures, and traumatic brain injury indicates that the frontal syndromes are not only frequent, but also likely the most common of not only neurological but also cognitive syndromes. The presentation can be dramatic and obvious but often also subtle, covert, and even frankly denied by the patient. A relatively frugal armamentarium for testing frontal syndromes is likely related to the limited understanding we have of the extent and nature of frontal syndromes. The relative paucity of available tests for the vast panoply of human cognitive brain disorders is sobering. One avenue to help remedy this may be the integration of the wealth of information concerning brain function and dysfunction from allied and overlapping disciplines such as psychiatry, psychology, speech and language, cognitive neuroscience, and behavioral neurology.

The long-time reliance on autopsy studies to determine brain behavior relationships was finally over with the advent of increasingly sophisticated cerebral computerized tomography (CT) scanning. This was followed closely by higher resolution neuroimaging including magnetic resonance imaging (MRI) for anatomical definition and positron emission computed tomographic (PET) studies for functional brain imaging starting in the 1980s. The convergence of clinical studies from neuropsychology, neurology, and neuroimaging culminated in a long overdue surge in frontal lobe research. Endeavors to promote brain and mind research continued, with the next decade (1990s) being declared the

decade of the brain [27], and the following decade beginning in 2000 led to the concept of the century of the mind with brain-mind institutes forming at major universities that garner a multidisciplinary approach, for best results. Two examples include the McGovern Institute for Brain Research MIT, Massachusetts, and Mind Brain Behavior Institute at Columbia University, New York, USA.

Approximately 90 % of the brain is involved in cognition, based on cerebral cyto-architectonics [28]. The most important and pervasive cognitive processes, frontal network syndromes, are ubiquitous in neurological and psychiatric disease; yet measurement remains poor with few available tests [24, 25, 29]. The most commonly clinically employed test, the Mini Mental State Examination (MMSE), does not even measure frontal systems [30]. Hence, there exists a dilemma between the need for accurate clinical frontal network system assessment and the current battery of tests available for this purpose. Clinical cerebrovascular, neurological decision making for example is severely constrained by a 4.5-h so-called thrombolytic therapy window [31]. In the setting of multiple concurrent tests including neuro-imaging, laboratory, and cardiac investigations, this does not leave more than a few minutes for clinical assessment of the patient. During a typical stroke approximately 2 million neurons and 14 billion synapses are lost each minute [32]. In this emergent setting, there is no place for formal neuropsychological assessment. It is also common experience in clinical practice that cognitive evaluation is challenging in the various stages of dementia. In some, the degree of cooperation or attention is limited to no more than a few minutes at best. Historically and philosophically, testing of the higher cortical brain functions has been approached differently by the three major disciplines (neurology, psychiatry, neuropsychology) concerned with assessment of behavioral and cognitive effects of brain lesions and conditions. Each has different "cultures" and approaches to this clinical challenge but because each has unique contributions, they complement each other. These comprise of (1) behavioral neurological approach comprising of a myriad of syndromes that are best described in ordinal and nominal data terms, (2) neuropsychiatric approach with syndromes described in terms of prespecified criteria (DSM-IV/V) and configured to nominal data, and (3) neuropsychological battery approach almost exclusively described according to numerical data and compared to normative data, less often ordinal and nominal data.

Time is brain (stroke) and time is limited (dementia for example). This necessitates a multi-tiered, time-based, cognitive testing approach. To benefit from all the varying clinical neuroscience approaches, the FNS testing methodology should be cross disciplinary while using a time-based battery of tests that range from minutes to several hours. The special relevance to FNS testing is advocated because this expansive cognitive network may be viewed as a supervisory and wide-ranging cognitive system (metacognition) that may be the most sensitive indicator of cognitive status.

To complicate matters however, cognitive reserve may mask brain pathology until late in certain brain disease processes. People with similar cognitive impairment may have markedly different Alzheimer's disease pathology for example, depending on their degree of brain and cognitive reserve. Because of the cognitive reserve hypothesis, now well buttressed by clinico-radiologic studies, clinical examination alone cannot discern cognitive impairment [33]. The cognitive reserve hypothesis proposes that people with similar cognitive impairments or even no impairment at all may nevertheless have rampant Alzheimer's pathology [34]. Hence clinical psychometric testing is unlikely to reliably diagnose many people that may benefit from specific disease therapies. Metabolic testing with positron emission tomography (PET) brain scanning is known to improve diagnosis and extend the window of AD diagnosis into the mild clinical and even preclinical phase. In addition to psychometric features, it is possible that certain behavioral neurological tests can diagnose disease earlier [35].

Another facet of complexity concerns the increasing number of classic dementia presentations being encountered that are caused by other treatable and at times completely reversible medical and neurological diseases. Examples of masqueraders of Alzheimer's disease for example include cognitive vascular disorders [36], cryptococcal meningitis [37, 38], hepatic encephalopathy [39], and masqueraders of frontotemporal lobe disorders such as Whipple's disease [40] and multiple sclerosis [41]. This is underscored by the recent revisions to the diagnostic criteria for Alzheimer's disease released in 2010, Alzheimer's Association International Conference on Alzheimer's Disease (AAICAD) [42].

Finally, Alzheimer's disease is today regarded as a vascular disease where clinically there is a continuum from stroke only to Alzheimer's disease only with the vast majority of people having features of both neuronal degeneration and vascular cognitive impairment [43].

As FNS is common to all these disease entities and as the most pervasive cognitive function in addition to its supervisory role, it makes sense to measure and monitor these, somewhat akin to fever and infection.

The absence of biomarkers makes the diagnosis of dementia challenging. At the time of writing four different frontotemporal lobe dementias (FTD) and four different Alzheimer's clinical subtypes are recognized, both of which have a frontal system variant manifesting with behavioral abnormalities [44]. Neuropathology is complex and many have mixed patterns. Clinical diagnostic accuracy is imperfect; yet treatment success as well as trials depend on it [45]. The dementia subtypes may also be viewed in terms of the predominant neurotransmitter disturbances such as anticholinergic deficiency in Alzheimer's, serotonin deficiency in FTD, and dopamine in cognitive vascular disorders (CVD). Specific replacement therapy is helpful in alleviating some symptoms [46, 47], whereas incorrect therapies may lead to deterioration [48]. Clinical cognitive evaluation is particularly challenging in the more advanced stages of dementia and sometimes attention span and motivation may be limited to a few minutes only. Cognitive evaluations in stroke patients have noted cognitive impairment as well as frontal network syndromes, being common in the weeks to months after stroke [49, 50]. Importantly, frontal network syndromes may be present regardless of the topography of the brain lesion. In addition to the frontal lobes this may be in the posterior cortex, subcortical, or subtentorial part of the brain. Even in brainstem and cerebellar stroke, up to half of patients may have some degree of frontal system impairment [51].

Although much progress has been made, even brief reflection of other cognitive functions of the frontal networks such as emotional intelligence, creativity, savant abilities, artistic ability, artistic appreciation, spirituality, religiosity, and the role of dreaming in maintaining optimum brain health provides sobering prospects of what is still unchartered territory. Currently we are armed with the most popular, simple bedside tests (MMSE, MOCA) that provide helpful guidance for distinguishing our most common dementia syndromes, but these as well as standard neuropsychological tests provide little, if any information, on the myriad of other frontal syndromes that have been documented [44, 52, 53]. Until now there has been a major focus on memory and subsequently executive function testing in common neurological conditions such as dementia, stroke, multiple sclerosis, and traumatic brain injury with other frontal network syndromes not formally tested [54]. For example, with frontotemporal lobe dementia or degeneration, it has been shown that an early presentation and useful way of monitoring the illness may be through evaluation of artwork by the patient [55].

This brings us to consider the "creative explosion" or "big bang of human evolution" that occurred within the last 30,000–40,000 years ago [56]. Convergent evidence from archeology, genetics, and evolutionary neuropsychology has forged a well-supported hypothesis that working memory (a core frontal system function) was the so-called "cognitive missing link" that enabled a cognitive fluidity and networking of the various intelligence domains (social, technical, natural history) of the human mind culminating in crossmodal connectivity and thence creativity. Although we do not have tests for a conundrum of frontal functions that we evolved with, we can at least test working memory, which is regarded as the "engine" of cognitive connectivity and executive function [57].

Sometimes we find simple tests that may discern and diagnose complex processes. The mirror neuron system (MNS) for example evolved at some stage in our primate history about 60 million years ago and can be affected by cerebral lesions. We can test for the MNS by documenting syndromes such as echopraxia, utilization behavior, and environmental dependency syndromes. These are not commonly employed tests, and yet they offer an important opportunity of how we can improve neurological evaluation and monitoring of complex FNS [58]. Together, the working memory circuit and the mirror neuron circuitry, both extensive frontoparietal cerebral circuits, are arguably the key circuits that made us human and that are both core frontal systems circuits that can be assessed clinically by relatively simple bedside tests.

Another important area of active research today is the role of sleep and particularly dreaming in FNS. Dreaming has been shown to improve memory, executive function, attention, depression, and creativity. Current hypotheses regard dreaming as a critical survival attribute particularly with regard to optimizing our polyadic relationships in society, seen by some as our biggest challenge as humans [59]. The adage, "we test for what we have tests for," is particularly pertinent in this discussion and serves as a reminder of the vast opportunities in cognitive neuroscience that await discovery. At the present time, with the tsunami of dementing illness upon us, coupled by the expense of cerebral assessments such as brain scans, any help we can muster from simple, quick, and reliable tools currently available and a more interdisciplinary assessment approach will be required. In the quest for simplicity, it may be useful to remember that there are three very frequent "pillars" of clinical frontal syndrome presentations. Although there is overlap amongst these core clinical syndromes, they frequently may contain all three in various combinations, or they may present in relative isolation:

Cognitive

Executive dysfunction disorders, inattention, working memory (predominantly assessed by neuropsychologists). Signature syndromes include small vessel cerebrovascular disease, multiple sclerosis, and traumatic brain injury.

Conative

Various forms of abulia, akinetic mutism, and amotivational syndromes that may initially come to the attention of psychiatrists and may at times be misdiagnosed as depression. Signature syndromes include extensive leukoaraiosis, traumatic brain injury, multiple sclerosis, and frontal lobe stroke syndromes.

Behavioral

Disorders with disinhibition, gambling tendencies, loss of empathy, social and emotional intelligence, facetious, and slovenly behavior are mostly the domain of behavioral neurology. Signature syndromes include frontotemporal lobe degenerations and dementia, traumatic brain injury, frontal lobe and subcortical strokes, and multiple sclerosis.

The book is organized into the approximate sequences of the evolution of the primate, the anthropoid, and later the hominin mind—from back to front. The visual cortical elaboration first differentiated within the primate brain, followed by temporal lobe object and facial recognition abilities, and then frontoparietal circuitry elaboration for visuomotor abilities in the arboreal environment. Left hemisphere specializations such as praxis evolved with stone tool making and episodic and working memory developments occurred with increasing environmental change the hominins faced. The frontal systems are considered last and followed the neuroimaging and cognitive therapy chapter because they are supervisory to all the other domains and are generally implicated to a greater or lesser degree, no matter where the cerebral lesion might be.

References

1. Groopman J. How doctors think. New York, NY: Mariner Books; 2008.
2. Seeley WW, Crawford RK, Zhou J, Miller BL, Greicius MD. Neurodegenerative diseases target large-scale human brain networks. Neuron. 2009;62:42–52.
3. Harlow JM. Passage of an iron rod through the head. Boston Med Surg J. 1848;39:389–93.
4. Harlow JM. Recovery from the passage of an iron bar through the head in publications of the Massachusetts Medical Society 1868;2:327–47.
5. Broca P. Nouvelle observation d'aphémie produite par une lésion de la moitié postérieure des deuxième et troisième circonvolution frontales gauches. Bull Soc Anat. 1861;36:398–407.
6. Tilney F. The brain from ape to man. New York, NY: Hoeber; 1928.
7. Penfield W. Mechanisms of voluntary movement. Brain. 1954;77:18.
8. Josephs KA, Hodges JR, Snowden JS, Mackenzie IR, Neumann M, Mann DM, et al. Neuropathological background of phenotypical variability in frontotemporal dementia. Acta Neuropathol. 2011;122:137–53.
9. Luria AR. Higher cortical functions in man. New York, NY: Basic Books; 1972.
10. Lhermitte F, Pillon B, Seradura M. Human autonomy and the frontal lobes. Part 1: imitation and utilization behavior. A neuropsychological study of 75 patients. Ann Neurol. 1986;19:326–34.
11. Lhermitte F. Human autonomy and the frontal lobes. Part II: patient behavior in complex and social situations: the "environmental dependency syndrome". Ann Neurol. 1986;19:335–43.
12. Heaton RK. Wisconsin card sorting test computer version 4. Lutz, FL: PAR Psychological Assessment Resources; 2003.
13. Trenerry MR, Crosson B, DeBoe J, Leber WR. Stroop neuropsychological screening test. Lutz, FL: Psychological Assessment Resources (PAR); 1989.
14. Bechara A. Iowa gambling test. Lutz, FL: Psychological Assessment Resources Inc; 2007.
15. Roth RM, Isquith PK, Gioia GA. BRIEF-A: behavior rating inventory of executive function adult version. Lutz, FL: PAR Neuropsychological Assessment Resources Inc; 2005.
16. Grace J, Malloy PF. Frontal Systems Behavior Scale. Lutz, FL: PAR Neuropsychological Assessment Resources Inc; 2001.
17. Delis DC, Kaplan E, Kramer JH. DKEFS. New York, NY: The Psychological Corporation; 2001.
18. Dubois B, Slachevsky A, Litvan I, FAB. The FAB. A frontal assessment battery at the beside. Neurology. 2000;55:1621–6.
19. Kertesz A, Davidson W, Fox H. Frontal behavioural inventory: diagnostic criteria for frontal lobe dementia. Can J Neurol Sci. 1997;24:29–36.
20. Reynolds CR. Comprehensive trail making test. Austin, TX: Pro-ed; 2002.
21. Culbertson WC, Zillmer EA. Tower of London. Toronto, ON: Multi Health Systems Inc; 2001.
22. Royall DR, Mahurin RK, Gray KF. Bedside assessment of executive cognitive impairment: the executive interview. J Am Geriatr Soc. 1992;40:1221–6.
23. Mesulam M-M. Large scale neurocognitive networks and distributed processing for attention, language and memory. Ann Neurol. 1990;28:597–613.
24. Kramer JH, Reed BR, Mungas D, Weiner MW, Chui HC. Executive dysfunction in subcortical ischemic vascular disease. J Neurol Neurosurg Psychiatry. 2002;72:217–20.
25. Tullberg M, Fletcher E, DeCarli C, Mungas D, Reed BR, Harvey DJ, et al. White matter lesions impair

frontal lobe function regardless of their location. Neurology. 2004;63(2):246–53.

26. Wolfe N, Linn R, Babikian VL, Knoefel JE, Albert ML. Frontal systems impairment following multiple lacunar infarcts. Arch Neurol. 1990;47:129–32.

27. The decade of the brain. The Library of Congress and National Institute of Mental Health. www.nimhinfo@nih.gov

28. Mesulam MM. Behavioral neuroanatomy: large scale networks, association cortex, frontal syndromes, the limbic system and hemispheric specialization. In: Mesulam MM, editor. Principles of behavioural and cognitive neurology. London: Oxford University Press; 2000. p. 1–120.

29. Royall DR, Lauterbach EC, Cummings JL, Reeve A, Rummans TA, Kaufer DI, et al. Executive control function: a review of its promise and challenges for clinical research. J Neuropsychiatry Clin Neurosci. 2002;14(4):377–405.

30. Folstein MF, Folstein SE, McHugh PR. "Mini-mental state". A practical method for grading cognitive state of patients for the clinician. J Psychiatr Res. 1975;12:189–98.

31. del Zoppo GJ, Saver JL, Jauch EC, Adams HP. Expansion of the time window for treatment of acute ischemic stroke with intravenous tissue plasminogen activator. A science advisory from the American Heart Association/American Stroke Association. Stroke. 2009;40:2945–8.

32. Saver JL. Comments, opinions and reviews. Time is brain—quantified. Stroke. 2006;37:263–6.

33. Stern Y. Cognitive reserve. Alzheimer Dis Assoc Disord. 2006;20:112–7.

34. Katzman R, Aronson M, Fuld P, Kawas C, Brown T, Morgenstern H, et al. Development of dementing illnesses in an 80 year old volunteer cohort. Ann Neurol. 1989;25:307–24.

35. Kemppainen NM, Aalto S, Karrasch M, Nagren K, Savisto N, Oikonen V, et al. Cognitive reserve hypothesis: Pittsburgh compound B and fluorodeoxyglucose position emission tomography in relation to education in mild Alzheimer's disease. Ann Neurol. 2008;63:112–8.

36. Knopman DS. The initial recognition and diagnosis of dementia. Am J Med. 1998;104:2S–12.

37. Ala TA, Doss RC, Sullivan CJ. Reversible dementia: a case of cryptococcal meningitis masquerading as Alzheimer's disease. J Alzheimers Dis. 2004;6:503–8.

38. Hoffmann M, Muniz J, Carroll E, De Villasante JM. Cryptococcal meningitis masquerading as Alzheimer's disease: complete neurological and cognitive recovery with treatment. J Alzheimers Dis. 2009;16:517–20.

39. Seiler N. Ammonia and Alzheimer's disease. Neurochem Int. 2002;41:189–207.

40. Benito-León J, Sedano LF, Louis ED. Isolated central nervous system Whipple's disease causing reversible frontotemporal-like dementia. Clin Neurol Neurosurg. 2008;110:747–9.

41. Stoquart-Elsankari S, Perin B, Lehmann P, Gondry-Jouet C, Godefroy O. Cognitive forms of multiple sclerosis: report of a dementia case. Clin Neurol Neurosurg. 2010;112:258–60.

42. AAICAD Conference Honolulu Hawaii, July 10-15, 2010

43. Viswanathan A, Rocca WA, Tzourio C. The vascular – dementia continuum. Neurology. 2009;72:368–74.

44. Snowden JS, Thompson JC, Stopford CL, Richardson AMT, Gerhard A, Neary D, et al. The clinical diagnosis of early onset dementias: diagnostic accuracy and cliniopathological relationships. Brain. 2011;134:2478–92.

45. Rascovsky K, Hodges JR, Knopman D, Mendez MF, Kramer JH, Neuhaus J, et al. Sensitivity of revised diagnostic criteria for the behavioral variant of frontotemporal dementia. Brain. 2011;134:2456–77.

46. Huey ED, Putnam KT, Grafman J. A systematic review of neurotransmitter deficits and treatments in frontotemporal dementia. Neurology. 2006;66:17–22.

47. Chollet F, Tardy J, Albucher JF, Thalamas C, Berard E, Lamy C, et al. Fluoxetine for motor recovery after acute ischaemic stroke (FLAME): a randomised placebo-controlled trial. Lancet Neurol. 2011;10:123–30.

48. Rafii MS, Aisen PS. Recent developments in Alzheimer's disease therapeutics. BMC Med. 2009;7:7.

49. Hoffmann M, Sacco RS, Mohr JP, Tatemichi TK. Higher cortical function deficits among acute stroke patients: The Stroke Data Bank experience. J Stroke Cerebrovasc Dis. 1997;6:114–20.

50. Hoffmann M. Higher cortical function deficits after stroke. An analysis of 1000 patients from a dedicated cognitive stroke registry. Neurorehabil Neural Repair. 2001;15:113–27.

51. Hoffmann M, Schmitt F. Metacognition in stroke: bedside assessment and relation to location, size and stroke severity. Cogn Behav Neurol. 2006;19(2):85–94.

52. Pendlebury ST, Markwick A, de Jager CA, Zamboni G, Wilcock GK, Rothwell PM. Differences in cognitive profile between TIA, stroke and elderly memory research subjects: a comparison of the MMSE and MoCA. Cerebrovasc Dis. 2012;34(1):48–54.

53. Freitas S, Simões MR, Alves L, Duro D, Santana I. Montreal Cognitive Assessment (MoCA): validation study for frontotemporal dementia. J Geriatr Psychiatry Neurol. 2012;25(3):146–54.

54. Bucker RL. Memory and executive function in ageing and AD: multiple factors that cause decline and reserve factors that compensate. Neuron. 2004;44:195–208.

55. Schott GD. Pictures as a neurological tool: lessons from enhanced and emergent artistry in brain disease. Brain. 2012;135:1947–63.

56. Mellars P. Major issues in the emergence of modern humans. Curr Anthropol. 1989;30:349–85.

57. Wynn T, Coolidge FL. The implications of the working memory model for the evolution of modern cognition. Int J Evol Biol. 2011;2011:741357. doi:10.4061/2011/741357.

58. Rizzolatti G, Fabbri-Destro M, Cattaneo L. Mirror neurons and their clinical relevance. Nat Clin Pract Neurol. 2009;5:24–34.

59. Franklin MS. The role of dreams in the evolution of the human mind. Evol Psychol. 2005;3:59–78.

Cognitive Archeology, Cognitive Neurology, and the Unraveling of the Connectome: Divulging Brain Function from Fractured Skulls and Fractured Minds

2

Evolutionary medicine may be deemed the most fundamental of the basic sciences and can provide insights into the neurobiological processes ranging from mitochondrial cytopathies to Alzheimer's disease [1]. The same principles are applicable to neurology. If we know how something is constructed we may sometimes have a better idea on how to fix it or gain better insights into the nature of brain syndromes. Notwithstanding the threefold enlargement of the human brain relative to great apes, the most impressive changes appear to have been in the various levels of connectivity. This has taken place both in terms of "hard-wired" fiber tracts and chemical- or state-dependent neurotransmitter proliferation and synaptic specialization. Hints that modern neurology has underestimated the importance of connectivity go back to a discourse between Charcot, a proponent of localizationism, and Brown-Sequard, who proposed distant effects of a lesion in 1875 [2]. More recently Lieberman suggested the minimal importance of Broca's area in motor aphasia but that subcortical damage (i.e., fiber networks) is a necessary component of aphasia and that aphasia never occurs in the absence of subcortical damage [3]. It has long been suggested that regional brain size increases were not the most important factor, but that the evolution of the cortical networks, particularly of the prefrontal cortex connecting all other brain regions, was the driver of human cognition and the mind [4]. Frontal dysfunction overlaps with vicissitudes of normal human function and a hallmark of the human brain is its connectome. Although humans have on average 2×10^{11} neurons, about one trillion glial cells, and 700 km of vessels, perhaps the most magnanimous vital statistic is the ~100,000 miles of axons comprising the neuropil. The defining feature of the human brain compared to other primates is a marked increase in connectivity and neuropil and increase in granular cortical areas [5, 6] (Fig. 2.1).

From Primates to Hominins

The primate brain had already increased in size relative to mammalian brain by the end of the Paleogene period (60–25 mya), by a factor of ~2, most likely due to frugivory and the challenging polyadic relationships consequent to the sociality adopted during this phase. With the globally, tropical conditions of the Paleocene-Eocene-Thermal-Maximum (PETM), ~55 mya dwindling, increasing global cooling, and aridity at ~34 mya were coincident with Antarctic glaciation. During this period, the ancient Tethys ocean, which had been associated with a warm equatorial ocean circulation, slowly became expunged, secondary to tectonic plate movements. With the obliteration of the Tethys ocean, the major ocean circulations changed from an equatorial to an interpolar, meridional type (Fig. 2.2). Cold dense water travelling northwards from Antarctica through the sea troughs and basins due to sea ice

© Springer International Publishing Switzerland 2016
M. Hoffmann, *Cognitive, Conative and Behavioral Neurology*,
DOI 10.1007/978-3-319-33181-2_2

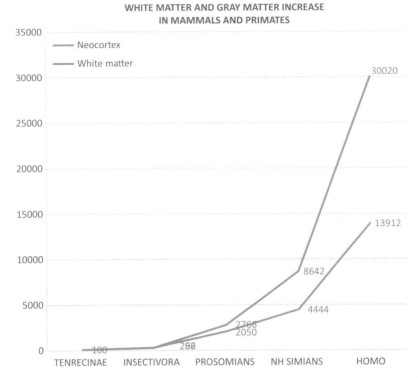

Fig. 2.1 The hallmark of human brain evolution was the profound allometric volume increase of white matter proliferation more so than the neocortical volume increase in *Homo sapiens*. Comparison in the figure is made to primates, nonhuman Simians, Prosimians, Insectivora, and the stem mammals here represented by the extant Tenrecinae, the Madagascar hedgehog. The Tenrecinae are regarded as representing stem Insectivora with relatively primitive brain morphology. The size of the various brain components is referred to the Tenrecinae base of 100 on the *Y*-axis. Figure compiled from the data of Stephan H, Baron G, Frahm H. Comparative Brain Research in Mammals, vol 1, Insectivora, Springer, New York

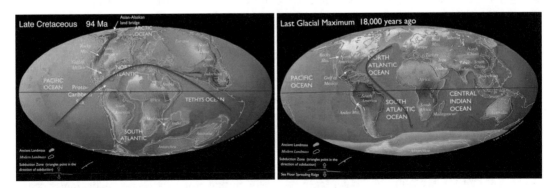

Fig. 2.2 Equatorial to interpolar circulation. The ancient Tethys ocean was associated with a warm equatorial ocean circulation. With tectonic plate movements gradually obliterating the Tethys ocean, the ocean circulations changed to an interpolar meridional type. Figure with permission from: Scotese CR. Atlas of Earth History, Volume 1, Paleogeography, PALEOMAP Project, Arlington, Texas, 2001, 52 pp and ocean circulation figure and adapted from Stow D. Vanished Ocean. How Tethys reshaped the World. Oxford University Press, Oxford, 2010

formation formed Antarctic bottom water. The resultant global cooling that commenced during these last 50 mya accelerated at about 34 mya and even more so ~6 mya, with the formation of the Arctic Pond [7]. These caused further challenges in terms of climate unpredictability and erratic food sources and corresponding frontal lobe specialization in earlier primates developed to reduce the number of errors incurred in these more challenging environments. Not all of hominin encephalization can be attributed to climate change. There were several drivers of increased brain size including predation, sociality, high-quality diet, chemical food processing (fire), and protolanguage [8]. These were consequent to the need for adaptation and flexible responses to environmental challenges and unpredictability. Climate however was a major factor that included global cooling and aridity episodes in relation to the four different Milankovitch cycles, with orbital forcing events that explain many of the Marine Isotope Stages. Also, marked and erratic fluctuations for example due to Heinrich and Dansgaard Oeschger events, precipitated by freshwater influx into the Atlantic Meridional Overturning Circulation (AMOC) down-welling in the northern Atlantic, are associated with whiplash-type climate changes from warm to cold spells [9].

East Africa and the Origins of the Modern Body, Brain, and Connectome

In addition to the climatic adversity, the change from arboreal to a more terrestrial habitat increased the predation risk particularly from the much larger than present-day apex predators such as Homotherium (preyed on pachyderms such as elephants and rhinoceros), Dinofelis, and Megantereon (both saber tooth cats). During this time (3.5–2.0 mya) the Australopithecines retained both arboreal and terrestrial features, remaining in an evolutionary stasis for about 1.5–2.0 mya. This changed with a sequence of three ice age (5–1 mya) periods precipitated by Arctic glaciation and a number of geological as well as orbital (Milankovitch cycling) forcing events. In

brief these included the closure of the Panama isthmus, Indonesian seaway closure, the opening of the Drake passage, African Rift valley formation (formation of complex topography), Himalayan formation, and Tibetan Uplift [10]. Together, these events conspired to produce the first hominin species, Homo erectus ~1.8 mya with a doubling in brain size, with obligate bipedalism, committed ground sleep (with increased slow-wave sleep, REM sleep, and dreaming), increased range of foraging, increasing sociality, the discovery of fire (~1.5 mya for protection and enabling chemical food processing), and the use of high-quality foods such as meat, fish, and underground storage organs. Ultimately predation was also considered a major factor in propelling early hominins towards sociality [11].

The next hominin, Homo heidelbergensis, represents a cognitive milestones with development of the modern human frontal lobe size. The archaic human skulls differed from those of early modern humans in having large brow ridges, but the slope of the inner frontal brain case does not differ suggesting that the shape of the frontal lobes reached a modern state in archaic humans ~300–600 kya as evidenced by the Bodo skull from Ethiopia 600 kya and the Kabwe skull from Zambia 300 kya (Homo heidelbergensis species). Its shape stabilized relatively early in our more recent cognitive evolution and this implies that the frontal lobes were well developed some time before one of the key technological revolutions that occurred in early humans [12] (Fig. 2.3). Blade technology and microlith bone tool evidence intimates that these inventions occurred by ~100 kya [13].

Several prefrontal subcomponents may be identified on a cytoarchitectonic and connectivity basis, each sub-serving different roles (Fig. 2.4). In an insightful evolutionary overview proposed by Passingham and Wise, the orbital prefrontal cortex (OPFC) assigns value to items; the caudal prefrontal cortex (PFC) and the frontal eye field areas enable searching for objects by mediating eye movement and attention. The medial prefrontal cortex (MPFC) is associated with choice of action influenced by prior experiences, by virtue of its connections with the hippocampus, amygdala, and the medial premotor cortex. The MPFC is able to

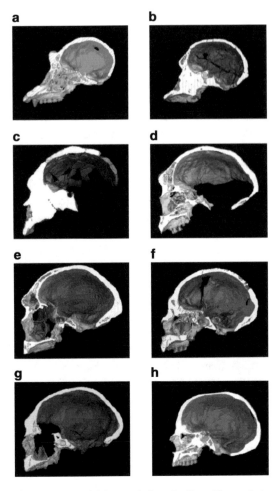

Fig. 2.3 Frontal lobe evolution timeline. The archaic human skulls differ from those of early modern humans in having large brow ridges, but the slope of the inner frontal brain case does not differ suggesting that the shape of the frontal lobes reached a modern state in archaic humans ~300–600 kya as evidenced by the Kabwe and Bodo skulls (a—Chimp, b—Australopithecine skull, c—Bodo, d—Kabwe, e—Homo heidelbergensis, f—Neanderthal, g—Neanderthal, h—Homo sapiens). Figure with permission: Bookstein F, Schaefter K, Prossinger H et al. Comparing Frontal Cranial Profiles in Archaic and Modern Homo by Morphometric Analysis. The Anatomical Record (New Anat) 1999;257:217–224

employ choice and action without external sensory stimuli and differs from OPFC in this regard in that the latter's action depends on external stimuli. The dorsolateral prefrontal cortex (DLPFC) has extensive connections to all other frontal areas as well as other brain regions including the OPFC, premotor cortex, posterior parietal cortex, and hippocampus. Together these mediate integration of the various forms of information and ultimately connection to the premotor cortex for action or inaction. The ventrolateral prefrontal cortex (VLPFC) receives input from the superior temporal (acoustic) and inferior temporal (visual) areas that allows actions based on combined acoustic and visual data. This region forms part of the mirror neuron circuitry, which allows decision making after a solitary event with overall error reduction [14]. The "apical" function of the frontopolar (FPC) region is associated with metacognitive capabilities, or self-reflection, as well as episodic memory, multitasking, relational integration, self-referential evaluation, and introspection [15]. The FPC matures late with dendritic spines developing up to adulthood, into the third and fourth decades [16, 17]. The progressive elaboration of the different prefrontal granular cortices and their connections was a key evolutionary feature amongst hominins, by allowing overall error reduction within their complex environment, allowing learning by imitation, rather than by trial and error and being able to simulate future scenarios by virtue of mental time travel and dreaming [18, 19].

At the microscopic level, proliferation of granular cortex (denoted by a distinct layer IV) in the prefrontal, parietal, and temporal cortices is a defining feature of modern humans. Granular prefrontal cortex, in particular, expanded in humans and is considered key for executive function abilities. Layer IV receives mostly afferents from thalamic neurons and is the site of intracortical connectivity. In addition, pyramidal cells in the human PFC are more branched and have much greater spine density when compared to the parietal, temporal, and occipital regions. Within layer III of the PFC, up to 61 spines per 10 μm have been recorded and when compared to the homologous area in macaques, the human layer III pyramidal cells are 23 times more spinous [20]. The neuronal discharge activity in the tertiary association areas, with cells that have many spines, such as the PFC, inferotemporal, and parietal cortex, is tonic as opposed to the phasic discharge typical of the primary areas such as visual area V1. This persistent firing has been interpreted as the neurobiological substrate of "holding" a memory or activity, during the so-

Fig. 2.4 Schematic representation of the evolutionary development of frontal lobe regions: caudal frontal (*purple*), ventrolateral frontal (*yellow*), dorsolateral frontal (*pink*) orbitofrontal (*orange*), frontopolar (*green*), and medial frontal (*blue*)

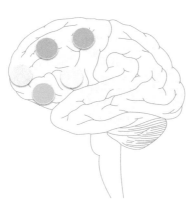
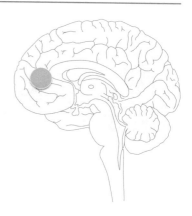

Table 2.1 Key stages or grade shifts in human cognitive evolution

- Increased sensorimotor integration
- Bipedality, meat diet
- Increased brain size
- Enhanced working memory—stage 7
- Mirror neuron circuit—stage 7
- Neurogenesis in dentate gyrus
- Autonoesis (subjective time awareness), mental time travel
- Neurobiological circuitry for higher consciousness, spirituality, language, and culture

called delay period, which in the PFC may be maintained even in the face of distractors. This process underlies working memory, which may constitute the core of frontal functions [21]. During hominin evolution, an increasing pyramidal cell complexity, consisting of highly branched and highly spinous pyramidal cells, occurred, particularly in the PFC [22].

Several key stages or grade shifts in human cognitive evolution may be defined (Table 2.1). Although archeological remnants of the earliest hominins, including the oldest of all, Sahelanthropus tchadensis, have been found as far afield as present-day Sahara desert and Australopithecines even further away in various parts of Southern Africa, general consensus of an East African Rift Valley origin remains (Fig. 2.5). The unique geological and ensuing climatic alterations in Eastern Africa resulted in a complex topographic and erratic climatic environment. Maslin et al. proposed their pulsed climate variability hypothesis that entails an overall extended aridity period of

East Africa but interspersed by briefer wet, humid periods that correlate with hominin speciation, brain growth, and out-of-Africa dispersals [23].

Hominin flaking of stone tools (Oldowan) becomes evident archeologically from 3.4 to 2.5 mya. Experimental evidence by Morgan et al. postulated that imitation of tool making by other hominins was a "low-fidelity social transmission" capability, leading to an almost one million year status quo of the stone flaking mechanism. Instruction by rudimentary protolanguage may have been an essential requirement for the next stage that led to blades or biface (Acheulean) technology [24]. This line of evidence also infers that cerebral reorganization occurred before an increase in brain size. This fits in well with the seven-stage extended mirror neuron system postulated by Arbib, whereby stages 1–3 were simple imitation procedures that are within chimpanzee capability, but by stage 4, complex imitation enabled the step to protosign (stage 5), protospeech (stage 6), and modern language (stage 7) [25]. Biface technology was associated with *Homo erectus* ~1.8 mya, by which time co-evolution of ground sleep, increased REM sleep, dreaming, slow-wave sleep, the use of fire and chemical food processing, and sociality with increased group size occurred. This stage also correlated with the first increase in brain size from about 450 cc to 800 cc. These "first" hominins were endowed with sufficient brain power to successfully conduct the first "out-of-Africa" expedition, reaching as far afield as modern-day China [26].

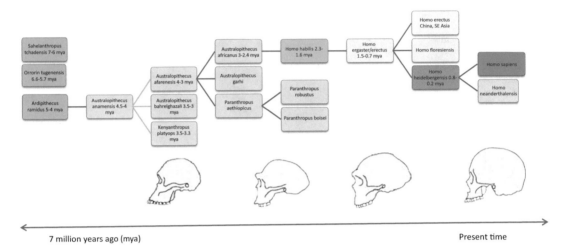

Fig. 2.5 Human evolution stages (mya): (1) "Missing link" genera and bipedality (5–7); (2) Australopithecines, large molars, and evolutionary stasis (4.5–2.6); (3) Homo habilis and Oldowan stone tools (2.5–1.9); (4) Homo ergaster and Archeulean biface blades (1.7–1.4); (5) Homo heidelbergensis and Schoeningen spears (0.7–0.6); (6) Homo sapiens, language, and art (0.05)

The next significant cognitive event is represented by *Homo heidelbergensis* with brain size (~1200 cc) close to modern humans, who occupied the time period 800–200 kya. They were represented in Africa by the *Homo rhodesiensis* and in Europe by *Homo heidelbergensis* and gave rise to both Neanderthals in Europe and modern humans in Africa. Due to further marked climactic fluctuations that resulted in erratic resources for hominins, several different granular frontal cortex subtypes developed that in essence culminated in error reduction by early hominins in their challenging environment. In addition, these new granular PFC areas enabled learning by imitation and mental time travel that in turn allowed imagining action, before engagement [27]. Examination of the *Homo rhodesiensis* skulls from Bodo and Kabwe revealed inner frontal brain cases to be not significantly different from modern human skulls. These findings imply that by that time the frontal lobes had already reached modern human proportions, well before the cultural evolution typically dated to 70–50 kya (Fig. 2.3) [12].

Southern Africa and the Origins of the Modern Mind: Do the Events Here Between 200 and 100 kya Explain the Sapient Paradox?

Archeological Evidence

Anatomically modern humans (AMH) evolved in eastern Africa. Did modern minds evolve in southern Africa? The term sapient paradox formulated by Renfrew refers to the unexplained time lag between AMH and the emergence of the modern mind ~50 kya [28]. Genetic, linguistic, and archaeological anatomical research has allowed a general consensus that not only did modern humans appeared in eastern Africa by 200 kya, but also the modern mind may have been honed in southern Africa. The timing and place of the cognitive and cultural explosion, evidenced by archaeological relics, that represent proxies, are uncertain. Intriguing research by Marean et al. provides tantalizing evidence for enhanced cognition appearing in the period 164–75 kya perhaps as a consequence of the extreme environmental challenges of cooling and aridity, posed by Marine Isotope Stages 6 (195–123 kya) and 5 (130–80 kya). The unique constellation of high-protein docosahexaenoic acid (DHA) from fish and shellfish consumption, yielded by the Agulhas current, temperate climate, as well as the immediately adjacent availability of complex carbohydrate (tubers), coincided with the substantiation of symbolic behavior, pigment use, and bladelet stone technology, found within the caves by the sea at the archeological research site, PP13B [29]. In a thought-provoking title "When the sea saved humanity," there was speculation that only a few thousand or as little as a few hundred modern humans survived these major ice age-induced adversity conditions, but which ultimately led to the formation of the modern human mind [30]. These fortuitous cave "condominiums" by the sea were not restricted to the extreme southern African coast as a similar find relates to the Sibudu cave about 1500 km to the north, on the Eastern South African coastline [31]. This approximately 2000 km stretch of coastline is characterized by multiple rocky outgrowths with intertidal pools that harbor shellfish and crustaceans in addition to fish.

Genetic Evidence

The molecular clock concept is based on the evolution of DNA sequences being relatively constant over time and has allowed the proposal of a "mitochondrial eve" emanating from Africa approximately 180 kya. Interestingly this occurred at a time of extremely low population density, probably, again, as a consequence of Marine Isotope Stage 6 (MIS 6). Subsequently, two modern human population groups appeared within Africa by approximately 130 kya that included the ancestral Khoe and San in southern Africa and one in central and eastern Africa. Mitochondrial DNA (mtDNA) analysis, specifically of the oldest mtDNA haplogroups (L0d and L0k), is found in particularly high frequencies in the southern African Khoe-San groups, in the time frame of ~100 kya. These represent the most penetrating mtDNA clades recorded in modern humans [32]. It is most likely that L0 is of southern Africa origin, from amongst the ances-

tral modern Khoe and San people. The evidence supports the L0 haplogroup originating in southern Africa, L1 in central Africa, and L2'6 in eastern Africa approximately 130 kya. This period coincided with the extreme aridity and cooler climate that precipitated the African "megadroughts" consequent to MIS 6 and MIS 5. It was during this period however that more extensive archaeological evidence of mode 3 or middle stone age tools, consisting of flake tools prepared from cores, appears (Fig. 2.6) [33].

Linguistics

The San have been termed "the oldest population on Earth" by Tishkoff et al. on the basis of their click language analysis. Although approximately 30 different click languages are spoken within the southern African region, only the hunter-gatherer Sandawe and Hadzabe people in eastern Africa speak a click language. It is presumed therefore that they migrated from their original south-western African habitat to eastern Africa [34]. Genetic analysis also provides insights into the subsequent modern human expansion from eastern Africa to the rest of the world (Fig. 2.7) [32, 35–37].

Additional Neurochemistry Evidence: Docosahexaenoic Acid and Synaptic Bandwidth

Humans have high-bandwidth synapses compared to rodents and DHA has many synaptic regulatory effects [38]. DHA availability has a profound effect on brain size and probably connectivity. We have a long association with DHA that goes back ~600 million years without any change in the molecule, whereas marked genetic changes took place during that time frame. Crawford expounded that DHA was probably more important than DNA and DHA dictated to DNA rather than the other way around [39]. Dietary omega-3 fatty acids have significant effects on neural gene expression. However the low availability of DHA in the land food chain supports the observation that terrestrial mammalian brains became relatively smaller with increasing body size. Mammals lack specific

Fig. 2.6 Mitochondrial
DNA analysis and human
evolution. Figure with
permission: Rito T,
Richards MB, Fernandes V
et al. The First Modern
Human Dispersals across
Africa. PLOS One
2013;8:e80031

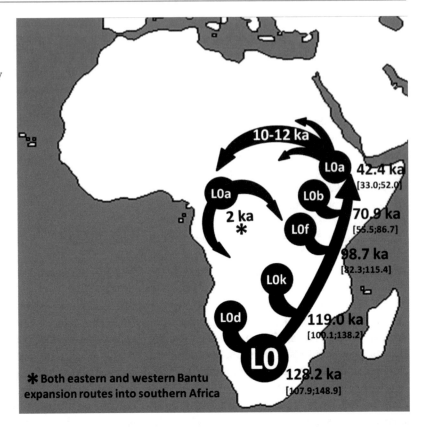

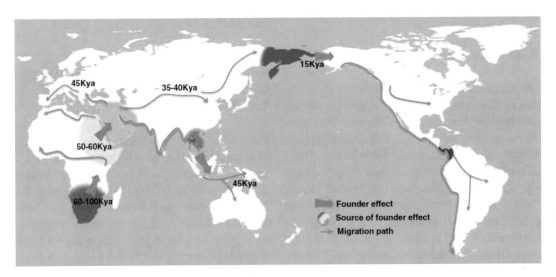

Fig. 2.7 Modern human dispersals since ~100,000 years beginning in southern Africa. Legend: Founder effects (new population established by a small group derived from a larger population) indicated by *broad arrows*, *small arrows* reflect migration routes, and colored areas reflect origins of the founder events. Figure with permission: Henn BM, Cavalli-Sforza LL, Feldman MW. The great human expansion. PNAS 2012;109:17758–17764

enzymes required for the synthesis of the precursors for omega 3-FA precursors. Lacustrine and marine environment food, rich in DHA together with micronutrients (iodine, zinc, copper, manganese, selenium), which counter peroxidation, may have been a key factor in hominin brain size increase as well as the surge in intra-connectivity [40].

Polyunsaturated fatty acids (PUFAs) and DHA in particular regulate neurogenesis, synaptic function, the function and structure of glial cells and endothelial cells, and brain inflammation. Unsurprisingly, PUFA metabolism impairment or dietary intake deficiency is associated with many neurological and psychiatric conditions influencing cognition and mood. The neurobiological mechanisms whereby these processes take place include regulation of membrane dynamics, receptor activation, cell signaling networks, modulation of brain endocannabinoids, synaptamine-mediated synaptogenesis, and neuronal growth and differentiation. In addition a DHA anti-amyloidogenic effect via the neuroprotectin D1 (NPD1), a DHA-derived mediator and regulator of brain glucose uptake, has recently been reported [41]. The protean DHA cellular functions provide insights into the importance of DHA in major human disease states such as dementia, depression, attention-deficit hyperactivity disorder, stroke, myocardial infarction, and even chronic headaches. For example, omega-3 fatty acids have antinociceptive properties that regulate several pain-related biochemical processes and omega 6 have pronociceptive properties. In a randomized trial, increasing omega-3 fatty acids and decreasing omega 6 fatty acids were effective in reducing chronic headaches and improving quality of life [42].

Was the Marked Increase in Brain Connectivity and Synaptic Bandwidth Due to a High-Intensity Sea Food and Tuber Diet of the Southern African Coast During MIS6?

Synaptic density increases from approximately 2000–5600 synapses overall per neuron in the brain of monkeys to 6800–10,000 in humans. In the prefrontal cortex this increase is even more dramatic where prefrontal pyramidal cells have an average of 15,100 spines, which represent a 72% increase, with respect to the macaque [43]. Human prefrontal pyramidal cells have on average 23× more spinous processes than the primary visual area, BA 17, which is regarded as a key factor in the high-level cognitive processing in humans [20]. Establishing the timing of the modern human mind more precisely than at about 50 kya seems less important than determining what circuitry became modern. With the modern frontal lobes placed by 600–300 kya (Kabwe and Bodo skull data), did subsequent high synaptic bandwidth develop and explain the sapient paradox? [38].

The Final Step in the Modernity of the Mind: The Parietal Lobes and Cerebellum Enlarge, Leading to Increased Intra-connectivity, the Modern Human Connectome, and Further Enhancement of Working Memory

Cerebellum enlargement occurred relatively late (35–11 kya) attributed to be a consequence of intra-connectivity with the frontal lobes via the frontopontocerebellar tracts. This reciprocal cerebellar enlargement was associated with connections to 14 different cortical areas that may have further promoted cortical efficiency. Physiologically, the cerebellum enables the enhancement and precision timing of neural responses in the domains of motor, sensory, and cognitive processing and signaling. This improved information processing ability would have been advantageous to modern humans during upper Paleolithic period, at a time of a marked increase in social and cultural complexity [44]. The neurobiological changes of the cerebellum included a cerebellum-to-cortex ratio significantly greater in modern humans, compared to Neanderthals. Other significant neuro-anatomical features included a high neuronal density, with 70–80% of the 100 billion neurons being located in the cerebellum, as well as a relative enlargement of the dentate cerebellar nuclei. All cortical regions project to the cerebellum via the pons and recipro-cally back to the cortical regions. The widely dis-tributed working memory circuitry enables refinement of cognitive, language, and motor action plans, improving efficiency. Vandervert et al. have even suggested that the cerebellar feed-back circuitry may have constituted the neurobio-logical basis of creativity and innovation [45]. The cerebellum has also been implicated in emo-tional processing, a key human attribute that developed in response to progressively more intricate sociality amongst AMH [46].

Parietal Globularization

Parietal lobe expansion also occurred very recently in human evolution (150–100 kya) with both vertical expansion and anterior widening, referred to as klinorhynchy, which was unique to modern humans [47]. Modern endocranial fea-tures derived from endocasts such as those from Jebel Irhoud (Morroco ~150 kya) and Skhul V (Near East ~120 kya) have modern human pro-files for example [48]. This morphological change enabled neural intra-connectivity and has been linked to significant modern human capa-bilities such as symbolic culture and exploration of new environs [49]. The neurobiological and neurophysiological parietal lobe functions that led to these capabilities were based on the func-tions that included analysis of number, space, time, and visuospatial and sensory inputs that allowed representation of the body in terms of both internal and external space representation. The cross-modal sensory function of the parietal lobes also allowed metaphor and language com-prehension. Other key functions included visuo-spatial working memory, motor planning (impaired in apraxia), hand–eye coordination, and three-dimensional spatial representations [50, 51]. These in turn empowered improved geographical navigation, more precise finger manipulation and control leading to improved toolmaking and manufacture of compound, and weapons ability. The appreciation of quantity, termed numerosity, is a function of the intrapari-etal sulcus (IPS) and finger counting to at least 5,

developed later. The evidence of stencils in Cosquer Cave, in France, may have represented an early numeric code dated at about 27 kya. The fact that most counting systems are decimal, fol-lowed by quinary (5) and vigesimal (20) systems, provides tantalizing evidence that human hands played a role [52]. Abnormalities seen with lesions in this region of the inferior parietal lobe are recognized clinically as Gerstmann's syn-drome (acalculia, finger anomia, right-left disori-entation, and dysgraphia). Number concepts could be much older, perhaps as much as 100 kya, as inferred by the Pinnacle Point and Blombos cave data from southern Africa, represented by strung beads as representing number concepts. These parietal functions, including the concept of symbolism, abstract thinking, metaphor, and lan-guage, are reflected in cave art, figurines, orna-ments, and more complex weaponry associated with modern humans in the Upper Paleolithic period (50–10 kya). Few if any were associated with the Neanderthals [53].

The superior parietal lobe that encompasses the IPS and precuneus, that mediate both numer-osity and prospective memory, in particular, may have been an important part of parietal lobe expansion. Functional imaging studies have shown the precuneus to have one of the highest metabolic activity regions of the brain, particu-larly during the resting state as part of the default mode network involving the parietal-medial pre-frontal network. A proposed function of this region is the integration of diverse neural net-works that underlies self-consciousness and auto-biographical memory [54].

What Explains the Relatively Abrupt Florescence of Advanced Behavior (the Modern Capacity for Culture) 50 kya?

All animals, excepting for apex predators, need to solve the predation challenge within their par-ticular environments. Primates solved this by functioning as a group, in essence sociality. The social brain hypothesis (SBH) has since revealed the close correlation of increasing group size with increasing brain size. With further group size increases, fission-fusion and geographically dispersed groups developed. Apart from group size, social complexity became even more significant with the increasing numbers of interacting dyads and triads [55]. With more dispersed groups, so-called virtual group members have to be factored into the social dynamics that primates have to consider that significantly increase the cognitive load of the evolving primates [56]. Brain circuitry deployed to cope with this increasing social complexity included using emotions to promote social bonds. Social grooming, being a very physical and face-to-face activity, was associated with strong emotions and release of oxytocin that promoted bonding. Others included theory of mind (TOM) and the related grades of intentionalities (grades 1–5), also termed mentalizing or mind reading. From these developed social emotions (shame, greet, guilt, remorse), different to the primary emotions, such as happiness, anger, fear, and sadness. Mammals in general are considered to have first-order intentionality, chimpanzees and australopithecines second order (regarded as sufficient for basic stone tool making), Neanderthals fourth order, and modern humans fifth order. With further group size increases, a shift from tactile grooming to vocal grooming is envisaged. This was achieved not only by early protolanguage, but also with the co-evolution of controlled use of fire and musicality. Early language would have enabled communication with a much larger number of individuals than one-on-one tactile grooming. Musicality, with its very emotive characteristics, likely evolved initially, akin to infant directed speech (IDS) type, chanting, dance, and probably preceded language and could well have amplified emotions in the social contexts. Subsequently during the period of Marine Isotope Stages (MIS), MIS6 (195–123 kya) and MIS5 (130–80 kya), evidence of composite tools, hafting, microliths, and use of ochre and shellfish signified cultural complexity. Larger populations and emotions were key in forming binding bonds, requiring more advanced TOM, in turn due to more dispersed populations in distance and time that tested social interaction [57].

With the exodus from Africa around ~70–50 kya (Fig. 2.7), Eurasian higher latitude occupation posed new challenges in the form of seasonality, variable day length, winter, and less concentrated food sources. With the resultant dispersed networks amongst the human bands, social bonding and networking helped mitigate these. Pearce et al. speculated that the smaller network size of Neanderthals, for example, in contrast to AMH, required less "cultural scaffolding" for social network maintenance [58]. More extreme winters and seasonality have selected for larger brains in both birds and primates, as well as within the hominin lineage. Behavioral flexibility, innovation, progressively larger group sizes, and social complexity placed further demands, promoting brain size increase as well as the need for high-quality foods and therefore meat acquisition. The encephalization, behavioral adaptations, and hominin speciation are in turn linked to the predominantly orbitally induced climatic fluctuations [59]. However there was a differential effect on encephalization imposed by the higher latitudes [8]. Neanderthals and AMH both had the largest brain sizes, but there were a number of differences. Some of the reasons proposed include relatively larger orbits and corresponding larger occipital lobes of Neanderthals due to decreased light, in the context of larger bodies compared to AMH. However this implied correspondingly more neural tissue dedicated to the somatic and visual function but within similarly sized brains [60, 61]. From this it follows that AMH had relatively larger brains when standardized for visual areas and body mass and less non-somatic neural areas, allowing AMH to have larger parietal lobes, enabling more connectivity, subserving social brain functions. Social cognition brain circuitry competed with that deployed for somatic and major organ systems [62] (Fig. 2.8). The orbitofrontal cortex size has been correlated with social cognitive competence, levels of intentionality, and mentalizing. AMH is postulated to have reached the fifth-order intentionality, whereas N reached up to the fourth order [63, 64].

With overall population size increasing with higher latitudes, AMH needed to fission into more groups compared to Neanderthals with the

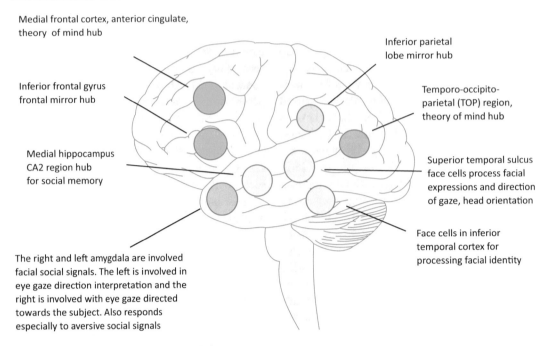

Medial frontal cortex, anterior cingulate, theory of mind hub

Inferior parietal lobe mirror hub

Inferior frontal gyrus frontal mirror hub

Temporo-occipito-parietal (TOP) region, theory of mind hub

Medial hippocampus CA2 region hub for social memory

Superior temporal sulcus face cells process facial expressions and direction of gaze, head orientation

The right and left amygdala are involved facial social signals. The left is involved in eye gaze direction interpretation and the right is involved with eye gaze directed towards the subject. Also responds especially to aversive social signals

Face cells in inferior temporal cortex for processing facial identity

Fig. 2.8 Schematic social brain circuitry hubs

average being 106 versus 152 for AMH [65]. The challenge of maintaining networks with geographical dispersion was resolved by periodic aggregations and by artifacts that served as proxies for face-to-face encounters (visual display hypothesis) as well as gift exchanges, all subsumed under the concept of cultural scaffolding [66]. Places of recurrent assemblages amongst these bands may have been centered around resource opportunities but developed into places of social gathering and promoted social cohesion. The larger more interconnected AMH groups not only impelled new innovations, but also led to a ratchet effect whereby knowledge accumulated, did not fizzle out, but was built upon and improved by previous advances [67].

Evolution of Modern Working Memory 200–50 kya: Sapient Paradox Candidates

Ultimately the augmentation or escalation of working memory capacity may have driven the appearance of modern executive function. Some researchers have envisaged the process on a

scale of 1–7, with chimpanzees, for example, acquiring level 2 and modern humans level 7 working memory [68, 69]. Possible contenders include:

1. A genetic event or epigenetic (methyl tagging modifying gene output) [70]
2. High-bandwidth synapses (high intake of aquatic food sources) [38]
3. Astroglial networks and gliotransmission [71]
4. Elaboration of prefrontal inhibitory functions (improved social interfacing) [72]
5. Prefrontal cortical inter- and intra-connectivity
6. Cerebellar and parietal lobe expansion and connectivity [46, 73]

These neurobiological processes in turn may have led to the enhancement of the subcomponents of the expansive circuitry that comprises the hypothetical WM model:

1. Phonologic storage capacity (syntax, recursion, and modern language)
2. Visuospatial sketchpad (simulation of the future, innovation)
3. Episodic buffer elaboration (creativity) [74, 75]

Converging Neurobiological and Clinical Evidence Supports a Connectomal Approach to Understanding Brain Function: The Role of Cognitive Neurology

At about 50 kya a major change occurred in the modern human mind. However, the neural hypothesis cannot be substantiated by fossil relics and needs to be inferred through behavioral correlations deduced from art and weapons, for example [76]. Can clinical lesion neurology provide complementary information and help perhaps as a more direct approach? Rather than being able to establish the timing, perhaps more important is what faculties evolved, other than the obvious one's, such as language. With the Geschwind Gastaut syndrome, for example, the philosophical pre-occupation and spirituality domains, hypergraphia, and other features that are "released" in the context of lesions or chronic seizures give insights into circuitry that has the temporal lobe as a hub. These now emerge with brain lesions or alterations that alter the balance of excitatory and inhibitory inputs to such circuits.

Neurobiological Evidence

The ultimate driver of brain size is frontal lobe white matter with frontal connectivity to all other brain regions evidenced by the marked neuropil increase by comparative extant primate studies [77]. Brain enlargement in humans is a function of a dramatic increase in white matter fiber tract connectivity. Evidence comes from comparative studies in primates and humans. Although the human brain has 8.6×10^{11} neurons and 10^{14} synapses, perhaps most impressive are the 150,000 km of fiber tracts [78]. Contrariwise, gray matter volume is not significantly different from extant primates. In anthropoid primates, the scaling of the white and gray matter of the cortex has been calculated at 3.5/3 [79]. This neocortical hyperscaling is attributed to the frontal lobe white matter increases. Hyperscaling of the brain refers to the relative increase in brain size with body size. Frontal lobe white matter growth is regarded as the primary constituent in brain size increase [80] and PFC white matter is relatively larger in humans compared to primates [81].

Clinical Lesion Insights: Diaschisis

With lesions, brain function may be increased rather than decreased due to the connectomal diaschisis concept, evidenced in cognitive syndrome examples such as field-dependent behavior, savant syndrome, Geschwind Gastaut syndrome, anxiety, depression, and many others. These syndromes may be viewed in terms of variations of frontal connectivity disruptions, or frontal network syndromes. Diaschisis refers to remote pathological or neurophysiological changes in the brain as a consequence of an injury in a different part of the brain. These so-called hodological effects, may be both excitatory or inhibitory. Different diaschisis subtypes include diaschisis at rest, functional, connectional, and connectomal. For example, functional diaschisis becomes evident only with activation or stimulation of a brain region that results in distant region effects, different to diaschisis at rest, and thus represents a dynamic type of diaschisis. More extensive changes in network connectivity, both intrahemispheric and interhemispheric, are also appreciated [82]. Connectomal diaschisis has been postulated to explain the more widespread network changes reported in the connectome measured by fMRI or MEG for example that can be appreciated after a remote lesion (Fig. 2.9) [83].

Clinico-Pathophysiological Insights: The Hub Failure Hypothesis

Computational neurology has given us major insights into brain connectomics. In essence brain networks are understood to comprise of small world networks that combine both robust, local connectivity and efficient long-distance connectivity. Extensively interconnected hubs within this system are then said to constitute a "rich club" or "connectivity backbone" [84]. Assessment of inter-regional connectivity may

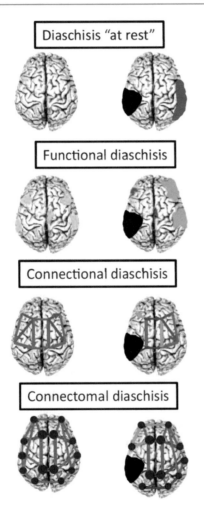

Fig. 2.9 Diaschisis classification. Figure with permission: Carrera E, Tononi G. Diaschisis; past, present, future. Brain 2014;137:2408–2422

aid the understanding of brain pathology as well as facilitating predictions about involvement of other brain regions. An important concept of brain networks is conceived under the concept of "centrality," a topological dimension, which refers to the number of connections to a given node. The term "rich club" denotes topologically central hub regions that are extensively interconnected and these central hub nodes populate the heteromodal tertiary association areas of the cortex. Damage to these central regions or to the connections between rich club components affects brain function more extensively than is the case if peripheral nodes are involved [85].

These highly interconnected hub nodes support integrative functions in the brain and are termed "functionally valuable" (Fig. 2.10). They have higher metabolic requirements and their connections are more expansive. Crossley et al. examined 26 different brain disorders and found that within the human brain networks, high-value hubs are preferentially affected by the majority and perhaps all brain pathologies. The ones significantly associated with hub pathology included frontotemporal lobe dementia, post-traumatic stress disorder, schizophrenia, temporal lobe epilepsy, juvenile myoclonic epilepsy, Aspergers, Alzheimer's disease, and progressive supranuclear palsy [86]. In addition, migraine may be similarly regarded as a hub failure process migraine [87]. Their liability to oxidative stress is a consequence of upgraded metabolism, attributed to increased neuronal spiking and synaptic activity [88, 89]. Complex networks are not limited to neuronal networks but also apply to genetic, immune, metabolic, and social networks, with which they interact. It may be readily appreciated that even focal pathology due to a brain tumor, discrete intracerebral hemorrhage, or bland infarct may have a much more widespread effect on the brain. Hub overload and failure occur for example in Alzheimer's disease, which specifically affects the hub regions, as well as in multiple sclerosis. A redistribution effect of the hubs is also evident. In traumatic brain injury for example, a hub redistribution effect has been noted from regions such as the precuneus and fusiform gyrus that has been postulated as a possible signature of TBI network reorganization [90, 91]. Resting-state network imaging has been instrumental, for example, in predicting recovery in people with minimally conscious state and vegetative states [92]. Importantly, hub failure, in many brain pathologies, present with cognitive impairment, in the domains of working memory (WM), executive function (EF), and attention [84]. Consistent with the hub failure hypothesis, from an evolutionary perspective therefore, the most recent expansion, that of the parietal lobe expansion, may have predisposed modern human to the ravages of Alzheimer's disease [93].

Fig. 2.10 A brain lesion (*black area*) and the effects on the network. Lesions of connector hubs (**b**) and community hubs (**c**), normal (**a**). Figure with permission: Carrera E, Tononi G. Diaschisis; past, present, future. Brain 2014;137:2408–2422

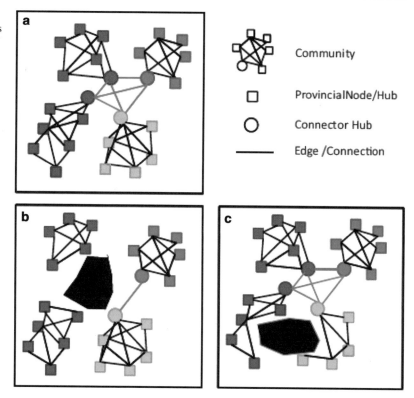

Clinical Insights from Clinical Cognitive and Behavioral Neurology

Can neurological lesion studies help by looking backwards through the course of development of the human mind? Can behavioral and cognitive neurology help solve the neural hypothesis of modern human behavior? The way the brain fragments due to lesions and the associated forms of the diaschisis phenomena permits some insights as to how it was constructed. Comparative primate studies show clear evidence of a number of markedly increased fiber tract systems in the human line and some that are unique to humans, such as the occipito-frontal fasciculus, which is the largest of all [94].

WM, EF, and attention are key processes that are invariably impaired in association with rich club failure. The important diagnostic implication is therefore the measurement of these core functions, as part of a routine in cognitive disorders in all people with cognitive or memory complaints, no matter how minor. Neuroimaging techniques such as functional imaging (PET brain, DTI) have already

yielded improved diagnostic capability in brain pathologies such as dementia, depression, and TBI. Resting-state network (RSN) analysis promises even more diagnostic accuracy, as it is integral to the connectomal understanding of the brain. This type of network imaging may also relate better to the ancestral evolution of the brain, in terms of its organizational changes of hyperconnectivity, dendritic arborization, and synaptic evolution, that took place aside from overall brain size increase.

Selected Examples of Cognitive Behavioral Syndromes That Reflect the Evolutionary Connectomal Neurobiology

Visual Dorsal and Ventral Radiations, Simultanagnosia, and Balint's Syndrome

Paleoneurological insights from basic lithic mode 1 (Oldowan) technology which began over about 3.4 mya (Dikika Ethiopia) are regarded as evidence of a climactic event in hominin evolution for which basic sensorimotor abilities sufficed, for this technology [95]. The later mode 2 (Archeulean) biface, hand-axes required more cognitive control for both spatial and visual object shape processing, subserved by the dorsal stream and the ventral radiations, respectively. Coordinating these two processes is not within ape capabilities and the stored information for such, contained within long-term procedural memory, only developed by ~1.8 mya. In addition, biface technology is also associated with the recruitment of the left hemisphere anterior intraparietal sulcus (IPS) and inferior frontal sulcus that are part of the mirror neuron system. This implies that social learning was a key element in acquiring the skills of biface technology [19].

Simultanagnosia (piece meal vision), optic apraxia (impaired reaching of objects under visual guidance), and optic ataxia (gaze direction to command) all co-occur in Balint's syndrome. These syndromes may result from bilateral parieto-occipital lesions such as eclampsia, stroke, and posterior reversible encephalopathy syndrome (PRES) that fragment the ventral and dorsal visual streams that permit object detection and manipulation under visual guidance.

Field-Dependent Behavior Is Important for Both Diagnostic and Therapeutic Reasons

Frontal brain lesions, especially infarcts, frequently cause field-dependent behavior (FDB) syndromes that may be regarded as an uncoupling of the mirror neuron system (MNS). The MNS is integral to our theory of mind, social cognition, intentionalities, and language. Hence, it presents an example of hodological system impairment, involving the mirror neuron system [96]. FDB is commonly associated with frontal pathology, attributed at least in part to the very extensive fronto-parietal connectivity [97]. The diagnosis of FDB is important, as the loss of personal autonomy in a person actively employed with relatively intact cognition may be associated with dire consequences. The superior aspects (supplementary motor areas) of the frontal lobes are involved with more elementary forms of FDB

such as imitation behavior (IB) and utilization behavior (UB) and the inferior frontal lobe (especially the right orbitofrontal region) aspects, implicated with more complex FDB such as environmental dependency syndrome (EDS) [98, 99]. IB and UB predominate in the early post-lesion period, with EDS being an FDB that arises later [100–102]. Apart from the diagnostic component, the therapeutic aspects of the mirror neuron system may be exploited, for example, in the setting of severe motor-deficit rehabilitation. Action observation engages similar circuits as if performing the action. Mirror visual feedback therapy is effective in stroke, phantom limb pain, and neuropsychiatric conditions [103, 104]. In addition the rehabilitative power of the MNS is supported by positive trials with gestural therapy and expressive aphasia [105].

Geschwind Gastaut Syndrome

So-called silent brain lesions, mostly those involving frontal lobes, temporal, and right parietal regions, are tertiary association cortical regions that present with more covert syndromes. Geschwind Gastaut syndrome (GGS) is one such entity defined by (1) a viscous personality, (2) metaphysical preoccupation, and (3) altered physiological drives. The "viscous personality" is regarded as the most fundamental component with circumstantiality in speech, over inclusive verbal discourse, excessively detailed information, prolongation of interpersonal encounters, hypergraphia, excessive drawing, or painting. Metaphysical preoccupations include intemperate moral and intellectual interests in religion and philosophy. Alterations in physiological drives such as hyposexuality, aggression, and fear constitute yet another facet to the syndrome [106]. Described to date, only in association with interictal epileptic syndromes and temporal lobe intracerebral hemorrhages, this syndrome allows a unique window into the diverse and manifold functions of the temporal lobes, particularly the right temporal lobe [107, 108]. The principal components of the GGS may reflect the assembly of brain circuitry development of the more recent human "cultural explosion" evolutionary development wherein thoughts of the afterlife, afforded by the enhancement of episodic memory and mental time travel, blossomed [109]. More importantly other diverse human cognitive developments (visual artistry, spiritual processing) are suggested by the unraveling of this part of the brain. It indicates how behavioral neurology contributes in a unique manner and how the assembly of the brain may have proceeded, something not amenable by fossilized evidence or interpretation of the neuroarcheological record.

Social and Empathic Dysfunction After Uncinate Fasciculus Lesions

Several large-scale fiber tracts, such as the arcuate fasciculus and the uncinate fasciculus (UF), have expanded markedly in hominin evolution (Fig. 2.11). The UF is a particularly extensive association fiber connecting the orbitofrontal cor-

tex (OFC), which mediates inhibitory control, to the anterior temporal lobe. This large tract is especially prone to shearing injury in traumatic brain injury (TBI), together with a propensity of the inferior frontal lobe and anterior temporal lobe cortical damage in TBI [110, 111]. The uncinate fasciculus is frequently impacted in common neurological disorders such as stroke, multiple sclerosis, and frontotemporal lobe syndromes. Reaching maturity in only third and fourth decades, this also renders it susceptible to a range of neuropsychiatric conditions in the young adult population [112].

The UF mediates choice, based on information gleaned from social and emotional input as well as episodic memory. The anterior temporal pole is a hub for memory storage of person-specific memory, theory of mind, and social memories. This readily explains the delusional misidentification syndromes (DMIS) where emotions and memory are disconnected. For example, a familiar person

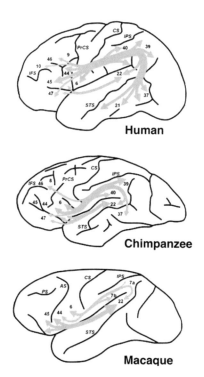

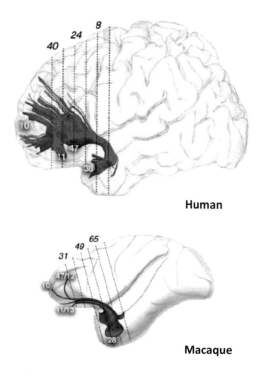

Fig. 2.11 White matter tract expansion: arcuate fasciculus and uncinate fasciculus evolution: Comparative human and primate analyses. Figure with permission: Rilling JK, Glasser MF, Preuss TM et al. Evolution of the arcuate fasciculus revealed with comparative DTI. Nature Neuroscience 2008;11:426–428 and Thiebaut de Schotten M, Dell'Acqua F, Valabregue R, Catani M. Monkey to human comparative anatomy of the frontal lobe association tracts. Cortex 2012; 48: 82–96

(Capgras syndrome) is regarded as a stranger yet the emotional reaction remains intact and vice versa (Fregoli's syndrome). A disruption of tracts between face-processing regions and emotive (limbic) areas allows one to perceive a familiar face but which is devoid of the expected emotional valence, and hence the person concludes that the "familiar" person must be a stranger or imposter [113]. Core UF-related functions pertain to episodic memory, linguistic, and social emotional domains. Others include generalized anxiety, schizophrenia, uncinate fits, and forme fruste Kluver Bucy syndromes [114]. Disruption of this tract due to relatively common diseases reveals features of the assembly of the social and emotional circuitry of the human brain.

Cerebellar Cognitive Affective Syndromes

Modern human brain size was substantiated by ~200 kya; yet archeological evidence supports modern cerebellum size only at ~28 kya. A relatively recent escalation in corticocerebellar connectivity occurred, particularly with the prefrontal cortex during the late Pleistocene period (2.5 mya to 12 kya). This has been correlated with improved cognitive efficiency, most likely mediated through enhanced working memory [44]. The clinical evidence of the cerebellar cognitive affective syndrome and cerebellar cognitive dysmetria attests to such neurobiological evolutionary developments. Neuroimaging evidence provides additional support with respect to the corticocerebellar, crossed diaschisis features, as well as the facial nucleus hypertrophy and corticobulbar tracts [115, 116].

Apraxia, Biface Technology, and Association with Language Evolution

Converging archeological, extant primate studies, human clinical and functional imaging points to a commonality and overlapping networks between language, praxis, and tool use. This leads to the supposition that the former may well have evolved from the praxis network. Both networks have an extensive left hemisphere circuitry. Gestural communication may have constituted an intervening state between stone tool production and the evolution of language by virtue of action recognition and imitation components [117]. Both praxis and language networks have key hub areas in the ventral premotor, superior temporal, and inferior temporal cortices, implying that they have commonality with respect to functional processes. Clinical evidence has noted the frequent appearance of both aphasias and apraxias with left hemisphere lesions. Language architecture in terms of syntax, semantic, and lexicon may also share neural architecture with praxis circuits. Broca's area (which contains motor programs that orchestrate speech muscles), the inferior parietal lobe (links concepts and words cross modally), and the superior temporal lobe (semantics) are linked by the mirror neuron circuitry. The cross-modal linking of visual (images), auditory (sounds), and motor maps thus occurred with synkinesis (movements of hands influencing mouth movements) located anteriorly and synesthesia in the posterior brain regions. Synkinesis may have been the mechanism whereby gestural protolanguage transformed into spoken language. The linking of these three principal regions together is the underlying concept of the synesthetic boostrapping hypothesis, proposed by Ramachandran [118].

The adjacent supramarginal gyrus is concerned with the production and imitation of skilled, complex movements. The origins date to the exquisite interaction between muscle and proprioception sense, together with vision, required for traversing the high canopies of the arboreal environment in early primates. A cross-modal abstraction developed between the branch visual image and the hand and arm proprioceptive, tactile muscle coordination. This enabled efficient and safe swinging through the trees and the basis of hand–eye coordination, the ancestral function associated with the supramarginal gyrus (SMG). When damaged, various forms of apraxia arise, which may be regarded as a mirror neuron system defect. A possible gene duplication of this region may have led to the angular gyrus (AG) as part of the inferior parietal lobe (IPL) which then became exapted for a different, cross-modal abstraction for word-object and concept linking. Clinical deficits of this region manifest with anomias and metaphor interpretation difficulty that may be tested with proverbs, for example [119].

The neurobiological basis of syntax (sentence construction) and its vast variability and meanings may be conceived of as an exaption of the basic stone knapping techniques. Using a stone tool, in the right hand of the knapper (subject), to strike (verb) another stone (object), usually held in the left hand, may constitute the basis of language syntax. Broca's region, already involved with motor programming, may also have been the subject of gene duplication evolutionary processes, with the new area becoming the new syntax area and no longer concerned with motor sequencing that was first used for tool manufacture and connected to the IPL (principally the supramarginal gyrus). This allowed hierarchical, multimodal tool assembly. The proposed duplication of this region then became concerned with syntactic language ability and linked to the MNS circuitry of the superior temporal lobe semantic abilities, Broca's area, and the inferior parietal lobe [120].

Diagnostic and Treatment Implications

From the foregoing it may be surmised that certain core cognitive domains should be tested, no matter what disease process is present. These include working memory, attention, executive function, and inhibition control. These are amenable to simple and relatively quick bedside testing [121]. Understanding these principles of the more recently derived "cultural circuits," mechanistically supports can drive the testing of new therapies such as melodic intonation therapy (MIT), gestural therapy in aphasia, and mirror visual feedback therapy (MVFB) for chronic pain and weakness. The diaschisis syndromes with hyper- and hypo-hodological effects present an opportunity for modulation and also a surrogate marker of restoration and improvement by

TMS and t-DCS [122]. These techniques use the diaschisis concepts to ameliorate deleterious excitatory and inhibitory effects. Enlisting more "ancient circuitry" for ameliorating brain disease such as music and visual art therapies (developed late in human evolution ~50 kya) seems promising in a variety of conditions that include hemiparesis, Parkinson's, and emotional disorders. Burgeoning interest and a surge in studies that modulate brain function in response to behavioral techniques such as meditation, cognitive behavioral therapies, and various art therapies have shown success in PTSD, TBI, Parkinson's disease, and dementias [123–125]. Meditation "brain builds" with white matter connectivity to the anterior cingulate gyrus increasing in efficiency and integrity as measured by diffusion tractography fractional anisotropy (FA) value increases [123]. In depression, increased activity in the amygdala and decreased activity in the PFC by f-MRI imaging may turn out to be a signature neuroimaging finding. This may be reversed by CBT [126]. For depression therapy, cognitive behavioral therapy (CBT) boasts success rates of 42–66 % versus pharmacotherapy results of 22–40 % [127]. A PET study revealed elevated right anterior insula activity, which responds better to drugs, and those with underactive insula responded better to CBT [125]. Furthermore, newer resting-state network imaging reveals abnormalities with depression and antidepressants, and shows promise in normalizing these networks, for example the default mode network [128].

Philosophical Outlook: Future Brain Development

During mammalian evolution, a shift from perceptual processing, that was predominantly chemosensory, to the primate visual predominant faculty occurred. The profound social challenges of the polyadic relationships that particularly encumber primates are due to the manifold social cues that are beset with ambiguity and variability. This places a premium on social cognitive computation. The departure from simple perception to the much more complex deductive-type thought processing has been conceptualized in the Inferential Brain Hypothesis [129] and the Reinterpretation Hypothesis [130], both of which expound the hominin capability of making deductions and extrapolating information from the environment. The proposed neurobiological circuitry of both includes the ventromedial PFC and amygdala that are integral to both chemosensory and social operations [131].

The two-way plasticity process between the brain and the environment, including of objects and materials, has yielded insights into treatment of neurodegenerative diseases. Delaying the onset and progression of dementias, for example by "environmental enrichment," is one example. The material engagement theory and Blind Man's Stick Hypothesis proposed by Malafouris are attempts to explore the sweeping and perspicacious interface between the human mind and the artifacts and materials that we shape and that in turn shape our brain. These cognitive prostheses or cognitive scaffolds indicate that the brain may be housed within the skull but the mind extends out into the material world and interacts with it in a two-way plasticity process termed metaplasticity [132, 133]. Gaining insights as to how the neurobiological circuitry evolved helps us to better understand the nature of our behavioral syndromes, their presentations, and their potential treatment.

However we may well be at the limits of both physical brain size and physiological function based on physics laws that impose restrictions on interneuronal communication and processing ability [134]. Cognitive performance correlates with brain network efficiency and resting-state networks such as the default mode network [135]. Appropriate optimism is in order here however, as the integrity of our networks can be improved by adhering to brain health regimens, and the more precise measurement of these networks can be accomplished by the various resting-state network evaluations and mathematical network analysis. The overall trend in the course of human evolution has been one of spiraling networking, first of our own brain (intracortical connectivity) and then to the interaction with other minds (intercortical connectivity). We are now in a brain-machine interface that continually expands this connectivity further. Even brain-to-brain communication, devoid of traditional sensory or motor networks, has been achieved [136]. For the practicing clinician today these findings underscore the important need to test for the core functions of attention, executive function, and working memory that reflect the chief hub failure manifestations, all of which are amenable to intervention.

References

1. Nesse RM, Dawkins R. Evolution: medicine's most basic science. In: Warrell DA, Cox TM, Firth JD, Benz JJ, editors. Oxford textbook of medicine. 5th ed. Oxford: Oxford University Press; 2010.
2. Brown-Sequard CE. Séance du 18 decembre. CR Soc Biol. 1875;424.
3. Lieberman P. Synapses, language and being human. Science. 2013;342:944–5.
4. Semendeferi K, Lu A, Schenker N, Damasio H. Humans and great apes share a large frontal cortex. Nat Neurosci. 2002;5:272–6.
5. Zilles K. Evolution of the human brain and comparative cyto and receptor architecture. In: Dehaene S, Duhamel JR, Hauser MD, Rizzolatti G, editors. From money brain to human brain. Cambridge, MA: Fyssen Foundation. MIT; 2005.
6. Elston GN, Benavides-Piccione R, Elston A, Zietsch B, Defelipe J, Manger P, et al. Specializations of the granular prefrontal cortex of primates: implications for cognitive processing. Anat Rec A Discov Mol Cell Evol Biol. 2006;288:26–35.
7. Bradley RS. Paleoclimatology. Reconstructing climates of the quarternary. 3rd ed. Amsterdam: Elsevier; 2015. and [Stow D. Vanished ocean. How Tethys reshaped the World. Oxford University Press, Oxford, 2010].
8. Schultz S, Nelson E, Dunbar RIM. Hominin cognitive evolution: identifying patterns and processes in the fossil and archaeological record. Philos Trans R Soc B. 2012;367:2130–40.
9. Kuhlbrodt T, Griesel A, Montoya M, Levermann A, Hofmann M, Rahmstorf S. On the driving processes of the Atlantic meridional overturning circulation. Rev Geophys. 2007;45, RG2001. doi:10.1029/200 4RG000166.
10. Raymo ME, Ruddimen WF, Froelich PN. Influence of the late Cenozoic mountain building on ocean geochemical cycles. Geology. 1988;16:649–53.
11. Klein RG. The human career. Human biological and cultural origins. 3rd ed. Chicago, IL: University of Chicago Press; 2009.

12. Bookstein F, Schaefter K, Prossinger H, Seidler H, Fieder M, Stringer C, et al. Comparing frontal cranial profiles in archaic and modern homo by morphometric analysis. Anat Rec (New Anat). 1999;257:217–24.

13. Mcbrearty S, Brooks AS. The revolution that wasn't: a new interpretation of the origin of modern human behavior. J Hum Evol. 2000;39(5):453–563.

14. Passingham RE, Wise SP. The neurobiology of the prefrontal cortex. Oxford: Oxford University Press; 2012.

15. Christoff K, Gabrieli JDE. The frontopolar cortex and human cognition: evidence for a rostrocaudal hierarchical organization within the human prefrontal cortex. Psychobiology. 2000;28:168–86.

16. Semendeferi K, Armstrong E, Schleicher A, Zilles K, Van Hoesen GW. Prefrontal cortex in humans and apes: a comparative study of area 10. Am J Phys Anthropol. 2001;114:224–41.

17. Jacobs B, Schall M, Prather M, Kapler E, Driscoll L, Baca S, et al. Regional dendritic and spine variation in human cerebral cortex: a quantitative Golgi study. Cereb Cortex. 2001;11:558–71.

18. Genovesio A, Wise SP, Passingham RE. Prefrontal-parietal function: from foraging to foresight. Trends Cogn Sci. 2014;18(2):72–81.

19. Wynn T. Archeology and cognitive evolution. Behav Brain Sci. 2002;25:389–438.

20. Elston GN. Cortex, cognition and the cell: new insights into the pyramidal neuron and prefrontal function. Cereb Cortex. 2003;13:1124–38.

21. Fuster JM. The prefrontal cortex: anatomy, physiology, and neuropsychology of the frontal lobe. Philadelphia, PA: Lippincott-Raven; 1997.

22. Goldman-Rakic PS. The prefrontal landscape: implications for functional architecture for understanding human mentation and the central executive. Philos Trans R Soc Lond Ser B. 1996;351:1445–53.

23. Maslin MA, Shultz S, Trauth MH. A synthesis of the theories and concepts of early human evolution. Philos Trans R Soc B. 2015;370:20140064.

24. Morgan TJ, Uomini NT, Rendell LE, Chouinard-Thuly L, Street SE, Lewis HM, et al. Experimental evidence for the co-evolution of hominin tool-making teaching and language. Nat Commun. 2015;6:6029. doi:10.1038/ncomms7029.

25. Arbib M. From mirror neurons to complex imitation in the evolution of language and tool use. Ann Rev Anthropol. 2011;40:257–73.

26. Klein RG. The human career. 3rd ed. London: University of Chicago Press; 2009.

27. Rowe JB, Owen AM, Johnsrude IS, Passingham RE. Imaging the mental components of a planning task. Neuropsychologia. 2001;39:315–27.

28. Renfrew C, Frith C, Malafouris L. The sapient mind. Archeology meets neuroscience. New York, NY: Oxford University Press; 2009.

29. Marean CW. Pinnacle point cave 13B (Western Cape Province, South Africa) in context: the cape floral kingdom, shellfish, and modern human origins. J Hum Evol. 2010;59(3-4):425–43. doi:10.1016/j.jhevol.2010.07.011.

30. Marean C. When the sea saved humanity. Sci Am. 2010;303:54–61.

31. Wadley L. Announcing a still bay industry at Sibudu Cave, South Afria. J Hum Evol. 2007;52:681–9.

32. Schlebusch CM, Lombard M, Soodyall H. MtDNA control region variation affirms diversity and deep sub-structure in populations from southern Africa. BMC Evol Biol. 2013;13:1–20.

33. Rito T, Richards MB, Fernandes V, Alshamali F, Cerny V, Pereira L, et al. The first modern human dispersals across Africa. PLoS One. 2013;8:e80031.

34. Pickrell JK, Patterson N, Barbieri C, Berthold F, Gerlach L, Güldemann T, et al. The genetic prehistory of southern Africa. Nat Commun. 2012. doi:10.1038/ncomms2140.

35. Henn BM, Cavalli-Sforza LL, Feldman MW. The great human expansion. Proc Natl Acad Sci U S A. 2012;109:17758–64.

36. Knight A, Underhill PA, Mortensen HM, Zhivotovsky LA, Lin AA, Henn BM, et al. African Y chromosome and mtDNA divergence provides insight into the history of click languages. Curr Biol. 2003;13:464–73.

37. Tishkoff SA, Gonder MK, Henn BM, Mortensen H, Knight A, Gignoux C, et al. History of click speaking populations of Africa inferred from mtDNA and Y chromosome genetic variation. Mol Biol Evol. 2007;24:2180–95.

38. Testa-Silva G, Verhoog MB, Linaro D, de Kock CPJ, Baayen JC, Meredith RM, et al. High bandwidth synaptic communication and frequency tracking in human neocortex. PLoS Biol. 2014;12(11):1002007. doi:10.1371/journal.pbio.1002007.

39. Crawford MA, Broadhurst CL, Guest M, Nagar M, Wang Y, Ghebremeskel K, et al. A quantum theory for the irreplaceable role of docosahexanoic acid in neural signaling throughout evolution. Prostaglandins Leukot Essent Fatty Acids. 2013;88:5–13.

40. Crawford MA. Docosahexaenoic acid in neural signaling systems. Nutr Health. 2006;18(3):263–76.

41. Bazinet RP, Laye S. Polyunsaturated fatty acids and their metabolites in brain function and disease. Nat Rev Neurosci. 2014;15:771–85.

42. Ramsden CE, Faurot KR, Zamora D, Suchindran CM, Macintosh BA, Gaylord S, et al. Targeted alteration of dietary n-3 and n-6 fatty acids for the treatment of chronic headaches: a randomized trial. Pain. 2013;154(11):2441–51. doi:10.1016/j.pain.2013.07.028.

43. Elston GN. Pyramidal cells of the frontal lobe: all the more spinous to think with. J Neurosci. 2000;20:RC95.

44. Weaver AH. Reciprocal evolution of the cerebellum and neocortex in fossil humans. Proc Natl Acad Sci U S A. 2005;102:3576–80.

45. Vandervert LR, Schimpf PH, Liu H. How working memory and the cerebellum collaborate to produce creativity and innovation. Creat Res J. 2007;19:1–18.

46. DeSmet HJ, Paquir P, Verhoeven J, Mariën P. The cerebellum: its role in language and related cognitive and affective functions. Brain Lang. 2013;127:334–42.

47. Bruner E. Geometric morphometrics and pale-neurology: brain shape evolution in the genus Homo. J Hum Evol. 2004;47:279–303.

48. Bruner E, Manzi G, Arsuaga JL. Encephalization and allometric trajectories in the genus Homo: evidence from the Neandertal and modern lineages. Proc Natl Acad Sci U S A. 2003;100:15335–40.

49. Bruner E, De La Cuétara JM, Holloway R. A bivariate approach to the variation of the parietal curvature in the genus homo. Anat Rec. 2011;294:1548–56.

50. Orban GA, Caruana F. The neural basis of human tool use. Front Psychol. 2014;5:310. doi:10.3389/fpsyg.2014.00310.

51. Bruner E. The evolution of the parietal cortical areas in the human genus: between structure and cognition. In: Broadfield D, Yuan M, Schick K, Toth N, editors. The human brain evolving. Gosport, IN: Stone Age Instiute Press; 2010.

52. Overmann KA. Finger counting in the upper Paleolithic. Rock Art Res. 2014;31:63–80.

53. Coolidge FN, Overmann KA. Numerosity, abstraction and the emergence of symboli thinking. Curr Anthropol. 2012;53:204–25.

54. Cavanna AE, Trimble MR. The precuneus: a review of its functional anatomy and behavioural correlates. Brain. 2006;129:564–83.

55. Dunbar RIM. Neocortex size as a constraint on group size in primates. J Hum Evol. 1992;20:469–93.

56. Barrett L, Henzi SP, Dunbar RIM. Primate cognition: from "what now" to "what if?". Trends Cogn Sci. 2003;7:494–7.

57. Richards MP, Pettit RB, Stiner MC, Trinkaus E. Stable isotope evidence for increasing dietary breadth in the European Mid-Upper Palaeolithic. Proc Natl Acad Sci U S A. 2001;98:6528–32.

58. Pearce E, Shuttleworth A, Grove M, Layton RH. The costs of being a high-latitude hominin. In: Dunbar RIM, Gamble C, Gowlett JAJ, editors. Lucy to language. Oxford: Oxford University Press; 2014.

59. Grove M. Amplitudes of orbitally induced climate cycles and patterns of hominin speciation. J Archael Sci. 2012;39:3085–94.

60. Pearce E, Bridge H. Is orbital volume associated with eyeball and visual cortical volume in humans? Ann Hum Biol. 2013;40(6):531–40.

61. Balzeau A, Holloway RL, Grimaud-Herve D. Variations and asymmetries in regional brain surface in the genus Homo. J Hum Evol. 2012;62:696–706.

62. Pearce E, Stringer C, Dunbar RIM. New insights into differences in brain organization between Neanderthals and anatomically modern humans. Proc R Soc Lond. 2013;280:20130168. doi:10.1098/rspb.2013.0168.

63. Powell JL, Lewis PA, Dunbar RIM, Garcia-Finana M, Roberts M. Orbital prefrontal cortex volume correlates with social cognitive competence. Neuropsychologica. 2010;48:3554–62.

64. Powell JL, Lewis PA, Roberts M, Dunbar RIM. Orbitofrontal prefrontal cortex volume predicts social network size: an imaging study of individual differences in humans. Proc R Soc Lond B. 2012;279:2157–62.

65. Pearce E. The effects of latitude on hominin social network maintenance. D. Phil. thesis, Department of Anthropology, University of Oxford; 2013.

66. McNabb J. The importance of conveying visual information in Archeulean society. The background to the visual display hypothesis. Hum Evol. 2011;1:1–23.

67. Dunbar RIM, Gamble C, Gowlett JAJ. Lucy to language. Oxford: Oxford University Press; 2014.

68. Read D, van der Leeuw S. Biology is only part of the story. In: Renfrew C, Frith C, Malafouris L, editors. The sapient mind. Archeology meets neuroscience. New York, NY: Oxford University Press; 2009.

69. Diamond A, Doar B. The performance of human infants on a measure of frontal cortex function, the delayed response task. Dev Psychobiol. 1989;22:271–94.

70. Carey N. The epigenetics revolution. New York, NY: Columbia University Press; 2012.

71. Squires L. Gliotransmission. In: Brady ST, Siegel GJ, Albers RW, Price DL, editors. Basic neurochemistry. 8th ed. New York, NY: Elsevier; 2012. p. 136–7.

72. Coolidge FL, Wynn T. Cognitive prerequisites for a language of diplomacy. In: Tallerman M, Gibson KR, editors. The Oxford companion to language evolution. Oxford: Oxford University Press; 2012.

73. Bruner E. Morphological differences in the parietal lobes within the human genus: a neurofunctional perspective. Curr Anthropol. 2010;51:S77–88.

74. Addis DR, Wond AT, Schacter DL. Remembering the past and imagining the future: common and distinct neural substrates during event e construction and elaboration. Neuropsychologica. 2007;45:1363–77.

75. Land MF. Do we have an internal model of the outside world? Philos Trans R Soc B. 2014;369:20130045.

76. Klein RG. The human career. Chicago, IL: Chicago University Press; 2009.

77. Smaers JB, Steele J, Caseb CR, Cowper A, Amunts K, Zilles K. Primate prefrontal cortex evolution: human brains are the extreme of a lateralized Ape trend. Brain Behav Evol. 2011;77:67–78.

78. Herculano-Hounzel S. The human brain in numbers: a linearly scaled up primate brain. Front Hum Neurosci. 2009;9:31.

79. Zhang K, Sejnowski TJ. A universal scaling law between gray matter and white matter of cerebral cortex. Proc Natl Acad Sci U S A. 2000;97:5621–6.

80. Smaers JB, Schleicher A, Zilles K, Vinicius L. Frontal white matter volume is associated with brain enlargement and higher structural connectivity in anthropoid primates. PLoS One. 2010;5:1–6.e9123.

81. Schoenemann PT, Sheehan MJ, Glotzer LD. Prefrontal white matter volume is disproportionately larger in humans than in other primates. Nat Neurosci. 2005;8(2):242–52.

82. Campo P, Garrido MI, Moran RJ, Maestu F, Garcia-Morales I, Gil-Nagel A, et al. Remote effects of hippocampal sclerosis on effective connectivity during working memory encoding: a case of connectional diaschisis? Cereb Cortex. 2012;22:1225–36.

83. Carrera E, Tononi G. Diaschisis: past, present and future. Brain. 2014;137:2408–22.

84. Stam CJ. Modern network science of neurological disorders. Nat Rev Neurosci. 2014;15:683–95.

85. Fornito A, Zalesky A, Breakspear M. The connectomics of brain disorders. Nat Rev Neurosci. 2015;16:159–72.

86. Crossley NA, Mechelli A, Scott J, Fox PT, McGuire P, et al. The hubs of the human connectome are generally implicated in the anatomy of brain disorders. Brain. 2014;137:2382–95.

87. Liu J, Zhao L, Li G, Xiong S, Nan J, Li J, et al. Hierarchical alteration of brain structural and functional networks in female migraine sufferers. PLoS One. 2012;7:51250.

88. Buckner RL, Sepulchre J, Talukdar T, Krienen FM, Liu H, Hedden T, et al. Cortical hubs revealed by intrinsic functional connectivity: mapping, assessment of stability and relation to Alzheimer's disease. J Neurosci. 2009;29:1860–73.

89. Saxena S, Caroni P. Selective neuronal vulnerability in neurodegenerative disease: from stressor thresholds to degeneration. Neuron. 2011;71:35–48.

90. Han K, Mac Donald CL, Johnson AM, Barnes Y, Wierzechowski L, Zonies D, et al. Disrupted modular organization of resting state cortical functional connectivity in US military personnel following concussive 'mild' blast related traumatic brain injury. Neuroimage. 2014;84:76–96.

91. Archard S, Delon-Martin C, Vértes PE, Renard F, Schenck M, Schneider F, et al. Hubs of brain functional networks are radically reorganized in comatose patients. Proc Natl Acad Sci U S A. 2012;109:20608–13.

92. Rosanova M, Gosseries O, Casarotto S, Boly M, Casali AG, Bruno MA, et al. Recovery of cortical effective connectivity and recovery of consciousness in vegetative patients. Brain. 2012;135:1308–20.

93. Bruner E, Jacobs HI. Alzheimer's disease: the downside of a highly evolved parietal lobe? J Alzheimers Dis. 2013;35(2):227–40.

94. Caverzasi E, Papinutto N, Amirbekian B, Berger MS, Henry RG. Q-Ball of inferior fronto-occipital fasciculus and beyond. PLoS One. 2014;9(6):100274. doi:10.1371/journal.pone.0100274.

95. Stout D, Passingham R, Frith C, Apel J, Chaminade T. Technology, expertise and social cognition in human evolution. Eur J Neurosci. 2011;33:1328–38.

96. Rizzolatti G, Fabbri-Destro M, Cattaneo L. Mirror neurons and their clinical relevance. Nat Clin Pract Neurol. 2009;5:24–34.

97. Lhermitte F. Human autonomy and the frontal lobes. Part II: patient behavior in complex and social situations: the "environmental dependency syndrome". Ann Neurol. 1986;19:335–43.

98. Besnard J, Allain P, Aubin G, Chauviré V, Etcharry-Bouyx F, Le Gall D. A contribution to the study of environmental dependency phenomena: the social hypothesis. Neuropsychologia. 2011;49:3279–94.

99. Bien N, Roebuck A, Goebel R, Sack AT. The brain's intention to imitate: the neurobiology of intentional

versus automatic imitation. Cereb Cortex. 2009;19:2338–51.

100. Ragno Paquier C, Assal F. A case of oral spelling behavior: another environmental dependency syndrome. Cogn Behav Neurol. 2007;20:235–7.

101. Volle E, Beato R, Levy R, Dubois B. Forced collectionism after orbitofrontal damage. Neurology. 2002;58:488–90.

102. Shin JS, Kim MS, Kim NS, Kim GH, Seo SW, Kim EJ, et al. Excessive TV watching in patients with frontotemporal dementia. Neurocase. 2013;19(5):489–96. doi:10.1080/13554794.2012.701638.

103. Garrison KA, Winstein CJ, Aziz-Zadeh L. The mirror neuron system: a neural substrate for methods in stroke rehabilitation. Neurorehabil Neural Repair. 2010;24:404–12.

104. Yavuzer G, Selles R, Sezer N, Sütbeyaz S, Bussmann JB, Köseoğlu F, et al. Mirror therapy improves hand function in subacute stroke: a randomized controlled trial. Arch Phys Med Rehabil. 2008;89:393–8.

105. Hanlon Brown RE, Brown JW, Gerstman LJ. Enhancement of naming in nonfluent aphasia through gesture. Brain Lang. 1990;38:298–314.

106. Bear DM, Fedio P. Quantitative analysis of interictal behavior in temporal lobe epilepsy. Arch Neurol. 1977;34:454.

107. Trimble M, Mendez MF, Cummings JL. Neuropsychiatric symptoms from the temporolimbic lobes. J Neuropsychiatry Clin Neurosci. 1997;9:429–38.

108. Hoffmann M. Isolated right temporal lobe stroke patients present with Geschwind Gastaut syndrome, frontal network syndrome and delusional misidentification syndromes. Behav Neurol. 2009;20:83–9.

109. Suddendorf T, Corballis M. The evolution of foresight: what is mental time travel and is it unique to humans. Behav Brain Sci. 2007;30:299–313.

110. Seo JP, Kim OL, Kim SH, Chang MC, Kim MS, Son SM, et al. Neural injury of uncinate fasciculus in patients with diffuse axonal injury. NeuroRehabilitation. 2012;30:323–8.

111. Ewing-Cobbs L, Prasad MR. Outcome after abusive head injury. In: Jenny C, editor. Child abuse and neglect: diagnosis, treatment, and evidence. St. Louis, MS: Elsevier Saunders, Inc; 2011.

112. Paus T, Keshavan M, Giedd JN. Why do many psychiatric disorders emerge during adolescence? Nat Rev Neurosci. 2008;9:947–57.

113. Hirstein W, Ramachandran VS. Capgras syndrome: a novel probe for understanding the neural representation of the identity and familiarity of persons. Proc R Soc B. 1997;264:437–44.

114. Von Der Heide R, Skipper LM, Kobusicky E, Olsen IR. Dissecting the uncinate fasciculus: disorders, controversies and a hypothesis. Brain. 2013;136:1692–707.

115. Ramnani N. The primate cortico-cerebellar system: anatomy and function. Nat Rev Neurosci. 2006;7:511–22.

116. Kamali A, Kramer LA, Frye RE, Butler IJ, Hasan KM. Diffusion tensor tractography of the human brain

cortico-ponto-cerebellar pathways: a quantitative preliminary study. J Magn Reson Imaging. 2010;32: 809–17.

117. Roby Brami A, Hermsdoerfer J, Roy AC, Jacobs S. A neuropsychological perspective on the link between language and praxis in modern humans. Philos Trans R Soc B. 2012;367:144–60.

118. Ramachandran VS. The tell tale brain. New York, NY: WW Norton and Company; 2011.

119. McGeoch PD, Brang D, Ramachandran VS. Apraxia, metaphor and mirror neurons. Med Hypotheses. 2007;69(6):1165–8.

120. Stout D, Toth N, Schick K, Chaminade T. Neural correlates of Early Stone Age toolmaking: technology, language and cognition in human evolution. Philos Trans R Soc B. 2008;363:1939–49.

121. Dubois B, Slachevsky A, Litvan I, Pillon B, The FAB. The frontal assessment battery at bedside. Neurology. 2000;55:1621–6.

122. Fregni F, Pascual-Leone A. Technology insight: non-invasive brain stimulation in neurology: perspectives on the therapeutic potential of rTMS and tDCS. Nat Clin Pract Neurol. 2007;3:383–93.

123. Tang Y-Y, Lu Q, Gen X, Stein EA, Yang Y, Posner MI. Short-term meditation induces white matter changes in the anterior cingulate. Proc Natl Acad Sci U S A. 2010;107:15649–52.

124. Koelsch S. Brain correlates of music-evoked emotions. Nat Rev Neurosci. 2014;15:170–80.

125. McGrath CL, Kelley ME, Holtzheimer PE, McGrath CL, Kelley ME, Holtzheimer PE. Toward a neuroimaging treatment selection biomarker for major depressive disorder. JAMA Psychiatry. 2013;70: 821–9.

126. Siegle GJ, Thompson W, Carter CS, Steinhauer SR, Thase ME. Biol Psychiatry. 2007;61(2):198–209.

127. DeRubeis RJ, Siegle GJ, Hollon SD. Cognitive therapy versus medication for depression: treatment outcomes and neural mechanisms. Nat Rev Neurosci. 2008;9:788–96.

128. Posner J, Hellerstein DJ, Gat I, Mechling A, Klahr K, Wang Z, et al. Antidepressants normalize the default mode network in patients with dysthymia. JAMA Psychiatry. 2013;70(4):373–82.

129. Koscik TR, Tranel D. Brain evolution and human neuropsychology: the inferential brain hypothesis. J Int Neuropsychol Soc. 2012;18:394–401.

130. Subiaul F, Barth J, Okamoto-Barth S, Povinelli DJ. Human cognitive specialization. In: Kaas J, editor. Evolution of nervous systems: a comprehensive review. New York, NY: Elsevier; 2007.

131. Gottfried JA, Zald DH. On the scent of human olfactory orbitofrontal cortex: metanalysis and comparison to non-human primates. Brain Res Rev. 2005;50:287–304.

132. Malafouris L. Beads for a plastic mind: the 'blind man's stick' (BMS) hypothesis and the active nature of material culture. Camb Archaeol J. 2008;18: 401–14.

133. Malafouris L. Between brains, bodies and things: tectonoetic awareness and the extended self. Philos Trans R Soc B. 2008;363:1993–2002.

134. Laughlin SB, Sejnowski TJ. Communication in neuronal networks. Science. 2003;301(5641):1870–4.

135. Van den Heuvel MP, Stam CJ, Kahn RS, Hulshoff Pol HE. Efficiency of functional brain networks and intellectual performance. J Neurosci. 2009;29: 7619–24.

136. Grau C, Ginhoux R, Riera A, Nguyen TL, Chauvat H, Berg M, et al. Conscious brain-to-brain communication humans using non invasive technologies. PLoS One. 2014;9:1–6.e105225.

Neurochemistry

3

The building blocks of all cells and the mechanisms by which they communicate with each other are central to the study of cognition. The complexity of the brain is readily apparent at the macroscale level and mesoscale level, with ~100 billion neurons and ~100,000 miles of connectivity fibers. However at the microscopic level there is an equally impressive intracellular and intercellular complexity. Cognitive and behavioral neurological syndromes may often be due to a chemical, a receptor, or other neurochemical perturbation. The brain functions as an electrochemical organ and many cognitive syndromes may be deciphered at the synaptic or intracellular level. A rudimentary knowledge of the basic cellular components, molecular constituents, neurotransmitters, receptors, and multiple levels of messenger systems is important in understanding and treating cognition. Neurochemistry is a mere subset of organic chemistry, which in turn is a part of chemistry in the wider context. A basic knowledge of neurochemistry and classification system of the various components is helpful in understanding many neurological and cognitive disorders such as seizures and autoimmune encephalopathy. This brief overview is not comprehensive and for more in-depth neurochemical insights the reader is referred to the excellent treatises by Brady, Nestler, Stahl, Harper, and Ganong that served as reference sources for this brief review [1–5].

Before listing the essential cellular, subcellular, and neurochemical constituents, a basic overview of the evolution of elements, minerals, organic chemistry, and neurochemistry is pertinent. The evolutionary origin of neurochemistry goes back over four billion years. From a chemically scientific point of view the process evolved from cosmochemistry to petrochemistry, geochemistry, organic chemistry, and finally neurochemistry (Fig. 3.1). The first stellar elements, hydrogen and helium, evolved to form the rest of the currently known heavier periodic table elements, the domain of cosmochemistry. Thereafter, in Earth evolutionary terms, it is petrochemistry, or the formation of rocks within the planetary system comprising of six main elements which make up 98 % of the Earth: silicon, oxygen, magnesium, calcium, aluminum, and iron. Of note is oxygen—over 99.9 % of oxygen is locked within the rocks as SiO_2. These six elements are also the components of Mercury, Venus, moon, and Mars [6].

The formation of the moon occurred early on in Earth's history, the first 70–110 million years with competing theories of origin. A giant impact occurring on a fast-spinning Earth is one possibility, also favored by the similarity in their isotopic composition [7, 8]. At this time the Earth day lasted about 5 h and subsequent to the origin of the moon, the Earth which initially revolved at a distance of about 15,000 miles as opposed to its current orbit about 239,000 miles from Earth. Consequences of this very much more proximate orbit were massive tides affecting the Earth. These did not involve oceans but the Earth's molten magma surface and estimates are that

M. Hoffmann, *Cognitive, Conative and Behavioral Neurology*,
DOI 10.1007/978-3-319-33181-2_3

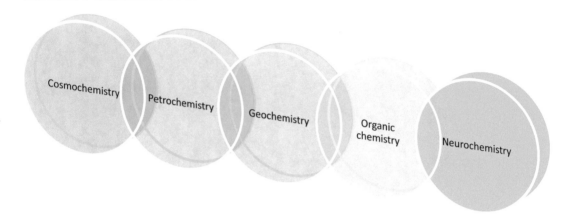

Fig. 3.1 The elements, minerals, organic molecules, and neurochemicals

they may have bulged about 1 mile outwards due to the moon's strong gravitational pull. The moon was very much larger and brighter moon appearing in the night sky which also featured intense volcanism during its earlier years [9, 10].

The phases of the Earth's evolution have been particularly well articulated by Hazen and a brief synopsis of his lucid account follows [6] (Fig. 3.2). During the black Earth, also called the Hadean period, about 4.4 billion years ago (bya), the Earth was reputedly black with a thin basalt veneer with volcanoes, meteors, and magma in a hostile-appearing world that had neither oxygen nor life enduring approximately 500 million years (Fig. 3.3). Hazen's theory of the co-evolution of the geosphere and biosphere posits that microbes were present in the seafloor of the Hadean-Archean Earth. In addition to being representative of the earliest living forms and the origin of life, these microbes also triggered mineral diversification by means of microbe fluid rock interfacing [11]. With gradual cooling the principal elements of the atmosphere, oxygen, nitrogen, sulfur, hydrogen, and carbon formed water, the oceans, and the blue earth. This is despite the fact that most of the Earth's water is held up in the mantle and core. The mantle transition zone, 250–400 miles deep, holds 9× the amount of the world's oceans and the lower mantle holds 16× the water of all oceans. Even more dramatic, the Earth's iron core has been estimated to store 80× the ocean's quantity of water. Zircon crystals last indefinitely and inherently tell time,

by virtue of containing uranium that has a half-life of ~4.5 billion years. The ratio of oxygen 18 to oxygen 16 within zircon is informative about the temperature during the formation of the crystal. The finding of 4.4 billion-year-old zirconium crystals in Western Australia, with relatively high oxygen 18 content indicating surface water interfacing, has yielded insightful evidence that about 150 million years after the Earth's formation there were oceans [12].

The gray Earth followed, due to largely granite formation and rise of the continents soon after the formation of the oceans. Some of the zircon crystals were also found to have quartz suggesting that granite crust formation occurred within the period ~4 bya. Granite formation required both surface basalt and the constant recycling of heat provided by the mantle and core. Granite is able to "float" upon the basalt layer being about 10 % less dense than basalt and the accumulation of layers of granite resulted in crust formation and some early mountain ranges such as the American Rockies. During this period a change from a basaltic crust dominated volcanoes to early granite crust formation first with scattered islands, eventually leading to large granite continents. This characterized the next billion years of what has been called vertical tectonics. Aided by asteroid collisions, these included volcanic ocean islands that arose from the oceans. This was followed later by the horizontal (lateral) plate tectonics characterized by mid-ocean ridge crust formation, moving continents,

Fig. 3.2 Evolution of the Earth. *bya* billion years ago, *mya* million years ago. Adapted from Hazen RM. The Story of the Earth Penguin 2012, New York

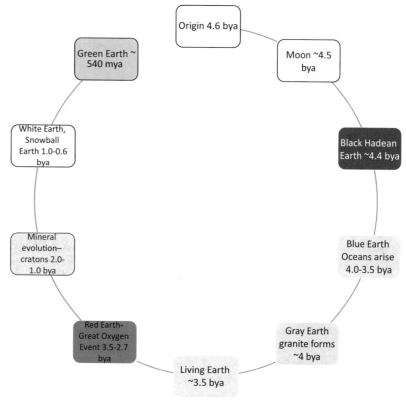

Fig. 3.3 Artistic rendition of likely appearance of the Hadean (*black*) Earth; giant moon, meteorites, and active volcanoes

and even higher mountain formations such as the Himalayas [6, 13].

Next came the living Earth with molecular evolution leading to the intricate processes of metabolism and genetics that together enabled perpetuation of these living forms during the period of 4.0–3.5 bya. First came the formation of RNA strands, which were able to fold and replicate. Organic chemical molecules first formed within deep sea hydrothermal origins in white and black smokers. Hazen proposed a co-evolution of the geosphere and biosphere whereby minerals change life and life forms change minerals. Rock cycles are not separate processes from the life cycles [14]. Thanks to the discovery of deep sea black smokers during the 1980s, this heralded the alternative and more plausible theory of life origins, first proposed by Miller in 1953—the primordial soup theory [15]. Mineral-rich fluids combine hot volcanic crust from ocean floor geysers with the hot water jets and ice-cold ocean water allowing mineral precipitation within which organic chemicals formed [16]. These hydrothermal vents (submarine hot springs) occur along mid-ocean ridges yielding mineral-rich waters into the ocean. Despite the absence of sunlight complex marine life accumulates here including clams, barnacles and anemones, limpets, and worms, all fueled by chemosynthesis [17]. Unicellular organisms appeared at 3.5 (bya) with the prokaryotes containing circular, anuclear free-floating DNA with 64 codons and 20 amino acids for protein assembly. These prokaryotes could aggregate into amorphous blue green algae and slime molds capable of cell-to-cell signaling using cyclic AMP ubiquitous in animal and plant cells today [18]. More complex cells evolved and subsequently multicellular life forms (metazoans) and eukaryote cells appeared ~1.7 bya. This symbiotic process involved the acquisition of mitochondria powerhouses which generated ATP. Eukaryotic cells were able to change their shape and allow movement in response to environmental stimuli. The basic cell machinery including ion channels, neurotransmitters, and receptors shared by all plant and animal life forms had evolved by this time [19].

Next came the red Earth, characterizing the first great oxygenation event. The photosynthetic organisms had secured the oxygenation of the atmosphere that occurred about 2.5 bya with oxygen rising to about 1 % of its current level. Dating rock strata to between 2.5 and 1.8 bya, extensive red and black iron oxide deposits have been found representing most of the world's iron ore deposits and manganese oxide deposits. It was during this time that the mineral explosion occurred. In Hazen's view, about 2/3 of the approximately four and a half thousand minerals formed as a consequence of the oxygen events underscoring the claim that most minerals on Earth are due to the evolution of life. The oxygen, associated weathering of iron containing granite and basalt turned into red soils, with the surface transforming from gray to rusty red and Earth would have resembled the red planet Mars at about 2 bya [11].

The term "Boring Billion" refers to the time of relatively less dramatic life and geological changes, but there was nevertheless change. This included the aggregation of all the original land mass on the Earth, into one big supercontinent that later broke up, as well as new rock formation. The first of these supercontinents was named Vaalbara about 3.3 bya, followed by Ur at about 3.1 bya; the earliest supercontinent was Kenorland 2.7 bya, followed by the supercraton Laurentia and then Rodinia during this time period. Cratons were large accumulations of rock that were destined to become the continents and moved around due to lateral plate tectonics [6].

At least three separate episodes of white Earth or snowball Earth with alternating periods of global ice formation with ice reaching all the way to the equators have been postulated. The global climate cycled between arctic and tropical (hothouse) cycles during the Sturtian glaciation 720 mya, Marinoan 650 mya, and Gaskiers 580 mya [20].

The time period of Green Earth at ~1.5 bya was associated with the supercontinent Rodinia rifting into a southern component called Gondwana and a northern Laurentia (current-day Greenland, North America, Europe). Subsequent tectonic movements caused a rearrangement into a supercontinent Pangea and superocean Panthalassa.

During this period the animal explosion occurred at about 530 mya, with early animals evolving, and soon thereafter the elaboration of protective shells evolved with subsequent specialization into armor, spines, claws, and teeth with overall increase in size. The biosphere was beginning and about 100 mya later some fish tackled the terrestrial environment [21].

Neurotransmitter Evolution

The major neurotransmitters such as acetylcholine, dopamine, serotonin, noradrenaline, adrenaline, and histamine have all been discovered in the three major living organisms, bacteria and fungi, plants, and animals. With their presence documented from protozoans and mammals, Roshchina suggested that the term biomediator as opposed to neurotransmitter is a more appropriate term. Their function in cells in altering electrical features and membrane permeability in both the microorganism and humans for example makes the study of the interaction between the two important in understanding the human conditions in both health and disease [22–26].

The Special Role of Dopamine in Humans, Cognition, and Neuropsychiatric Disease

Dopamine underlies core components of advanced cognition including working memory, cognitive flexibility, motor planning, temporal processing speed, abstraction, and creativity. This may in turn be due to the unique evolutionary history of dopamine in humans. Dopamine has been considered to be the key neurotransmitter with advanced intellect. Comparative analyses show that advanced intelligence in divergent nonhuman animal lineages, such as birds, with differing brain size and structure have been correlated with a dopaminergic "predominant" brain. One point of view posits that dopamine became the dominant neurotransmitter in the hominin line, having undergone a major expansion amongst primates, and is particularly concentrated in the prefrontal cortex. Furthermore, humans show a doubling of the caudate nucleus (gray matter with a particular predominance of dopamine) size relative to the chimpanzee caudate nucleus dopamine concentration. Furthermore there is a paucity of evidence for the expansion of any of the other neurotransmitter systems within the hominin line. Furthermore, the theory emphasizes the importance of increased shellfish and meat consumption which provided the precursors to dopamine in the form of tyrosine increase, iodine from seafood and meat, and its importance for increased thyroid hormone production as well as essential fatty acids [27].

Working memory ability is central to advanced human (abstraction, creativity) and the neurobiological substrate of such activity resides within the extensive frontoparietal working memory circuitry and the cellular basis of working memory. The latter is represented by neurons that are able to respond to anticipation and also maintain persistent firing within a delay period. Termed memory cells, these connect our past and future across the temporal divide of action or sometimes inaction. The metabolically active dorsolateral prefrontal cortex (DLPFC) is particularly susceptible to neuromodulation mediated by various neurotransmitter systems but in particular to dopamine. Lesion studies in primates have revealed that after catecholamine depletion in the DLPFC of monkeys, their failure in spatial working memory abilities was as severe as that after removal of the DLPFC itself. Furthermore, an inverted U-shaped response has been found with respect to dopamine. Excess dopamine (stress for example) with D1 receptor stimulation and suboptimal D1 stimulation with too little dopamine both resulted in decreased DLPFC neuronal firing [28].

A Brief Overview of Neurotransmitters, Intracellular (Cellular Machinery), Intercellular (Neurotransmitters), Receptors, and Astroglial Networks

Neurotransmission is electrochemical in nature and has both rapid effects and slower effects. Rapid effects are mediated by neurotransmitters and ion channels. Slower effects are mediated by neuromodulators and altering gene expression such as immediate early genes such as c-Fos and c-Jun and late genes. Signal transduction occurs via a four-step process. The second messenger systems in turn regulate third and fourth messenger systems that regulate both gene expression and accompanying protein synthesis. There is a major intracellular amplification network that subserves a marked complexity in chemical

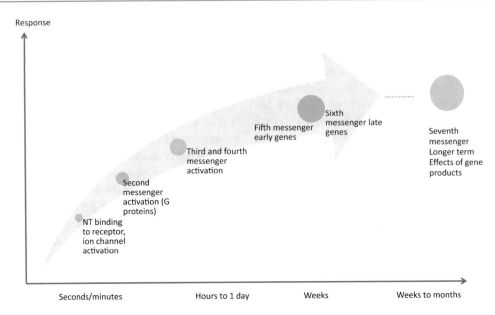

Fig. 3.4 Intracellular messenger cascades

transmission of a signal from the first messenger to the second, third, fourth, fifth, and sixth messengers (Fig. 3.4). In addition to the three-dimensional signaling cascades, a fourth dimension in terms of time occurs through these systems with more immediate responses within seconds, minutes, and hours to others that last days to weeks (late gene activation).

Intercellular messaging may be associated with synaptic activation by amino acid neurotransmitters such as glutamate, GABA, monamines including dopamine, and serotonin, nucleosides (adenosine), as well as peptides. Some neurotransmitters do not use a synapse such as diffusible gases (nitric oxide, carbon monoxide) and lipid neurotransmitters (endocannabinoids) that may act as retrograde neurotransmitters. Both fast and slow transmission is possible. Ultrafast transmission within milliseconds is relayed via ligand-gated ion channels (ionotropic) such as glutamate and GABA synapses. Slower transmission measured in seconds to minutes occurs via receptors linked to guanine nucleotide-binding proteins (G proteins) that in turn cause alpha subunit dissociation from the beta gamma complexes with subsequent downstream activation kinases, phosphatases, DAG, cGMP, cAMP, and Ca++. In general the second

messenger systems mediate slow transmission and may module the much faster glutamate and GABA transmission. The second messenger systems and those beyond markedly increase both flexibility and the complexity of the electrochemical signaling process. Second messenger system involves G proteins classified as heterotrimeric and small G proteins. The G protein structure consists of alpha, beta, and gamma subunits that dissociate upon neurotransmitter binding to a receptor and link the neurotransmitter directly to a particular ion channel (Fig. 3.5).

G proteins are implicated in a number of neurological, cognitive, and neuropsychiatric disorders and over 35 different types been recognized to date. Currently they are categorized into four main families:

Gs family	Adenylyl cyclase stimulation
Gi family	Inhibits adenylyl cyclase, activates K+, and inhibits Ca++ channels
Gq family	Activates PI-PLC
G12 family	Activates proteins Rho-GEFs

In addition small G proteins may activate or inactivate certain proteins behaving like molecular switches, akin to micro-RNA and the epigenetic

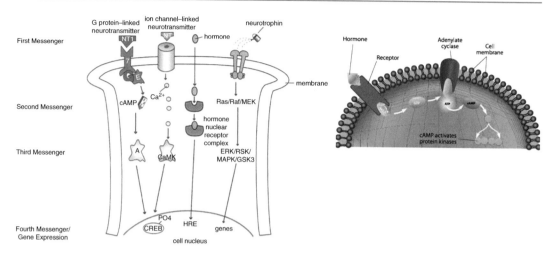

Fig. 3.5 The principal cellular signal transduction schematic cascades that enable neurons to respond to signals in manifold ways. All link initially to a specific receptor with subsequent activations of second, third, and fourth messenger molecules. These include the G-protein-, ion channel-, hormone-, and neurotrophin-linked systems.

Left figure with permission from Stahl S. Essential Psychopharmacology. Neuroscientific Basis and Practical Applications (3rd ed). Cambridge University Press, New York, 2008. Right figure: G protein and 2nd messenger activation and recycling. Figure credit: Shutterstock.com

gene "switches." An abbreviated list of some G proteins includes the following:

Ras	Signal transduction, growth factor control
Rab	Vesicle trafficking
Rho	Actin microfilament assembly
ARF	Assembly of Golgi complex
Ran	RNA and protein trafficking between nucleus and cytoplasm

To date, small G proteins have been implicated in the neurological conditions of tuberous sclerosis (mTORC 1 gene activation by the small G protein Rheb) and hereditary sensory motor neuropathy (gene mutation in small G protein, Rab7).

Influencing gene expression within the cell nucleus is important so that the organism may adapt to environmental changes and needs. Several mechanisms exist whereby this is transacted including chromatin structure changes, DNA-to-RNA transcription, and protein translation. Some of the gene expression regulation mechanisms by chromatin are achieved by histone acetylation, methylation, phosphorylation, acetylation, ubiquitinylation, and glycosylation. Histone acetylation, for example, is implicated in several cognitive processes such as learning, memory, antidepressant action, and various neuropsychiatric diseases.

For example curcumin is a histone acetyltranferase (HAT) inhibitor mediating anti-inflammatory and anti-neoplastic effects.

Micro RNAs have a regulatory function when binding to particular m-RNAs and influencing their translation inhibiting their translation. To date hundreds of mi-RNAs have been reported in mammalian cells and signify a new dimension of gene regulation identified in mammalian cells. Transcription factors are targeted by the intracellular signaling pathways. For example they may affect the nucleus directly as occurs with steroid hormone receptors via NF-kappa B that is cytoplasmic in response to ligand binding. Cyclic AMP response element-binding (CREB) protein, for example, is regulated by C-AMP and calcium activation, and is targeted by phosphokinase A (PKA) that after translocating to the nucleus is involved in encoding neurotransmitters, neuropeptides, and transcription factors. CREB represents a form of neural plasticity mechanism involved in learning and memory and long-term potentiation. The AP1 transcription factors include two major families, Fos and Jun (bZIP proteins), that are transient effects, activated within minutes, on gene expressions—termed immediate early genes. In contrast late-response genes are activated and inactivated over a period of hours.

Neurotransmitter Classification

Monoamines
Acetylcholine
Serotonin

Catecholamines
Dopamine
Norepinephrine
Epinephrine

Other amines
Histamine
Tyramine
Octopamine
Tryptamine
Melatonin
Agmatine

Amino Acids
Inhibitory
Gamma aminobutyric acid
Glycine
Excitatory
Glutamate
Aspartic acid
Gamma-hydroxy-butyrate
D-Serine

Neuropeptides
Opioids
Opiocortins
Enkephalins (Met-enkephalin, Leu-enkephalin
Dynorphin
Kyotorphin
Nociceptin
FMRFamide

Tachykinins
Substance P
Physalaemin
Kassinin
Uperolein
Eledoisin
Bombesin
Substance K

Neurokinin A
Neurokinin B

Other peptides
Bradykinin
Carnosine
Calcitonin G-related peptide
CART (cocaine and amphetamine-related script)
Neuropeptide Y
Neurotensin
Delta sleep factor
Galanin
Orexin
Melanocyte-concentrating hormone

Secretins
Secretin
Glucagon
Vasoactive intestinal peptide
Gastric inhibitory peptide

Insulins
Insulin
Insulin-like growth factors I and II

Somatostatins
Somatostatins
Pancreatic polypeptide

Gastrins
Gastrin
Cholecystokinin

Purine and Pyrimidine
Adenosine
ATP, ADP, AMP
Uridine

Gases (gasotransmitters)
Nitric oxide
Carbon monoxide
Hydrogen sulfide

Cannabinoids (lipids)
Anandamide
Palmitoylethanolamide (PEA)

Other
Brain natriuretic peptide

Heart
Atrial natriuretic factor

Hypothalamic releasing hormones
CRH, LHRH, TRH, GHRH, GnRH

Neurohypophyseal hormones
Vasopressin
Oxytocin
Neurophysins

Pituitary peptides
ACTH
TSH
GH
Lipotrophin
Alpha-MSH
Oxytocin
Vasopressin
Prolactin

Selected Overview of Some of the More Important Neurotransmitters

Glutamate

Glutamate is the major excitatory cerebral neurotransmitter accounting for approximately 75 % of excitatory transmission. The intimate astroglial neural network converts the spent glutamine into glutamate with reuptake postsynaptically. NMDA receptors allow large amounts of Ca influx that require both glycine facilitation and unblocking of magnesium at the receptor site. Whereas NMDA receptors are exclusive to neurons, kainate and AMPA receptors are found on neurons and glia. Glutaminergic receptors include:

Metabotropic

Ionotropic

AMPA—4 subtypes
Kainate 5 kainate
NMDA (NRI, NR2A, NR2B, NR2C, NR2D, NR3A, NR3B)

Excessive excitotoxicity that may follow stroke, recurrent seizures, and traumatic brain injury may destroy neurons. A high concentration of NMDA receptors within the hippocampal formation plays a role in LTP, learning, and memory. For example, overexpressed NR2B in the Doogie mouse forebrain results in an enhancement of hippocampal CA1- LTP [29]. More recently autoimmune syndromes affecting glutamate have been recognized.

The AMPA receptor is involved in limbic encephalitis which may be reversed by treatment with IVIG, adriamycin, and cyclophosphamide. Antibodies to Glu R1/R2 AMPAR (the predominant subtype in the hippocampus) can test positive and then become negative after 1-year treatment for example [30].

GABA

GABA is the major inhibitory neurotransmitter within the mammalian nervous system.

GABA receptors include:

GABA-A
GABA-B
GABA-C

The GABA-A subtypes include:

Alpha (1–6 subtypes)
Beta (1–3)
Gamma (1–3)
Delta
Epsilon
Phi

These are targeted by drugs such as alcohol, barbiturates, benzodiazepines, flumazenil, propofol, zolpidem, zaleplon, and eszopiclone.

GABA-B receptor may be blocked by baclofen.

Valproate, tiagabine, and vigabatrin inhibit GAT, a GABA transporter, and so potentiate GABAergic transmission.

GABA-C receptors are located in the retina.

Dopamine (DA)

Dopamine nuclei are located in midbrain nuclei and influence the nigrostriatal, mesocortical, and mesolimbic systems. Projection to the N accumbens and limbic subcortical regions subserve reward, addiction, and working memory for example.

Dopamine receptor (D1–5) activation is mediated by heterotrimeric G proteins.

D2 overstimulation is implicated in schizophrenia, D3 receptors are predominant in the nucleus accumbens, and D4 with the frontal cortex have particular affinity for atypical antipsychotics that include clozapine for example.

Acetylcholine (Ach)

Located at the neuromuscular junction, autonomic ganglia, the basal forebrain, and pontomesencephalic cholinergic nuclei. Ach functions include regulation of the sleep–wake cycle, memory, and learning. Both nicotinic and muscarinic receptors are found in the brain. M1-like receptors (M1 and M3) predominate in the brain, especially the frontal regions. M2 receptors occur in the heart and M2-like receptor (M2, M4) activation induces hyperpolarization and function as presynaptic autoreceptors that result in an overall decrease in cholinergic activity. M5 are located on blood vessels. Muscarinic receptors are coupled to G proteins, potassium channels, adenylyl cyclase, and phospholipase C. Inactivation by cholinesterases has clinical relevance in that glial cell-mediated BuChE increases with dementia severity and inhibition by rivastigmine for example affects both AChE and BuChE.

Serotonin

Located in the cerebral cortex, raphe nucleus, hypothalamus, and limbic system cerebellum, the highest concentration is in GIT enterochromaffin cells and myenteric plexus. Serotonin receptors also occur on platelets. Receptors are both presynaptic and postsynaptic, mostly mediated by G proteins, adenylyl cyclase, or phospholipase C and include:

5HT 1A, 1B, 1D, 1E, 1F
5HT2A, 2B, 2C
5HT3
5HT4
5HT5 A, B
5HT6
5HT7

From a clinical point of view, euphoria and hallucinations may be induced by activation of cerebral 5 HT2 receptors; 5HT2A mediates platelet aggregation, smooth muscle contraction, and 5HT3 activation in the area postrema, related to vomiting; and 5HT4 in GIT promotes peristalsis and secretion. The receptors 5HT 6 and 5HT 7 in limbic system and antidepressant medications have a high affinity for 5HT6 receptors.

Noradrenaline and Adrenaline

Noradrenaline is widely distributed in the nervous system, present at sympathetic post-ganglionic endings. Both noradrenaline and adrenaline are also secreted by the adrenal medulla. The receptors include:

Alpha 1 (A,B,D)
Alpha 2 (A,B,C)
Beta 1, 2, 3

Cell bodies producing NA are found in the midbrain locus coeruleus and other pontine and medullary cell bodies, which are regulated by alpha 2 autoreceptors, presynaptically located. There is some correlation with lowered NA and the behavioral disturbance of Alzheimer's disease.

Histamine

Histamine receptors include:

H1
H2
H3

These receptors are found in the forebrain thalamus and tegmentum, mediate histamine's excitatory effects, and may be countered by tricyclic antidepressants, chlorpromazine and diphenhydramine.

Orexins

Orexins A and B emanate from cell bodies in the lateral hypothalamus and bind to the G protein-linked receptors OX1 and OX2. They are involved in maintenance of arousal and deficiency is associated with the narcolepsy syndrome by influencing cholinergic and monoaminergic neurons. Orexins are also involved in lipid metabolism, brown fat activation, and mood.

Gasotransmitters

Nitric Oxide (NO)

NO mediates the signal from postsynaptic neurons communicating with presynaptic endings in memory formation with long-term potentiation (LTP) and long-term depression (LTD). NO occurs in three different forms:

N Nos—neuronal
E Nos—endothelial
I Nos—inducible

N Nos is found in neocortex and hippocampus and may be co-expressed with neurotransmitters. E Nos has neuroprotective functions within the endothelium and is concerned with vascular autoregulation. I Nos is associated with cytokine activation and inflammation [31].

Hydrogen Sulfide (H_2S)

Found in blood vessels, it activates K_{ATP} channels controlling potassium out of smooth muscle cells, which in turn affects calcium entry into the muscle cells allowing muscular dilatation. It may be that while NO is concerned with large vessel vasodilatation, the function of H_2S is in small vessel vasodilatation. Hence it is important in myocardial infarction, stroke, and hypertension. Furthermore, as a neuromodulator it plays a role in neural circuitry responsiveness to cognition, learning, and memory. Because it promotes glutathione production, an antioxidant and neuroprotective agent, H_2S is being investigated in regard to suspended animation, termed H_2S hibernation, that may have clinical use in acute stroke and heart attack neuroprotection. Garlic contains sulfide compounds that are converted into H_2S within red blood cell membranes [32].

Cannabinoids

The lipid neurotransmitters, the endocannabinoids, anandamide, and 2-arachidonoyl glycerol (2-AG), are the ligands for the CB1 receptor found in cerebral cortex, particularly in the pain circuitry. Stimulation of the CB1 receptor by anandamide and tetrahydrocannabinol (THC) for example mediates euphoria analgesia and has anxiolytic effects. THC targets the CB1 receptor in the nucleus accumbens to trigger dopamine release. The CB2 receptor endogenous ligand is palmitoylethanolamide (PEA) with functions including pain reduction. Cannabinoids have been implicated in the "runner's high." Intense exercise such as prolonged running is associated with euphoria, analgesia, and anxiolysis associated with increased blood levels of the opioid β-endorphin and anandamide (endocannabinoid). The anxiolytic effect appears to be mediated by the CB1 receptors within the forebrain via GABAergic neurons. The analgesic effect is mediated by peripherally located CB1 and CB2 receptors [33].

Agmatine

One of the most recently discovered cerebral neurotransmitters, agmatine, is a polyamine that may bind to a number of different receptors and mediates anxiolytic, anticonvulsant, and antinociceptive properties as well as ischemia protection. It also has a modulatory function in memory and learning [34].

Ion Channel Overview

Ion channels are critical to the timing and ion flow within nerve impulses, cardiac conducting tissues, and heart beat for example. When impaired, we use the term channelopathies. Examples include seizure conditions, migraine, myotonia, episodic ataxia, and, in the cardiac arena, the prolonged QT interval. Most anti-seizure agents act on the sodium channels. Glial cells also have neurotransmitter receptors, ion channels, and gap junctions. Astrocytes in particular have high concentrations of excitatory amino acid transporters (glutamate) for potassium uptake, being linked to each other by gap junctions. Intracellular calcium waves are mechanism whereby astrocytes communicate with each other, in turn related to neuronal activity.

Ion channels may be classified into two basic types:

- Voltage-gated like ion channel group
- Ligand-gated ion channel group

Ion Channel Classification

Calcium—Ca^{++}_v
L channels
T channels
P/Q, N, and R channels

Potassium
Six transmembrane domain channels
K^+_v delayed rectifiers
K^+_v A-type channels
K_{Ca} (calcium-activated K channels)
Sk channels (subtype of K_{Ca} channels)
2 transmembrane domain channels
K_{ir}—inward rectifier K channels
G protein-coupled K_{ir} 3 channels (GIRKs)
K_{ir} (ATP) channels
CNG—Cyclic nucleotide-modulated ion channels

TRP—Transient receptor potential channels
K_{2p}—2 pore K channels (leak channels)

Chloride
Ligand-gated Cl⁻ channels (GABA, glycine)
CLC channels
CFTR channel

Sodium—$Na^\pm_v$
Uniform across animal species

Calcium Ion Channel

Different calcium channel subtypes include:

L channels (L—long open time)
The antihypertensive and anti-anginal compounds such as verapamil, diltiazem, and nifedipine target these channels.
T channels (T—tiny/transient currents)
Used by thalamic neurons that generate cortical rhythms of absence-type seizures blocked by ethosuximide
P/Q, N, and R channels

These are concerned with hormone and neurotransmitter release, the latter at the neuromuscular junction for example. Impairments in these channels occur in familial hemiplegic migraine and episodic ataxia, for example, and poisoning by certain spider and sea snail toxins may block them.

Sodium Channels

These channels are targeted by some of the most common neurological drugs such as the antiepileptics, carbamazepine and phenytoin. Conditions such as periodic paralyses, myotonias, and prolonged QT interval syndrome are associated with sodium channel malfunction. Impairment of their functioning may be caused by cocaine and poisoning by tetrodotoxin from octopus and puffer fish.

Potassium Channels

These are concerned neuronal stabilizing effects. For example the HERG K$^+$ channel dysfunction may cause cardiac dysrhythmias due to inability to repolarize the cardiac ventricles. A neurological disease associated with these channels is episodic ataxia.

Chloride Channels

In general these channels have a modulating effect on both neuronal and cardiac myocyte excitability. The major inhibitory neurotransmitters, glycine and GABA, act upon ligand-gated chloride channels [35].

Neurotransmitter Syndromes May Be Recognized as Syndromes of Cognitive Impairment and May Present as Neurotransmitter Deficiency Syndromes or of Hyperactivity

Parkinson's syndrome is perhaps the best known dopamine deficiency syndrome. However the manifestations of dopamine deficiency may also present with conditions such as apathy or abulia. Encephalitis lethargica with antibodies to midbrain and basal ganglia neurons may result in a profound dopamine deficiency state that can respond dramatically to dopaminergic therapy [36].

Syndromes of excess neurotransmitter states include:

Excess

Serotonin Syndrome/Toxidrome

Cognitive impairment (encephalopathy, headache, coma), autonomic activation (tachycardia, hypertension, hyperthermia), and somatic effects (tremor, myoclonus) occur in response to excess serotonin usually due to co-administration of antidepressant serotonergic stimulating drugs in association with other centrally acting medications.

Neuroleptic Malignant Syndrome

This is due to a sudden and significant reduction in dopamine levels such as after withdrawal from anti-Parkinson's mediation with levodopa; the blockage by antipsychotics such as haloperidol and phenothiazines such as chlorpromazine or the atypical antipsychotics such as clozapine, risperidone, and quetiapine; as well as antiemetic treatment by metoclopramide. The condition is caused by dopamine blockage of D2 receptors in the basal ganglia; it also presents with encephalopathy, tremor, rigidity, muscle cramps, hyperpyrexia, and autonomic instability. The main distinguishing features from the serotonin syndrome include the muscular rigidity, rhabdomyolysis, bradykinesia, elevated CPK, and leukocytosis.

Anticholinergic Toxidrome

Presenting symptoms and syndromes include encephalopathy, hallucinations, seizures, myoclonus, autonomic activation, pyrexia with dry skin, and dilated pupils. Causative medications include atropine, benztropine, antihistamines, antidepressants, antipsychotics, and anti-Parkinson's medications.

Cholinergic Toxidrome

This syndrome is characterized by encephalopathy, bradycardia and bronchospasm/bronchorrhea, miosis, and hypothermia. Typical offending agents include organophosphate poisoning, mushroom poisoning, and carbamates.

Malignant Hyperthermia

Encephalopathy, hyperpyrexia, autonomic activation, and muscular rigidity in genetically susceptible individuals (ryanodine receptor gene) in response to neuromuscular blocking agent, succinylcholine.

Deficiency

Serotonin deficiency syndromes (depression)

Dopamine deficiency syndromes (Parkinson's, akinetic mutism, abulia)

Cholinergic deficiency syndromes (Alzheimer's disease)

Norepinephrine deficiency syndromes (impaired attention and concentration, depressed mood)

Less profound hyperdopaminergic syndromes may be recognized and may reflect a relative serotonin underactivation and dopamine DA overactivation within the left hemisphere and ventral prefrontal cortex. These include a spectrum of conditions ranging from the impulsive conditions such as attention-deficit hyperactivity disorders, mania, addictions, and schizophrenia to the compulsions, obsessive compulsive disorder, and autism spectrum conditions. Dopamine excess has been implicated in a number of these exclusively human conditions.

References

1. Brady ST, Siegel GJ, Albers RW, Price DL, editors. Basic neurochemistry. 8th ed. Elsevier: Amsterdam; 2012.
2. Nestler EJ, Hyman SE, Malenka RC. Molecular neuropharmacology. A foundation for clinical neuroscience. 2nd ed. New York, NY: McGraw Hill; 2009.
3. Stahl S. Stahl's essential pyschopharmacology. 3rd ed. Cambridge: Cambridge University Press; 2008.
4. Rodwell VW, Bender D, Botham KM, Kennelly PJ, Weil PA. Harper's illustrated biochemistry. 30th ed. New York, NY: Lange Publications, McGraw Hill; 2015.
5. Barrett KE, Barman SM, Boitano S, Brooks HL. Ganong's review of medical physiology. 25th ed. New York, NY: Lange McGraw Hill; 2015.
6. Hazen RM. The story of the earth. The first 45 billion years, from stardust to living planet. New York, NY: Penguin; 2013.
7. Ćuk M, Stewart ST. Making the Moon from a fast-spinning Earth: a giant impact followed by resonant despinning. Science. 2012;338(6110):1047–52.
8. Mastrobuono-Battisti A, Perets HB, Raymond SN. A primordial origin for the compositional similarity between the Earth and the Moon. Nature. 2015;520(7546):212–5.
9. Lunine JI. Physical conditions on the early Earth. Philos Trans R Soc Lond B Biol Sci. 2006;361(1474):1721–31.
10. Halliday AN. A young Moon-forming giant impact at 70–110 million years accompanied by late-stage mixing, core formation and degassing of the Earth. Philos Trans A Math Phys Eng Sci. 2008;366(1883):4163–81.
11. Grosch EG, Hazen RM. Microbes, mineral evolution, and the rise of microcontinents-origin and coevolution of life with early earth. Astrobiology. 2015;15(10):922–39.
12. Watson EB, Harrison TM. Zircon thermometer reveals minimum melting conditions on earliest Earth. Science. 2005;308(5723):841–4.
13. Trail D, Tailby ND, Sochko M, Ackerson MR. Possible biosphere-lithosphere interactions preserved in igneous zircon and implications for Hadean earth. Astrobiology. 2015;15(7):575–86.
14. Mulkidjanian AY. On the origin of life in the zinc world: 1. Photosynthesizing, porous edifices built of hydrothermally precipitated zinc sulfide as cradles of life on Earth. Biol Direct. 2009;4:26. doi:10.1186/1745-6150-4-26.
15. Miller SL. A production of amino acids under possible primitive earth conditions. Science. 1953;117(3046):528–9.
16. Kreysing M, Keil L, Lanzmich S, Braun D. Heat flux across an open pore enables the continuous replication and selection of oligonucleotides towards increasing length. Nat Chem. 2015;7:203–8. doi:10.1038/nchem.2155.
17. Stow D. Vanished ocean. Oxford: Oxford University Press; 2010.
18. Fuqua C, White D. Prokaryotic intercellular signaling – mechanistic diversity and unified themes. In: Fairweather I, editor. Cell signalling in prokaryotes and lower Metazoa 2004. Dordrecht: Kluwer Academic Publishers; 2004.
19. Caveney S, Cladman W, Verellen LA, Donly C. Ancestry of neuronal monoamine transporters in the Metazoa. J Exp Biol. 2006;209:4858–68.
20. Hoffman PF, Kaufman AJ, Halverson GP, Schrag DP. A neoproterozoic snowball earth. Science. 1998;281(5381):1342–6.
21. Mitchell RN, Kilian TM, Evans DA. Supercontinent cycles and the calculation of absolute palaeolongitude in deep time. Nature. 2012;482(7384):208–11.
22. Roshchina VV. Evolutionary considerations of neurotransmitters in microbial, plant, and animal cells. In: Lyte M, Freestone PPE, editors. Microbial endocrinology, interkingdom signaling in infectious disease and health. New York, NY: Springer; 2010.
23. Kruk ZL, Pycock CJ. Neurotransmitters and drugs. New York, NY: Chapman and Hall; 1990.
24. Roshchina VV. Neurotransmitters in plant life. Plymouth: Science Publication; 2001. p. 283.
25. Kuklin AI, Conger BV. Catecholamines in plants. J Plant Growth Regul. 1995;14:91–7.
26. Kulma A, Szopa J. Catecholamines are active compounds in plant. Plant Sci. 2007;172:433–40.
27. Previc F. The dopaminergic mind in human evolution and history. New York, NY: Cambridge University Press; 2009.
28. Arnsten AFT. The neurobiology of thought. Cereb Cortex. 2013;23:2269–81.
29. Yang Q, Liao ZH, Xiao YX, Lin QS, Zhu YS, Li ST. Hippocampal synaptic metaplasticity requires the activation of NR2B-containing NMDA receptors. Brain Res Bull. 2011;84(2):137–43.
30. Baraller L, Galiano R, Garcia Escrig M, Martinez B, Sevilla T, Blasco R, et al. Reversible paraneoplastic

limbic encephalitis associated with antibodies to the AMPA receptor. Neurology. 2010;74:265–7.

31. Benarroch EE. Nitric oxide: a pleiotropic signal in the nervous system. Neurology. 2011;77(16):1568–76.

32. Yang G et al. H_2S as a physiologic vasorelaxant: hypertension in mice with deletion of cystathionine gamma-lyase. Science. 2008;322:587–90.

33. Fuss J, Steinle J, Bindila L, Auer MK, Kirchherr H, Lutz B, Gass P. A runner's high depends on cannabinoid receptors in mice. Proc Natl Acad Sci U S A. 2015;pii:201514996.

34. Uzbay TI. The pharmacological importance of agmatine in the brain. Neurosci Biobehav Rev. 2012;36(1):502–19.

35. Wu DC, Wang YT. Allosteric potentiation of glycine receptor: chloride currents by glutamate. Nat Neurosci. 2010;10:1038.

36. Dale RC, Church AJ, Surtees RA, Lees AJ, Adcock JE, Harding B, et al. Encephalitis lethargica syndrome: 20 new cases and evidence of basal ganglia autoimmunity. Brain. 2004;127:21–33.

Vision: Elementary and Complex Visual Processing

4

Vision directs our attention and attention, in turn, is nature's answer to massive sensory input that our brains are subjected to. Our senses gather information continuously and hence we are subject to sensory overload. The human sensory system receives ~11 million bits of information per second; yet it has been estimated that we can only process ~16–50 bits per second (0.0002 %). Countless decisions are being made by our brain on a continuous basis, mostly at a nonconscious level from which is derived the very rough estimate that we process only 5 % at the conscious level and 95 % at the nonconscious level [1, 2]. At another level and another approximation, about half of the cortex is concerned with vision and vision usually overrides other senses when there are conflicting inputs. The evolution of our visual system, its predominance over the other senses, and the ability to process most sensory information nonconsciously can only be explained and understood in terms of paleoneurological or evolutionary biological and neuroscientific process.

Evolutionary Insights

Vision is our major sensory apparatus. This obviously is not the case for all animals, let alone mammals. The process whereby this evolved takes us back to about 340 million years ago (mya) when early reptiles (stem amniotes) evolved from amphibians and subsequently diverged into two major clades: the Sauropsids, which led to present-day birds and reptiles, and the Synapsids, with the only surviving clade being mammals. The predominant sense of early mammals was olfaction. They were nocturnal and vision was generally poor as it was unimportant. The development of small ear bones allowed transmission of air pressure oscillations to the cochlea organ, which enabled extension of hearing into higher frequencies not undetectable by their reptilian counterparts and predators.

Overall brain size increased together with the various sensory faculties which diversified markedly during mammalian evolution in the period of 300–200 mya to fill an immense variety of ecological niches. The drivers for increased encephalization and origin of the neocortex included innovations of senses such as olfaction and hearing, miniaturization, parental care, endothermy, elevated metabolism, and nocturnality. The Synapsida mammalian lineage diverged from the tetrapods during the Carboniferous period about 300 mya at which time they possessed relatively low-resolution olfactory capability, vision, and hearing, but had basic tactile sensitivity and relatively basic motor coordination. The first surge in encephalization is exemplified by Morganucodon that possessed an EQ of ~0.32, about 50 % larger compared to the basal cynodonts. This was largely due to olfactory bulb, the olfactory cortex (pyriform cortex), thalamus, and cerebellum. During this time improved high-frequency hearing developed with the reduction of the middle ear ossicles

M. Hoffmann, *Cognitive, Conative and Behavioral Neurology*,
DOI 10.1007/978-3-319-33181-2_4

which however remained still attached to the lower jaw and hair was a sensory tactile organ, only later developed further and exapted for fur and insulation to facilitate thermoregulation. The second surge in encephalization, represented by Hadrocodium, with an EQ of ~0.5, had further expanded olfactory bulbs, olfactory cortex, and detached middle ear ossicles (from the lower jaw) that are now detached from the jaw and suspended beneath the cranium. The animals known as the crown Mammalia signified the third pulse, predominantly in olfaction enhancement together with nasal turbinate development that enabled a tenfold enlargement of the olfactory epithelium. Visual systems evolved much later among mammals [3].

Brain Evolution Occurs in a Mosaic Manner

The earlier mammals exploited nocturnal niches and evolved newer somatosensory, auditory, and visual areas complemented by agranular parts PFC. Later in evolution granular PF evolved in early primates as they foraged in the tree canopies fruits, nectar, tender leaves, flowers, and insects. Subsequently during anthropoid evolution and diurnal foraging, foveal vision and trichromacy developed. These were aided by granular PFC development such as the ventral, dorsal, and frontopolar that facilitated foraging and reduced predator risk (Fig. 4.1).

The Brain Has a Two-Tier System

An initially evolved more fundamental, unconscious system, essential for survival and processing of large amounts of information processing and the more basic sensory processing common to all vertebrates, involved the superior colliculus and pulvinar. The pulvinar is almost nonexistent in the rat, is present in cats but small in size, and in humans comprises about 40 % of the thalamus, larger than any other thalamic nuclei. One of the functions of the SC was to trigger freezing in relation to fast-advancing objects. Bears and dogs for example have problems identifying still objects. The subsequent conscious component addition, typical of mammals with the lateral geniculate nucleus (LGN) visual, improved perception and although slower could override the more elementary SCP circuitry. The LGN has particularly extensive cortical connections allowing better identification of predators and enabling decisions whether to flee, attac,k and also warn conspecifics as Vervet monkeys do for example. Primates evolved their visual apparatus further and incorporated an extensive fear module (amygdala) into their vision circuitry (Fig. 4.2).

With primates the principal sensory–environment interface is vision and in humans vision-related cortex occupies approximately ½ of the cortical area. In general better vision facilitates food source location, avoiding danger and facilitating associated with longer life expectancy. The vertebrate visual apparatus that included the

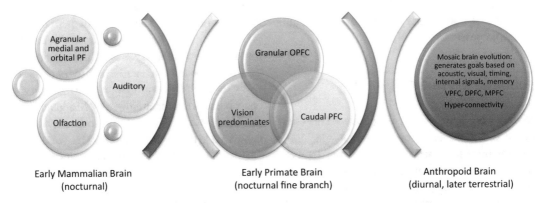

Fig. 4.1 Conceptual sensory, visual and prefrontal cortex evolution. Adapted from Passingham RE, Wise SP. Neurobiology of the Prefrontal Cortex. Oxford University Press, Oxford 2012

a Visual pathways

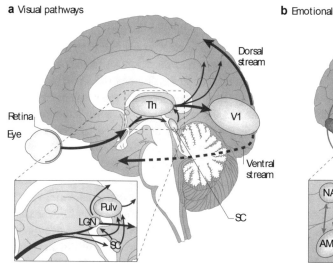

b Emotional pathways

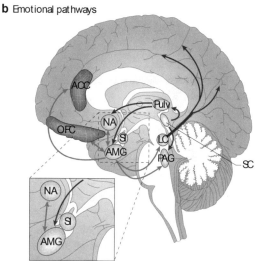

Fig. 4.2 (**a**) Vision and (**b**) emotion. Cortical and subcortical pathways. Figure with permission: Tamietto M, de Gelder B. Neural bases of the non-conscious perception of emotional signals. Nature Neuroscience Reviews 2010;11:697–709

superior colliculus in the midbrain and pulvinar allowed shifting of attention to the most critical or important and unexpected environmental events without the need of any assessment or conscious awareness that only slowed this system. Freezing is a common reflex in mammals in response to objects that suddenly present in the peripheral vision. The principal function of the SC therefore is predator detection and evasion and connections via the substantia nigra allow rapid interruption of their activity to focus on suddenly appearing objects. However in addition to the extensive posterior cortex dedicated to vision, the retina forms the first part of the process. Retinal cones function during a typical day (photopic vision), process to light. Retinal rods allow processing of light when light levels are low (scotopic vision) and cones, a third cell type, the ip-RGCs (intrinsically photosensitive retinal ganglion cells) situated in the ganglion cell layer that relay the pupillary light reflex, circadian rhythm control, and aspects of behavior. In addition connections have been reported to the suprachiasmatic nucleus and thalamus and be involved in conscious sight as well as mesopic vision (between photopia and scotopic vision).

In general the earlier mammals such as the insectivores and marsupials have two major types

of vision cortex, primary (striate) visual cortex, V1 (BA 17), and association (extrastriate) cortex, V2 (BA 18). Rodents have expanded to 3–4 major visual cortical areas (up to 12 extrastriate cortical areas), including areas V2–V5. The diurnal lifestyle of squirrels furnished them with 95 % cones in the retina, increased density of ganglion cells projecting to both the LGN and superior colliculus, and further increased extrastriate areas with both striate and extrastriate cortex 4–8× times larger than rodents with seven major visual areas [4, 5].

Feline and primate visual cortex shows the greatest elaboration with up to 15–19 visual cortical areas in cats and 25 or more in primates, identified to date [6]. In particular V5 or MT (middle temporal area) is characteristic for primates. In general the earlier more caudal visual areas of V1–V5 connect to both the dorsal and ventral streams although V4 is mostly associated with the ventral stream and V3 has V3 d and V3 v components. In primates MT/V5 is the most prominent extrastriate region after V2 and is responsible for both local motion cues as well as more global motion cues. MT seems to be unique to primates but this remains conjectural. The posterior medial lateral sulcus PMLS complex of cats may be a homologue of the primate MT, as both have similar architectonic appearances and are involved in motion processing. In primates area V8 is involved in color processing and V7 remains conjectural [7].

The different visual competencies of New- and Old-World monkeys form the basis of the snake detection theory, proposed by Isbell. Her hypothesis posits that snakes were responsible for the principal selective pressure on expansion of primate visual circuitry and the differing trichromatic and dichromatic capabilities of primates. Venomous snakes appeared about 60 mya at a similar time that mammals such as primates, rats, and bats generally had much swifter locomotive abilities. At this stage Madagascar had already separated from Africa and South America was separating from Africa from about 150 to 40 mya. Hence, Malagasy prosimians had not been exposed to venomous snakes and remained with relatively poor vision. The ancestors of modern platyrrhines arrived in South America about 35 mya at a time when there were no venomous snakes on that continent. They arrived much later by about 23–10 mya and rattlesnakes only 3 mya presumably via Asia. All primates in this group have dichromatic (or red green color blind) vision except for the Howler monkey. Contrariwise, all catarrhines (Old-world, African monkeys) as well as humans have excellent trichromatic vision. Malagasy prosimians (lemurs) have excellent olfaction and small brains and vision is relatively poor. In contrast catarrhine monkeys (Old World) have poor olfaction, good vision, and large brains. The platyrrhine primates of the New World have intermediate vision. This variable evolutionary subjection to venomous snakes is presumed to be the cause of differences in the visual abilities amongst primates [8].

A Central Function of Unconscious Brain Processing Is Filling in of Blanks Due to Partial Information

Relatively limited information is processed by the brain when surveying the environment—a large part of the percept is by inference. A cen-

tral function of higher visual functions concerns predator detection and various aspects of camouflage such as color and physical are both camouflage and partitioning from the environmental structures. For example particular patterns that have certain repetitiveness excite motion-detecting neurons, a valuable and specific evolutionary asset [9, 10]. In addition the brain has to actively "fill in" data to accommodate the blind spot due to the origin of the optic nerve fibers, devoid of retinal tissue. Microsaccades and macrosaccades help solve this problem. Microsaccades refer to eyes moving several times per second during eye gazing and enable filling in the percept but in a nonconscious manner. Macrosaccades occur about 100,000 times per day—similar to the heartbeat frequency. Our vision percept is therefore constructed by the unconscious components of our minds representing an interpretation to maximize overall likelihood of survival. Aside from the visual circuitry, this applies also to hearing and memory [11].

Blindsight, Change Blindness, Inattention Blindness, and Bistable Figure Perceptual Change

In addition to the active filling in to account for the blind spot, some other examples include the clinical phenomenon of blindsight, change blindness, and inattention blindness. In addition there can be a perception that can change without there being any change in the sensory stimulation.

Evolutionary Explanations of Blindsight: Subcortical Components Enable Blindsight

People with primary visual (V1) lesions retain some degree of visual capacity but do have perceptual awareness or the acknowledgement of the percept. For example they may be able to navigate a route, avoiding specifically placed obstacles, and be correct in facial expression identification (angry or happy faces) better than by chance alone, approximately two-thirds of the time. In type 1 blindsight, there is no awareness of perceptual stimulus but can predict better than by chance alone features of a stimulus such as movement or location. In type 2 blindsight some awareness of the stimulus is retained evidenced by the tracking motion of the eyes for example but still devoid of a visual percept [12–14].

In people that have sustained left visuospatial neglect due to right parietal damage, blindsight may be demonstrated in a different manner. For example Marshall and Halligan devised an experiment consisting of an intact house and a house on fire. Most prefer the house not on fire implying nonconscious processing of the information (Fig. 4.3) [15].

Perception of an image can also change without there being any alteration in the sensory stimulus. This can be demonstrated by ambiguous figures such as the Rubin figure and Necker cube and underscores the role of the prefrontal cortex and attentional systems in image perception. The normal alternating perception of face and vase

Fig. 4.3 Blindsight: Unconscious processing, spatial neglect example, and a preference to live in the lower house. Figure with permission: Marshall JC, Halligan PW. Blindsight and insight into visuospatial neglect. Nature 1988;336:766–767

for example can become impaired due to frontal lobe lesions, alternations impaired due to frontal lobe lesions [16].

The extensive occipital networks to all other brain regions emphasize the extensive visual networks and the concept of not only lesion-site impairment but also more remote impairment. These in turn can manifest as hypofunction, hyperfunction, or dysfunction that refers to both location and the hodological aspects. The most extensive occipital radiations are depicted in Fig. 4.4.

Neurophysiological Aspects of Vision Processing

The V1 (BA 17) area integrates the visual fields and enables binocular vision. With areas further along the ventral and dorsal streams processing from more basic (form, shape) to intermediate (depth) and higher level vision (objects, faces) processing occurs. Cytoarchitectonic and more recently functional MRI has allowed an appreciation of the magnitude of different visual areas within the human brain currently in the order of 30.

The current understanding of the different visual processing areas is summarized in Fig. 4.5 [17].

Basic, Intermediate, and Higher Visual Processing

Shape, color, depth, and motion are evaluated in differing downstream ventral and dorsal stream cortical regions. For example basic visual perception depends on separation of the background and foreground of a scene, the principle of good continuation (contour saliency) and co-linearity of lines (linked rather than branching for example). For example a smooth line within a complex background is more likely to be discerned. Furthermore top-down effects on perception are important, as the expectation of what one expects to see influences ultimately what is seen [18]. Features assembled in the basic visual processing of V1 include combination of visual information enabling binocular vision, contrast, brightness, light–dark contrasting stimuli,

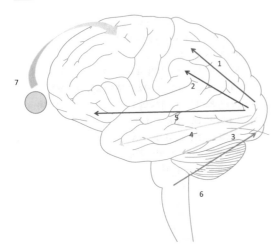

Fig. 4.4 Visual radiations: (*1*) dorsal occipito-parietal (where), (*2*) occipito-parietal (when), (*3*) occipito-temporal (what), (*4*) occipito-temporal amygdala, (*5*) occipito-frontal (via extreme and external capsule), (*6*) brainstem (peduncular hallucinations) (remote), (*7*) ocular (Charles Bonnet syndrome)

and line orientation. Lesions here can cause field defects and cortical blindness. Brightness and color also depend on context (Fig. 4.6) ([17] and [www.lottolab.org]). The BA 18 region or V2 is relatively large, has connections to the frontal eye field, and is involved in eye movements as well as auditory and sensorimotor cortices. The BA 19 or V3 region subserves predominantly spatial orientation, stereoscopy, motion, and depth perception. V3a is situated superiorly to V3, being particularly sensitive to contrast and motion in humans, that differs from the function in monkeys. V4 subserves color (V4v, ventral component) and size evaluation (V4d, dorsal component) [19]. The V5 designated area is at the temporo-parieto-occipital junction and is especially critical for motion detection including direction and speed of the object or stimulus. Two tracts subserving this function have been reported: a direct one from the retina via thalamic nuclei and the

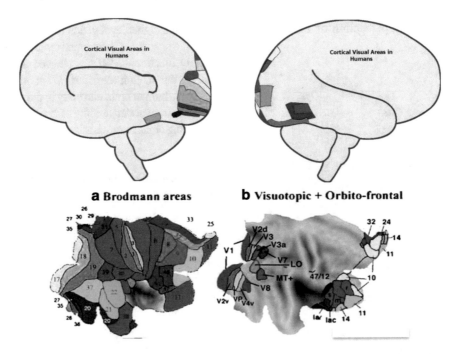

Fig. 4.5 Visual areas V1–V8. *Upper figures* legend: V1/2/3 subserve binocular vision, line orientation, contrast, and brightness. V1—*blue*, V2—*lilac*, V3—*yellow*, V4 (color vision)—*dark blue*, V5 (motion processing)—*light blue*, V6 (three-dimensional vision)—*dark pink*, LOC—object shape and FFA (fusiform face area)—*dark peach*, PPA (parahippocampal place area for places more than objects)—*brown*, EBA (extrastriate body is for body parts)—*dark lilac*. Adapted with permission from Kandel ER, Schwartz JH, Jessell TM, Siegelbaum SA, Hudspeth AJ (Eds). Principles of Neural Science, 5th ed, McGraw Hill, 2013, New York and Piech V. *Lower figures* (**a**) and (**b**): Visual cortical areas (V1–V8). The PALS B12 atlas multiple coordinate system demonstration with inflated and flat maps correlated with the Brodmann areas (*lower left figure* (**a**)) and in relation to the visual orbitofrontal cortical areas (*lower right figure* (**b**)). Figure with permission: Van Essen DC. A Population-Average, landmark and surface-based (PALS) atlas of human cerebral cortex. Neuroimage 2005;28:635–662

Fig. 4.6 Illusions: Brightness perception of colors depends on surrounding context. The Helmholtz-Kohlrausch effect. The four *blue*-appearing tiles on the superior surface of the cube and the six *yellow*-appearing tiles on the superior surface (*left figures*) are in fact all the same *grey* in color (*right figures*). Figure with permission: Image by R. Beau Lotto at www.lottolab and Corney D, Haynes JD, Rees G, Lotto RB. The Brightness of Colour. PLoS One. March 2009 Vol 4, e5091

indirect, major one, traversing V1 [20, 21]. V6 is situated caudally within the parietal lobes and has been linked to three-dimensional vision, self-motion analysis, and visual guidance during arm reaching and finger pointing in humans [22, 23]. Area V7, a previously identified visual map, lies within the posterior intraparietal sulcus, shares foveal representations with the intraparietal sulcus, and is involved in saccadic eye and reaching movements [24]. V8 is associated with color interpretation separate from V4 [25].

Higher Level Processing

The inferior temporal lobe encodes complex visual stimuli and mediates object recognition, faces, and hands. An important neurophysiological process, perceptual constancy, is required for object recognition. Under this concept is included size constancy (the object at different distances perceived to be the same size), position constancy (objects the same no matter where they are in the visual field), and viewpoint invariance (three-dimensional objects viewed from differing angles) [17].

Change Blindness and Inattention Blindness

Attention is important in visual processing which may be both voluntary and involuntary. This fails in people with left parietal lesions for example that demonstrate left hemineglect that

may be trimodal (visual, auditory, tactile). In viewing a picture or scene for example, attention is generally paid to a few select objects reliant on our saccadic input that allows foveal concentration. The image on the fovea shifts with each saccade. Considerable changes in successive presentation of a picture for example frequently go unnoticed because of this fragmented attentional aspect. With changes involving a face for example, functional imaging reveals activity in the fusiform gyrus, which may or may not have entered conscious awareness. However if awareness of the changes has indeed taken place, the additional functional activity occurs in the prefrontal and parietal cortex (Fig. 4.7) [26]. A related but different condition, known as inattention blindness, occurs with a failure to recognize an event or object that is generally in plain sight and attributed to attention focused elsewhere. First described by Mack and Rock in 1992 it is particularly well demonstrated by the Invisible Gorilla Test of Chabris and Simon (http://www.theinvisiblegorilla.com/gorilla_experiment.html) [27, 28].

Perception can also change without any change in sensory stimulation which is attributed to normal prefrontal cortical functioning. This may be demonstrated by ambiguous or bistable figure tests such as the Rubin Vase and Necker Cube (Fig. 4.8). The alternating perception of such figures may not occur with various prefrontal lesions [16]. This top-down influence of attention on perception may not

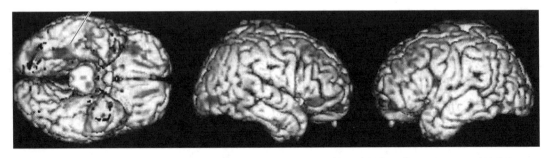

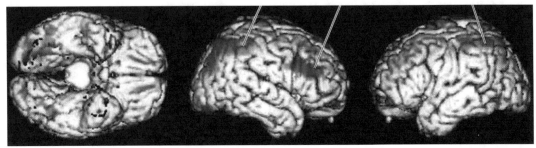

Fig. 4.7 Brain activation and change blindness. Unconscious detection in fusiform face area (*upper panel*) and conscious report increase in frontal and parietal region (*lower panel*). Figure with permission: Beck DM, Rees G, Frith CD, Lavie N. Neural correlates of change and change blindness. Nature Neuroscience 2001;4:645–650

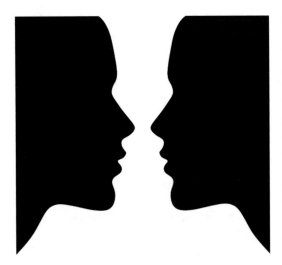

Fig. 4.8 Bistable figures: The Rubin Vase. Image credit: Shutterstock.com

occur with certain strong emotional stimuli. For example a percept of a frightening face triggers a fear response in the person's autonomic nervous system that may not be accompanied by a conscious perception of the stimulus [29].

Grouping

This principle can be viewed as nature's mechanism to counter camouflage. Grouping is a temporally based process, with temporal structure (referring to the timing of changes) or grouping according to the same global pattern, even in the absence of the changes being synchronous. First described by gestalt psychologists about 100 years ago, this can be explained more precisely in neurophysiological terms; time-based neural spiking dominates spike synchrony [30]. A possible neurophysiological explanation of grouping has been proposed by Singer and Gray. Different neuronal spikes fire in an oscillatory manner (in the cat) in response to the various components of an image such as a camouflaged animal at a rate of 40 Hz. The spike trains become synchronous on recognition of the whole image or animal in this instance [31]. Context is critical for perception and the same principles apply to the haptic (tactile) processes and grouping operates not only in the visual but also in the haptic domain [32] (Fig. 4.9).

Fig. 4.9 Grouping and camouflage—the snow leopard. With permission: Shutterstock.com. Image credit Dennis W. Donohue

Peak Shift Effect and Supernormal Stimulus

These are regarded as one of the foundations of the newly conceived neuroesthetics discipline. However it was Tinbergen who first discovered this phenomenon in herring gulls (Larus argentatus) whereby he noted that the young chicks peck a red patch of the mother's lower beak to elicit feeding [33]. Today it is employed by artists for example in drawing caricatures whereby a particular feature of an object or face (bushy eyebrows, shape of the nose) is amplified. Other fundamental processes and hypothesis that help us understand art and neuroesthetics and the contribution of Zeki's laws of constancy and abstraction and Ramachandran's eight laws of the artful brain are discussed in the section of Art and the brain.

From an evolutionary point of view this may represent a mechanism whereby our vision sense fast tracks information to aid in the survival (predation, rapid recognition), food sourcing, and procreation of our species. Our lexicon of shapes and images within the inferior temporal lobes have presumably been fashioned in our evolutionary development to prefer or favor certain shapes and images. These may be relevant to our preference for particular facial features and body shapes. This may also be an explanation pertaining to the meaning conveyed by abstract art. In Ramachandran's view, successful artists are able to construct ultranormal stimuli that have a greater excitatory effect on some of our visual circuitry by "tapping into the figural primitives of our perceptual grammar" [34].

Overview of Symptoms and Syndrome Presentation of Elementary and Complex Visual Disorders

Disorders in this sensory realm may manifest with syndromes that result from an inhibitory effect of the cortical function by lesions or electrical mechanisms (hypofunction), stimulatory effects (hyperfunction), as well as remote effects (diaschisis mechanisms). Recent functional neuroimaging and lesion study analyses have allowed a hypothetical classification of the various diaschisis mechanisms, namely diachisis at rest, functional diaschisis, and connectional and connectomal diaschisis [35]. In pathophysiological terms, therefore, the presentations of human brain lesions may be grouped into topological hypo- and hyperfunction and connectomal (or hodological) hypo- and hyperfunction as recently proposed by ffytche et al. [36]. These processes may not be present in isolation but with a particular syndrome, more than one pathophysiological, i.e., hypo- or hyperfunction with or without hodological hypo- and hyperfunction may be implicated. Examples of hypofunction include scotomas, field defects, and cortical blindness. Hyperfunction effects include hallucinations and remote effects, presumably diaschisis effects, include diverse syndromes such as post-traumatic stress syndrome, Charles Bonnet syndrome, and peduncular hallucinosis. Migraine with aura is a good example of the hypofunction of a scotoma being closely followed by the hyperfunction of the scintillating teichopsias.

A Classification of the Principal Circuits Relating Visual Disorders, Topological, Hodological, and Ancient Circuitry Phenomena (Fig. 4.4)

1. V1/V2/V3 lesions and related syndromes including evolutionary ancient syndromes
2. Ventral stream (what)-related syndromes
3. Dorsal stream (where)-related syndromes
4. Occipito-parietal (when) stream-related syndromes
5. Occipito-temporal amygdala and related syndromes
6. Occipito-inferior frontal and related syndromes
7. Lesions outside the cortex affecting vision—Charles Bonnet (retinal disease and peduncular hallucinosis (brainstem lesion))

Visual Disorders Associated with the More Elementary Cortical Areas, V1, V2, and V3 for Example

Heminanopias, scotomas visual hallucinations: Represent basic loss of vision deficits, a type of hypofunction-deficit syndrome that includes homonymous heminanopias, quadrantanopias, and scotomas. Elementary visual hallucinations or nonformed hallucinations may pertain to colors, lights, and geometric figures or shapes and may be without particular structure, termed phosphenes, which can originate from the retina or visual cor-

tex, in the absence of light stimulation and due to electrical, mechanical stimulation or radiation. They may be associated with neurological diseases such as stroke and multiple sclerosis but also in visual deprivation, sometimes with a sequence of different colors appearing (prisoner's cinema), meditation, and stimulating drug use and in astronauts [37]. If with some kind of structure they are termed photopsias also. Fortification spectra or teichopsia is a common form occurring in migraine for example with scintillating, jagged lines that were likened to a fortress [38] (Table 4.1).

Agnosias (apperceptive and associative): A general definition of agnosias is an acquired disorder whereby in the context of normal visual perception and language there is a lack of knowledge or meaning of the visual stimulus. Perception and recognition may be considered to lie on a continuum and agnosia exists somewhere between these two poles, that is, beyond the initial perceptual process but prior to the recognition achieved through multimodality memory processes. Nonverbal and verbal recognition is important in diagnosing agnosias as is copying the object in question and matching to the real object. Apperceptive visual agnosia and visual associative agnosia are clinical deficits that demonstrate the impairments related to grouping and line continuity. In apperceptive agnosia (also referred to as visual form agnosia), the person is able to verbally describe simple shapes such as a square, circle, triangle, and the number 4 for example, but unable to copy them as the appreciation of the

Table 4.1 Syndromes related to primary occipital cortical regions BA 17–19 (V!–V3)

1. Hemianopias, scotomas
2. Simple hallucinations, phosphenes, photopsias
3. Visual apperceptive agnosia (visual form agnosia)
4. Visual associative agnosia
5. Perceptual categorization defect
6. Cortical blindness
7. Anton's and inverse Anton's syndrome
8. Blindsight
9. Riddoch syndrome
10. Scieropia
11. Astereopsis

object parts or wherein basic sensory representations of objects cannot be assimilated. The deficiency appears to be at a more elementary level whereby there is difficulty in identifying the continuity of lines and the grouping of and integration of the subcomponents of the figure or object. This is well demonstrated by the image of 7415 Landis et al. [39] and is a function of the posterior to middle portions of the inferior temporal lobe (Fig. 4.10).

In associative agnosia, the person is able to copy a line drawing of a cup, pen, or ring but is unable to identify the object or assign meaning to the particular object. The more basic visual sensory data is registered by the processing of the symbolic meaning of the object that is impaired and is a function of the more mid to anterior aspects of the inferior temporal lobe [40]. A similar disorder, anomic aphasia, may be confused with visual associative agnosia, which may be differentiated by employing both verbal and nonverbal testing. In the former, when shown the picture of a key, they are able to state that it is a device to open doors for example, but unable to name it. This can be tested by picture-matching tasks, such as a key with the door and its lock, which would be done correctly. In the latter, by requesting a verbal description of the object in front of them, there is ability to describe the basic shapes and contours but cannot associate name of the object insofar as the symbolism it represents, such as a key for example. The approximate lesion locations of the syndromes of cortical blindness, apperceptive agnosia, and associative agnosia are depicted in Fig. 4.11.

Perceptual categorization deficit: Refers to those people who have difficulty in recognizing scenes or objects that are presented in unusual angles, views, or illumination. The key features of a particular object may not be seen and the object therefore not recognized. It has been postulated that this is a disorder of visual imagination rather than due to the perception process. Responsible lesions are right more than left inferior posterior parietal lobes [41].

Cortical blindness, Anton's syndrome, inverse Anton's syndrome, and blindsight: Extensive bilateral damage to the visual cortex and contiguous optic radiations severely affects vision to the point of blindness, termed cortical as opposed to retinal or ocular blindness. This may be associated with active denial of blindness, in which case it is referred to as Anton's syndrome. Pathophysiologically this may represent a diaschisis or disconnection between the primary visual area and parietal lobe (the site of body schema representations). Inverse Anton syndrome, which is defined as a "denial of seeing," is proposed to be a disorder of diaschisis or disconnection of the frontoparietal attentional network from primary visual cortex [42].

As discussed above, blindsight may occur in a clinically blind person from occipital lobe lesions, and retains (more than by chance alone) some capability of navigating obstacles in their path or depicting a light pointed at them in their blind hemifield for example. Pathophysiologically, this is explained by retina sending signals via the superior collicular nuclei and pulvinar to the intact parietal cortex (Fig. 4.11).

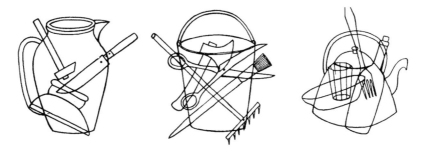

Fig. 4.10 Apperceptive agnosia may be tested by the Poppelreuter Ghent Figure Tests. Figure credit: Della Sala S. Poppelreuter-Ghent's Overlapping Figures Test: Its Sensitivity to Age, and Its Clinical Use. Archives of Clinical Neuropsychology 1995;10:511–534

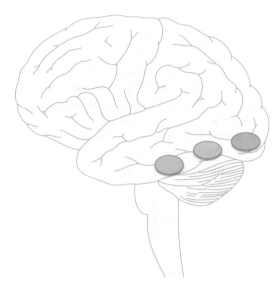

Fig. 4.11 Schematic, approximate inferior temporal cortical lesion areas associated with visual cortical blindness (*pink*), apperceptive agnosia (*blue*), and visual associative agnosia (*green*)

Riddoch syndrome and gnosanopsia. Agnosopsia and gnosopsia: The Riddoch syndrome can be considered a variant of the cortical blindness group of disorders, which includes Anton's, inverse Anton's, and blindsight. As the V5 motion cortex is functional it is presumed that this is activated via the more direct connections from the eye via the pulvinar and lateral geniculate nucleus of the thalamus [43].

The concept of the Riddoch phenomenon or syndrome includes these varying manifestations, first described by a British neurologist during the World War 1 syndrome; George Riddoch attributed this syndrome due to the limited shielding of the occiput by British helmets at the time and the pervasive trench warfare that frequently only exposed the head and cranial trauma accounted for about 25 % penetration injury sustained by their forces. There seemed to be a predilection for the occipital region and cerebellar regions. These wounds typically did not cause death and another factor that advanced since to explain this type of injury was the relatively low-velocity bullet used during that time. His index case involved an officer leading his men in battle when he sustained a bullet injury that traversed his occipital lobe, leaving him dazed, likely concussed but able to continue the fight. With a subsequent collapse he awoke only 11 days later at which time he was disorientated for place, perceiving shapes and colors but still able to identify movement without being able to decipher what was moving. Hence there is awareness of movement but without the percept of the moving object termed gnosanopsia [44]. Awareness and visual discrimination may therefore be uncoupled by a lesion in the V1 region. Agnosopsia refers to lack of awareness but retained discrimination. A subsequent report over 80 years later by Zeki and ffytche had additional observations with fMRI in that a difference in V5 area cerebral activity with relatively less activity was seen in agnosopsia compared to gnosopsia.

Hence they proposed three variations within this syndrome: being able to discriminate and also being aware, termed gnosopsia (gnosis in Greek refers to knowledge and opsia to vision); being aware but unable to discriminate an item in the blind visual field (gnosanopsia); and being able to discriminate but no consciousness awareness (agnosopsia). The latter is closest to what is generally referred to as blindsight [43].

Scieropia occurs with bilateral occipital lesions and is the symptom of an overall dimming or decreased illumination of the environment, or general darkening. It was first described by Wapner whose patient's vision was explained as a twilight vision [45].

Stereopsis and astereopsis: Binocular vision enables stereopsis or depth perception and good visual acuity and ocular alignment must first be determined before making a diagnosis of astereopsis. Dating back to our primate past, stereopsis was critical in judging distance and depth perception in our challenging arboreal environment. Today it remains a very important attribute in anyone operating an aircraft and vehicle and in numerous technical occupations. Insights into the neurophysiology of stereopsis have been gleaned from lesion studies and functional imaging study. A commonly used test, the random dot stereogram test, and fMRI supported the dorsal parieto-occipital cortex and superior parietal lobule in processing stereopsis with a right hemisphere dominance [46]. This was corroborated by a regional blood flow study, using PET brain imaging and Julesz-type random dot stereograms that activated both middle temporal lobe regions, BA 18, 19, and 7 and again with a right hemisphere predominance [47]. The intraparietal sulcus appears to be an important hub for this faculty as demonstrated in another functional MR study. The human middle temporal lobe V5+, together with the IPS and ventral occipital region, forms part of a right hemisphere-dominant network for both stereopsis [48, 49]. Less commonly bilateral lesions of the superior visual cortex cause astereopsis but the overwhelming evidence so far is that this may be a right hemisphere-dominant function [50]. Binocular depth perception involves neural circuitry of the nuclei, posterior parietal lobe, and associated white matter connectivity between these. Leukoaraiosis secondary to cerebrovascular disease can impair this circuitry and can produce stereoblindness and impair stereoacuity. This has been shown to occur more often with right hemisphere as compared to left hemisphere involvement and may also be due to subcortical and parietal involvement [51]. Testing proceeds with testing of ocular alignment first by the red-green glasses and Randot and then cortical association areas concerned with stereopsis by the random dot stereograms or the Four Dot test. Stereopsis can be a critical ability depending on occupations as well as pastimes as illustrated by the recently reported windsurfer illusion, involving stereopsis ability [52].

Ventral Stream Disorders

These involve alterations of objects, color, text, faces, and places. Object alterations may take the form of hallucinations or illusions or hypofunction, such as object agnosia (Table 4.2).

Color may be seen without sensory input, termed chromatopsia, and may be brighter than normal, termed umbropsia or seeing a particular color, such as red, without specific vision stimulation, termed erythropsia. Hyperchromatopsia and achromatopsia represent hyperfunction and hypofunction effects of the color areas of the visual cortex and its connection (V4 and perhaps V8). Common acquired disorders of color perception that may involve part (hemifield, quad-

Table 4.2 Ventral stream disorders

| Achromatopsia |
| Color anomia |
| Color agnosia |
| Hyperchromatopsia (increased activity in color cortex) |
| Color hallucinations/illusions |
| Prosopagnosia |
| Facial hallucination/illusion |
| Facial intermetamorphosis |
| Prosopometamorphopsia |
| Object agnosia |
| Object hallucination/illusion |
| Object anomia |
| Optic aphasia |
| Object agnosia |
| Environmental agnosia |
| Landscape hallucinations |
| Micropsia, macropsia |
| Pelopsia (appear closer), telopsia (further away) |
| Synesthesia (colored hearing, colored music, colored grapheme) |
| Pareidolia—Images of faces and objects seen in visual patterns such as clouds, plants thought to be due to a hyperconnection of visual areas |
| Lilliputian hallucinations |

rant) or all of the visual field, with normal visual acuity, are termed central achromatopsia. Clinical presentation may be partial with a fading of colors, often described by patients as everything appearing in shades of gray to complete loss where only black and white are seen. Causative lesions usually involve the bilateral occipito-temporal cortices of the lingual or fusiform gyri, due to posterior cerebral artery infarcts. They are often associated with visual agnosia, prosopagnosia, visual field, and pure alexia. Hemiachromatopsia may occur due to a unilateral occipito-temporal lesion and if due to right occipito-temporal lesion that may be the sole cognitive deficit without other demonstrative deficit [53]. It is not known how, but the left and right posterior hemispheres process color in different ways. Neuroimaging by functional MR scanning has implicated the causative lesion to be the lingual gyrus, middle third, or associated white matter adjacent to the posterior part of the lateral ventricle, termed V4 and V8 [25]. Testing of achromatopsia can easily be performed moving a red sphere from the hemifield ispilateral to the lesion to the contralateral side and a typical result is that the hemifield color may change from red to grayish in the process. Color agnosia, a related problem that may be misdiagnosed as achromatopsia, refers to an inability to name colors as well as being unable to point to colors when referred to their given names. The causative lesions are usually left hemisphere and associated deficits are a right homonymous hemianopia, alexia, and intact color vision in the left hemifield. Color anomia represents a two-way deficit of both naming a color and being unable to point to a color given its name. To distinguish these conditions a number of tasks are required including color matching, pointing to the color and asking the name of the color, saying a color name and asking to point to it, and Ishihara plate testing. Color anomia, color dysphasia, and color agnosia may be viewed as being on a continuum with progressively more impairment.

Prosopagnosia: Refers to a specific inability of recognizing familiar faces with retained ability to recognize generic aspects of faces. Hence such patients may not be able to recognize their own face. Similarly, agnosias can relate to non-living items such as cars and different species of birds. There may however be a nonconscious recognition of familiar faces identified by skin conductance response testing. This underscores the premise that neural circuitry responsible for processing emotionally relevant signs is separate from those that process basic features and facts related to a stimulus [54].

Prosopometamorphopsias refer to different type of illusions of faces. Lesion studies have correlated these to the lateral occipital region. Hallucinations of faces without alteration of the basic features have been linked to abnormalities in the fusiform face area or occipito-temporal cortical area [55]. In a stimulation study during assessment for surgical epileptic candidates, right ventral prefrontal cortical stimulation resulted in both face illusions and hallucinations and was regarded as part of face processing circuitry [56].

Environmental agnosia: An inability to recognize familiar places with relatively intact perception and memory and attention. It has been attributed to right medial temporo-occipital lesions, rarely due to left TO lesions. It represents a category-specific agnosia akin to face agnosia or prosopagnosia [57]. Object shape is generally deciphered within the lateral occipital cortex according to texture, color, and configuration. However Cavina-Pratesi et al. found that the areas specific for color, texture, and shape were separate from those that processed color, texture, and shape together, such as occurs for the more complex perceptual stimuli that pertain to places and faces [58].

Visual object agnosia may be both general and category-specific agnosia. With general visual object agnosia, this refers to an inability to recognize an object at a generic level such as inability to identify the percept as a face at all or that a bird is a bird. Optic aphasia may be associated with visual naming deficit in the setting of normal tactile and auditory naming ability. The causative lesions in the medial occipito-temporal cortices are more extensive than with prosopagnosia for example. Category-specific visual object agnosia refers to difficulty with seeing a particular object within a generic category that may be with respect to living things (birds, plants) or inanimate (cars and tools).

Visual text hallucinations: Described in people with Lewy body disease and various ocular abnormalities, hallucinations of text may occur in up to 25% of people with visual hallucinations. These may consist of letters (orthographic hallucinations) or words and be nonsensical or convey some meaning by way of meaningful phrases or sentences (syntacto-semantic visual hallucinations). These need to be distinguished from similar schizophreniform presentations and responsible lesions have been bilateral occipito-temporal infarcts [59, 60].

Thalamic dazzle syndrome: First described by Cummings and Gittinger in 1981 it comprised of a delayed (3 months) light intolerance after a right thalamic and occipital stroke. It was considered to be analogous to the Dejerine–Roussy central pain syndrome that may occur after thalamic infarction [61]. A subsequent report involved a monocular dazzle syndrome after bilateral paramedian thalamic infarcts by Du Pusquier et al. Although concomitant dysphasia, vertical diplopia, right visual field loss, and right-sided long tract signs accompanied the deficit, the dazzle syndrome was the most distressing to the patient [62].

Synesthesia: The definition of this syndrome is the experience of a sensation in one domain by stimulation in another sensory domain. Examples include colored hearing, colored music, and colored graphemes. Neurophysiologically there is a proposed hyperconnectivity to parietal lobe and temporal centers which may account for synesthesia of color with numbers and sounds. For example grapheme-color synesthesia has a prevalence of 1.2 % in the general population, but synesthesia in general may be acquired such as secondary to multiple sclerosis, stroke, and sensory deprivation [63].

Visuospatial agnosia refers to a difficulty with locating a room in one's own familiar house which would be an extreme example but more often locating a building or street in the city may be another manifestation. A number of different mechanisms may be responsible such as memory abnormalities, agnosias, and neglect. A visuospatial agnosia is usually due to bilateral occipito-temporal and occipito-parietal areas. Geographical disorientation and planotopagnosia is part of this syndrome with difficulty in finding countries, cities, or states on a map, due to similar lesion locations. Functional magnetic resonance imaging studies implicate the parahippocampal regions more in the place identification in comparison to face recognition. From functional imaging studies (fMRI) it appears that a specific area within the parahippocampus (parahippocampal place area) corresponds more to places more than faces. Disorders of spatial analysis can be tested by the judgment of line orientation test [64].

Metamorphopsias or *dysmetropsia* refers to a group of illusory conditions with respect to size and shape that include visual aberrations of micropsia (smaller), macropsia (larger), teleopsia (far away), and pelopsia (very close). These are often associated with seizure conditions but also retinal pathologies, and neurovirological, neoplastic, and vascular causes affecting the occipito-temporal region areas BA 18 and 19. These are

also known as the Todd syndrome or Alice in Wonderland syndrome [65, 66]. These may be due to a process known as constancy whereby the impairment in object constancy is due to temporal lobe integration of the distance of the object with respect to visual angle of the retina [67]. Lilliputian hallucinations refer to the experience of seeing small people often with a particular outfit or costume. It has been associated with migraine, schizophrenia, encephalitis, and certain medications such as amantadine. The proposed mechanism is one of impairment of the visual constancy mechanism [68, 69].

Pareidolia is an illusory perception of non-existent objects or faces, which may take the form of pareidolia seen in environmental patterns for example trees, flowers, or clouds. It has been attributed to hyperconnectivity within visual cortical areas. Administration of a specific pareidolia test has been reported with an increased incidence of Parkinson's disease as opposed to the general population [70, 71].

Dorsal Stream Syndromes

Simultanagnosia and Balint's syndrome: Whereas visual agnosia may be regarded as the prototypical disorder of the ventral stream, simultanagnosia is the prototypical disorder of the dorsal stream. Simultanagnosia may also be regarded as the central component of Balint's syndrome. Visual acuity and identification of even very small objects such as small print or ants on the floor are possible. However vision is restricted to shaft vision. Also referred to as piecemeal vision, it refers to an impairment in the ability to maintain attention over the different parts of a visual field that results in a particular object becoming fragmented. Basic visual function is preserved and the preservation of macular vision allows excellent visual acuity but with extreme restriction of the visual field [72] (Fig. 4.12).

Dorsal and ventral simultanagnosia: With dorsal simultanagnosia, attention is markedly restricted in the visual domain and this precludes the ability to see more than one component of an object or

Fig. 4.12 The brain has a
two-tier system: the
superior colliculus/pulvinar
system unconscious circuit
and the lateral geniculate
nucleus conscious pathway.
Image credit: Shutterstock.
com

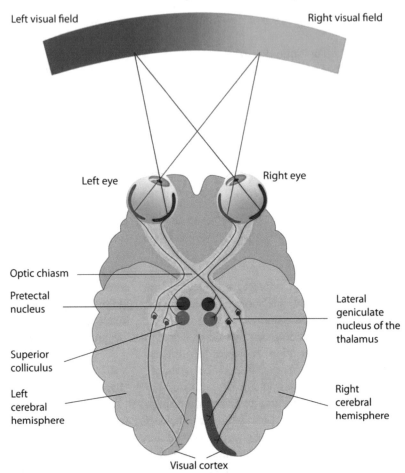

an image at a time. In the case of ventral simultanagnosia an object can be recognized but with a limitation of the number per unit time. For example when testing with a complex picture, for example the Cookie Theft Picture test which is commonly employed for such purposes, the interpretation of the number of people present is slow and often inadequate [73, 74] (Table 4.3).

Balint's syndrome comprises of simultanagnosia—not being able to perceive the visual field in holistic manner, optic ataxia—impaired target pointing using visual guidance and optic apraxia, an impairment of being able to direct gaze at a new stimulus, also viewed as visual scanning deficit. Although macular vision is intact there if an inability to stabilize the focus and sequence of movements resulting in the fragmented vision. The optic ataxia component is associated with normal

pointing ability to personal body parts because of retained somatosensory (tactile, auditory) input. Only visual guidance is affected. With optic apraxia, the normal saccadic process is affected and consequently the impaired ability to correctly orient to novel stimuli the visual fields, especially the periphery. Usual associated lesions include an inferior quadratic field cuts. Neuropathologically, the responsible lesions involve the parts of the parietal lobe that subserve visual attention in the spatial domain and typical lesion locations associated with Balint's syndrome include the more dorsal occipito-parietal regions whereas visual agnosias are typically associated with the more ventral occipito-parietal region [75]. A comprehensive review of current theories of simultanagnosia by Dalrymple et al. suggests that causative mechanisms are likely to be related to both

Table 4.3 Dorsal stream syndromes

Dorsal and ventral simultanagnosia
Optic ataxia
Oculomotor apraxia
Balint's syndrome
Akinetopsia motion vision
Cinematographic vision
Motion hallucination/illusion
Palinopsias
Polyopia
Entomopia
Visual perseveration
Visual fading
Visual extinction
Trailing phenomenon
Delayed palinopsia (object returns)
Illusory visual spread (object pattern spreads)
Positive afterimages
Negative hallucination
Visual allesthesia
Inverted vision
Visual vestibular disorders
Autoscopy/heautoscopy
Extracampine hallucinations
Oculogravic and oculogyric illusion
Zingerle's automatosis
Paroxysmal perceptual alteration (PPA)

object- and space-based attention processing abnormalities [76]. Pathological processes commonly causing simultanagnosia include eclampsia, posterior reversible encephalopathy syndrome (PRES), the primary form of Call Fleming syndrome, and bilateral watershed infarctions [77].

Akinetopsias and Zeitraffer phenomenon: These are defects in motion processing that encompass impaired smooth pursuit eye movement, problems with appreciating the movement of objects, and deficits such as cinematic vision. The neurophysiology and neural circuit implicate the parieto-occipito-temporal regions as well as the parietal insula and cerebellum. Perception of visual movement is dependent on temporal intervals, spatial displacements, and movement sensations within the visual realm. The cortical area referred to as V5/MT is specific for movement appreciation or processing of motion. Both functional MR studies and detailed case reports of

lesion studies have attested to this [78]. Akinetopsia is associated with pathologies such as stroke, epileptic seizures, and the focal, posterior cortical atrophy syndrome of Benson [79, 80]. The Zeitraffer phenomenon is a particular subtype in which there is an alteration in the perception of the speed of the object in motion [81]. A commonly employed neuropsychological test for ascertaining this faculty is by the Random Dot test [82].

Palinopsia group of disorders, polyopias, entomopias, delayed palinopsia, and illusory visual spread: Palinopsia may be defined as persistence of the visual appreciation of the image of an object after it has disappeared or removed or the recurrence of it when it is no longer present. A number of different palinopsias may be differentiated such as visual perseveration, illusory visual spread (design or motif of object spreads to the adjacent areas), delayed palinopsia (image of item returns after a pause), and trailing phenomena (stationary items, images, or objects trailing a moving item). Neurophysiologically this has been attributed to an impairment of the coordinate systems that are a function of the parietal lobe and the occipito-parietal connections. Whereas diplopia is usually due to oculomotor nerve dysfunction or their brainstem connections and less commonly neuromuscular lesions of the ocular muscles, polyopia, seeing many images of an object frequently in rows or particularly numerous typically over a hundred is termed entomopia. Typical causative lesions include migraine, head trauma, stroke, substance abuse related, as well as medications such as topiramate. When there are multiple representations of the object the term polyopia is used and when hundreds this is referred to as entomopia [83–86].

Visual alloesthesia: Refers to the condition whereby the visual field is tilted, inverted, or rotated and ascribed to impairments of the circuitry connecting the visual cortex, parietal cortex, and vestibular cortex [87, 88].

This is different from visual alloesthesia in which a sensory stimulus, applied to one side of the body, is reported to be felt or perceived on the contralateral side. Most commonly appreciated with left hemisensory tactile neglect due to right hemisphere stroke, for example, a touch stimulus of the left hand is perceived on the right unaffected hand [89, 90].

Autoscopic hallucinations: Derived from Greek, autos (self) and skope (looking at), it refers to an out-of-body experience with the projection of one's body or face above or from an external point of view typically lasting seconds to minutes. They may take a number of different forms and Brugger et al. described at least six different forms. These include the feeling of a presence of another person in the immediate extracorporeal area, inner heautoscopy (one's inner organs are the subject of the visual hallucination), negative heautoscopy (unable to perceive one's body either when seen in a mirror), the autoscopic hallucination (exact visual percept of one's face or trunk) of oneself, occasional perception of only one's face or trunk, the out-of-body experience, and heautoscopy also defined as heautoscopic proper referring to seeing one's self [91]. These experiences have been correlated with the dysfunction of the TPO and impaired integration of tactile, vestibular, visual, or proprioceptive tactile and vestibular stimuli [92]. Related disorders of parieto-temporal dysfunction, whether due to hypo- or hyperfunction or by hodological effects, include oculogravic illusion, oculogyral illusions, caloric stimulation-related visual hallucinations, Zingerle's automatosis, and neuroleptic induced oculogyricc crises, also termed paroxysmal perceptual alterations [93–95]. Oculogravic refers to the sensation of the displacement of items with a change in gravity (described with respect to diving aeroplane for example), oculogyral (sensation of movement of a light in a darkroom after being rotated). In Zingerle's automatosis there is the sensation of a visual hallucination in association with vestibular or motor impairment and when a text or other visual pattern is briefly intensified also referred to as paroxysmal perceptual alteration 5 after neuroleptic medication [96, 97].

Visual axis perception disorders such as angular, horizontal, and vertical appreciation has been described with anterior parietal lobe impairment or disconnection from the anterior parietal lobe and cortical vestibular regions or vestibular nuclei [98].

Perky effect and eidetic imagery: Visual mental imagery may interfere with actual visual perception—termed the Perky effect. It may lead to a decrease in perception and considered a connectivity impairment between the visual cortex and memory. Eidetic (photographic memory) refers to the ability to accurately recall objects, sounds, or images with exceptional accuracy. Eidetic imagery may represent the opposite to the Perky effect in a sense, with a hyperconnectivity between these two visual cortex and memory areas [99–102].

Occipito-Parietal (When) Pathway

The estimation of the motion of objects, whether occurring simultaneously or not, and the overall temporal order of events have all been attributed to the right parietal cortex as they are impaired with lesions in that region. This has also been shown to occur in response to transcranial magnetic stimulation over the right parietal cortex. Differentiation of two events from the same spatial coordinate but that differ in the temporal (time) domain forms the basis of visual neglect syndromes. Furthermore the right parietal lobe is dominant over both visual fields which differs from the parietal lobe control over spatial attention whereby the contralateral field is controlled by the one hemisphere. Hence the right parietal lobe is likely the site of the "when" pathway and is dominant in that it subserves both visual fields (Fig. 4.13) [103–105].

Disorders of the Occipito-Temporal Cortex Including the Amygdala (Fig. 4.14)

The amygdala conveys critical information concerning social knowledge and facial emotions in response to visual information such as fear and aggression [106, 107]. Some relevant syndromes that are related to the occipito-temporal, most importantly the amygdaloid complexes and their circuitry, include the following:

Kakopsia, kalopsia: These refer to unpleasant and pleasant subjective experiences, respectively, in response to visual perceptions. These are derived from the Greek words kakos (unpleasant or bad, and opsis, seeing). The term kalopsia derives from the Greek terms kalos, seeing items as beautiful or comforting, and opsis, seeing. A likely explanation may be due to hyperfunction on a connectivity, diaschisis, or hodological

Fig. 4.13 Simultanagnosia—piecemeal vision

means of the links between the emotional brain networks (particularly the amygdala) and the visual cortex [108, 109].

Post-traumatic stress disorder (PTSD): In PTSD, the person relives previously stressful events, experiences flashbacks, and often has a multi-sensory modality experience with strong emotional, vivid memory, recurring dreams, visual hallucinations, hyperarousal, and autonomic components. It may be regarded as a occipito-amygdaloid hyperconnectivity syndrome.

Kluver-Bucy syndrome: This may present as a visual agnosia with objects not recognizable. It is associated with hypersexuality, a kind of automatic assessment of objects, and placing them in their mouth. Although associated with anterior temporal lobe lesions, it represents a syndrome with a prominent part of which is visual agnosia.

Disorders of the Occipito-Frontal Fasciculus via the Extreme and External Capsule (Figs. 4.15 and 4.16)

Delusional misidentification syndromes or content-specific delusions refers to a number of differing delusions usually associated with right frontal lesions (see frontal lobe chapter). With the more common types such as Capgras and

Fig. 4.14 Right parietal lobe function—event timing. Adapted from Battelli L, Pascual-Leone A, Cavanagh P. The "when" pathway of the right parietal lobe. Trends in Cognitive Sciences 2007;11:204–210

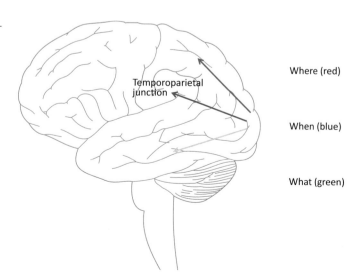

Temporoparietal junction

Where (red)

When (blue)

What (green)

Fig. 4.15 Occipito-temporal pathway. Figure with permission: Catani M. Jones DK, Donato R, ffytche DH. Occipito-temporal connections in the human brain. Brain 2003;126:2093–2107

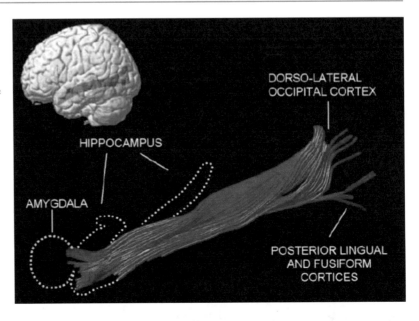

Fig. 4.16 Inferior fronto-occipital fasciculus. Figure credit: Caverzasi E, Papinutto N, Amirbekian B, Berger MS, Henry RG. Q-Ball of Inferior Fronto-Occipital Fasciculus and Beyond. PLOS One 2014;9:6, e100274

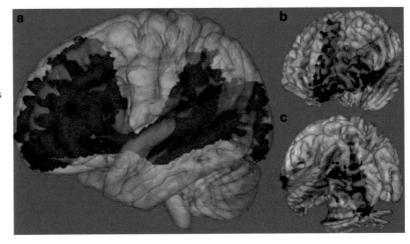

Fregoli's syndromes, the network for facial recognition remains intact but the circuitry mediating the emotional associations and responses is impaired. Autonomic monitoring during familiar face presentations is also abnormal. In a strict definition it is not so much a disorder of thinking but a processing abnormality in the perception domain. Capgras syndrome can in a sense be viewed as opposite to prosopagnosia [110].

The topographical or hodological pathophysiological features of the visual imagery or visual memory-related syndromes such as anorexia and Charcot Willbrand syndromes are less certain.

Cerebral lesions due to strokes, tumors, or infections may target neural circuitry that mediates dreaming. The resultant clinical syndromes may take the form of Charcot's variant with the specific faces, movement, or color during dreams or Willbrand's variant or global anorexia with complete absence of dreaming. Anoneirognosis refers to a reality-dream/reality confusional state. Etiopathogenetic mechanisms have included carbon monoxide poisoning, dementias, stroke, cerebral trauma, and hydrocephalus and usually are associated with bilateral occipito-temporal lesions [111, 112].

Remote Effects: Tracts 6 and 7 (Fig. 4.4)

Charles Bonnet syndrome (CBS): This refers to complex visual hallucinations in association with ocular visual loss. It may represent a type of release phenomenon or denervation hypersensitivity, or a hyperactivity, on a hodological basis from a remote anatomical area. The development of complex visual hallucinations, typically Lilliputian type, consists of people, characters, and sometimes objects, diminutive in appearance. Most commonly they are of people, faces in both color, or movement, lasting 3–5 min, and occur about three times per day and emerge over 6–12 months. People are insightful that these hallucinations are not real and the hallucinatory experience is limited to visual domain and does not occur in the other sensory systems such as auditory, tactile, smell, or taste [113].

Peduncular hallucinations: Refer to visual hallucinations secondary to brainstem lesions such as infarcts. They are perceived as realistic in contrast to those of CBS and may be very vivid and colorful and typically occur at nighttime. They have been attributed to a disruption of the ascending reticular activating system and has similarities to the type of hallucination that occurs in normal REM sleep [114].

Clinical Bedside and Metric Testing

Bedside

Establishing basic visual processing and normal function is essential before progressing to more complex visual disorder assessment. Visual acuity is tested by Snellen tests or similar charts and visual fields are assessed by confrontation testing or computerized perimetry.

Color blindness and color vision can be tested with a variety of simple tests such as with Ishihara plates, the six-color items in the Boston Diagnostic Aphasia Test, or the Western Aphasia subtest items. Thereafter the patient is presented with drawings or photographs of objects, faces, animals, colors, and complex scenes. These will identify patients with simultanagnosia and prosopagnosia. Complex pictures such as the WAB picture or Boston Aphasia Diagnostic Cookie Theft Picture Test are useful for simultanagnosia testing [74]. Differentiation between visual form agnosia and associative visual agnosia can be achieved by instructing the person to copy drawings of simple objects. This will be possible only in those with visual agnosia but if the person has visual form agnosia further differentiation between associative visual agnosia from naming impairment is then required. The latter can identify the object for example by pantomime. The person with visual agnosia can identify the object through other sensory modality such as tactile or auditory route. The cueing conditions of the Boston Naming test provide a mechanism for distinguishing between a naming impairment and visual associative agnosia.

Metric Tests

There are several dozen metric tests useful for discerning complex visual disorders and their associated hodological syndromes. A number of the more commonly used tests are presented below. Further orientation with respect to these metric tests may be found within Lezak's 5th edition of Neuropsychological compendium [115]:

1. Visual acuity by Snellen or similar chart
2. Visual field testing by confrontation or computerized charting
3. Boston Naming Test
4. Pyramid and Palms Test
5. Ishihara plates
6. Line drawings
7. Famous faces
8. Objects and colors
9. Visual Object and Space Perception battery (VOSP)
10. Birmingham Object Recognition Battery
11. Benton's Judgment of Line Orientation Test
12. Poppelreuter Overlapping Figures Test
13. Hooper Visual Organization Test
14. Gollin Figures
15. Test of Facial Recognition
16. Color—Farnsworth's Dichotomous Test for Color Blindness (D-15)

Fig. 4.17 Visual object and space perception battery subtest images. Figures with permission: Warrington E, James M. VOSP. Harcourt Assessment. The Psychological Corporation. The Thames Valley Test Company, London, 1991

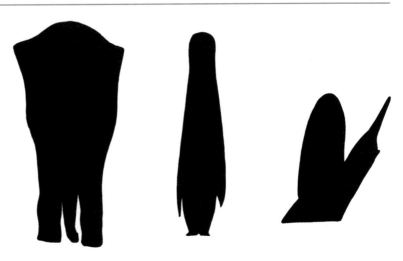

17. Color to Figure Matching Test
18. Line Bisection Test
19. Familiar and unfamiliar face tests such as the Benton and Van Allen's facial recognition test can be employed for objective assessment

The VOSP test for example is particularly useful in that it tests for a number of different complex visual processing modalities both for the ventral object identification and dorsal visuospatial circuitry, consisting of eight subtests with normative data. For example, object perception tests include views of letters, objects, and animals that are presented in incomplete forms. The screening test is used for evaluation of whether the candidate may be able to proceed to the subsequent eight subtests that include:

Shape detection screening test whereby 50 % of cards have degraded letter "X"s

1. Incomplete letters
2. Silhouettes
3. Object decision test
4. Progressive silhouettes
5. Dot counting tests
6. Position discrimination test
7. Number location test
8. Cube analysis test

The incomplete letters test includes 20 letters that are variously concealed ranging from 30 to 70 % with the mean score in a control group

19.3 ± 0.8. The silhouettes test consists of 15 silhouettes of objects and 15 of animals, which are presented in various angles and orientations that challenge the observer. Mean values for over 50-year age group are 23 ± 4 out of a total of 30. The object decision test has 20 cards with 4 black shapes per card, only one of which is a real object silhouetted with the normative range 17.7 ± 1.9.

The progressive silhouettes include two objects, a handgun and trumpet that progressively become more revealing, each having ten sequential images. Mean values are 10.8 ± 2.5 (total 20 images). The dot counting test presents arrays of dots that vary between five and nine in a random pattern with the maximum score of 10 and the control group mean of 9.9 ± 0.2. The position discrimination tests attempt to discern whether a single black dot on a white card is in the center or off center. Two adjacent cards are presented, one of which has centered dot, one being off center. Of 20 stimuli cards, normal controls are 19.6 ± 0.9. The number location test uses two cards, one with numbers randomly placed and a second card with a single dot placed on a white blank card. The task is to decipher which number corresponds spatially to the dot and ten stimuli cards with normal mean values being 9.4 ± 1.1. The cube analysis test consists of 3D representations of cubes that range from 5 to 12 in number, some of which are obscured. There are ten stimuli cards and the normal mean range is 9.2 ± 1.2 (Fig. 4.17) [116].

Selected Pathological Processes Specific to the Visual Cortex and Their Radiations and Treatment Implications

The treatment of elementary and complex visual processed is tightly linked to the underlying pathophysiology of the etiological process. For example even occipital cerebral infarcts causing homonymous field defects improve with time. Key however is the effective secondary cerebrovascular prophylactic therapy. A recent case report of a patient with a Riddoch syndrome due to posterior circulation steno-occlusive disease for example improved to an extent at least, without further strokes occurring, after successful interventional angioplasty and stenting [117]. Some syndromes may linger for months and years, and others for hours, days, or weeks. With eclampsia, a fairly common condition in some countries and in the Call Fleming syndrome and other conditions of secondary posterior circulation vasospastic disorders, the resultant visual agnosias may resolve completely within hours, days, or weeks. In others, due to progressive neurodegenerative disease the approach remains one of symptomatic treatment and brain health and fitness approaches to slow the march of dementias. With Lewy body disease treatment with acetylcholinesterase inhibitors may provide significant symptomatic relief of the visual hallucinations. With occipital seizures, antiseizure therapy may effectively treat the condition.

Posterior Reversible Encephalopathy Syndrome (PRES)

Improved multimodality MR imaging has helped with the identification of this syndrome that represents an acute neurological syndrome presenting with seizures, headaches, a variety of visual disorders, and autonomic instability. Radiologically the lesions involve the bilateral parieto-occipital cortices associated with both cytotoxic and vasogenic edema and at times hemorrhagic

lesions. The vasospastic nature of the posterior intracranial circulation is easily recognized by MR angiography. Resolution typically occurs over days to weeks and the more common associated pathophysiological triggers include chemotherapeutic agent administration, autoimmune conditions, eclampsia, hypertensive encephalopathy, human immunodeficiency virus infection, subarachnoid hemorrhage nephrotic syndrome radiological procedures, and immunosuppressive agents. Characteristic radiographic findings include bilateral regions of subcortical vasogenic edema that resolves within days or weeks. The underlying pathophysiology has been attributed to endothelial dysfunction [118]. When no obvious cause can be discerned, the primary form, termed Call Fleming syndrome, is appropriate [119, 120] (Fig. 4.18).

Eclampsia

Women presenting with eclampsia often complain of poor vision or blindness and depending on whether it is tested for simultanagnosia is common in patients with eclampsia. In a cognitive neurological study of 30 women with eclampsia using the Cookie Theft Picture Test and a simple scoring system to detect simultanagnosia, the cognitive bedside test performed well against multimodality MRI and diffusion-weighted imaging of parieto-occipital lesions. The validated CTPT-designed simultanagnosia detection was abnormal in 96.7 % of the eclampsia patients. Specifically other neurological scales such as the Canadian Neurological Scale were invariably normal [74].

Benson's Syndrome or Posterior Cortical Atrophy Syndrome (PCAS)

This is one of the rarer focal atrophy syndromes that presents with complex visuospatial disorders in addition to other cognitive impairments such as apraxia, agraphia, and motor planning disorders. These patients typically present with visual agnosia, alexias, Balint's syndrome, and prosopagnosia. Two different types may be differentiated: a

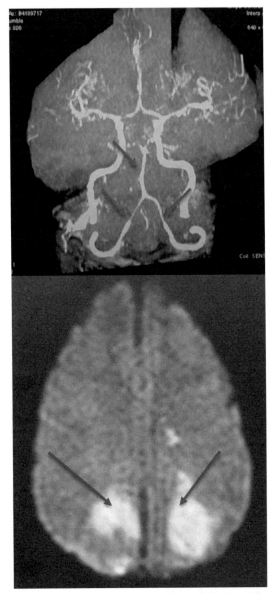

Fig. 4.18 Posterior reversible encephalopathy syndrome and clinically simultanagnosia, showing spasm and irregularity (reversible) of the blood vessels (*upper image arrows*) with bilateral occipital lesions (*lower image arrows*)

ventral visual stream or occipito-temporal subtype presenting with alexias, prosopagnosia, and visual agnosia, and a less frequent occipital predominant entity that presents with more elementary visual disturbances. The pathology has long been attributed to Alzheimer's disease process but the

pathological findings of Lewy body disease cerebrovascular, prionopathies, and frontotemporal lobe dementias have also been documented. For example Caroppo et al. identified a mutation in the GRN gene in a person with PCAS [121]. Diagnosis may be facilitated by metabolic positron emission tomography (PET) scanning which can reveal impairments in terms of hypometabolism in the occipital areas that antedate the characteristic occipital atrophy (Fig. 4.19).

Post-traumatic Stress Syndrome (PTSD)

The symptoms flashbacks, vivid memories of traumatic events, and recurring dreams are part of a wider syndrome that include dyssomnias, irritability, anger outbursts, impaired concentration, exaggerated startle responses, and hypervigilance. Evolutionary Insights regard this syndrome as an exaggerated and persistent stress response as proposed by the smoke detector principle [122]. From a neuropathological perspective, the RGS2 protein that decreases G protein signaling, SNP FK506, SNP ADCYAPR1, and perturbations in the PACAP-PAC1 pathway, all involved in the abnormal stress responses have been implicated [123]. Aside from the pharmacotherapeutic agent, prazosin (blocks noradrenaline release in the brain) medication has also demonstrated benefit [124, 125].

Progressive Lewy Body Disease

Diffuse Lewy body dementia (DLDB) and Alzheimer's disease (AD) share features at a clinical, neuroimaging, pathological, and pharmacotherapeutic level. Parkinson's disease (PD), Parkinson's with dementia, DLBD, and AD may be viewed as a pathological spectrum with loss of cholinergic and dopaminergic neurons. In addition, overlap with cognitive vascular disorders, AD, and frontotemporal lobe disorders occurs.

The core features entail Parkinsonism, fluctuating attention, alertness, and cognition and visual hallucinations. Other symptom complexes may include syncopal events, recurrent, autonomic

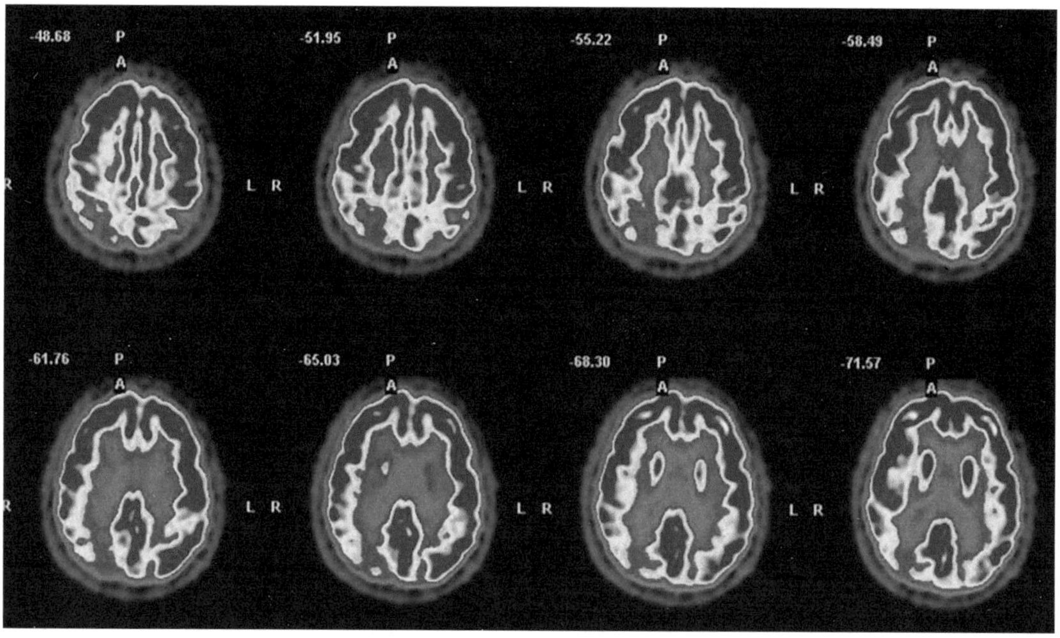

Fig. 4.19 PET brain scan of posterior cortical atrophy syndrome showing bilateral posterior hypometabolism

dysfunction, other sensory modality, hallucinations, delusions, and depression. Supportive investigative findings include REM sleep behavior disorder, severe neuroleptic sensitivity, and PET brain imaging revealing low dopamine transporter activity in the basal nuclei, preserved medial temporal volume by MR imaging, occipital lobe hypometabolism, EEG slow-wave activity, and transient temporal lobe sharp waves. Hallucinations are a hallmark feature of DLDB that may respond to anticholinesterase therapy [126, 127]. A specific 5HT-2A antagonist has been implicated in the visual hallucinations of DLDB [128]. Because of the diagnostic challenges that are both clinical and radiological, a number of neuroimaging metabolic features by PET brain scanning have been identified that include:

1. Global glucose hypometabolism including the occipital cortex
2. Hypometabolism of the lateral occipital cortex (most sensitive)
3. Preservation of the posterior cingulate also called the posterior cingulate island sign (PCIS), associated with the highest specificity [129]

Hallucinations occur in 50–75 % of those with DLBD and associated with a cortical Ach deficiency. Treatment with acetylcholinesterase inhibitors is usually effective.

Cerebral Leukodystrophies with Parieto-Occipital Disease (Fig. 4.20)

The majority of the leukodystrophies appear to have a posterior predominance of the white matter disease by neuroimaging. In the late-onset globoid leukodystrophy for example, 100 % of patients in a neuroradiological series had parieto-occipital involvement [130]. In a series of seven children with metachromatic leukodystrophy a posterior white matter predominance was noted in all cases [131]. Similarly cerebrotendinous xanthomatosis, a type of autosomal recessive metabolic leukodystrophy associated with a deficiency of an enzyme concerned with cholesterol catabolism, has white matter neuroimaging features with a predilection for posterior predominance [132]. Even in the newly discovered forms

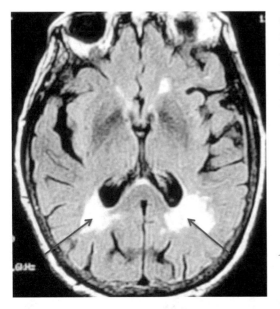

Fig. 4.20 Posterior leukodystrophy and visual perceptual problems

of leukodystrophy such as the autosomal dominant, adult leukodystrophy the extensive T2 white matter hyperintensities that reflect the white matter involvement showed a preference for involvement of the posterior cortical regions [133]. In multiple sclerosis, a wide variation of white matter lesions is possible and depending on the particular distribution of the leukoaraiosis, differing cognitive syndromes may be appreciated. For example, a patient presenting with predominantly visual form agnosia with the addition of ideomotor apraxia and dysgraphia had extensive leukoaraiosis in both occipito-temporal regions and other areas of the brain [134].

Neurotoxicological Syndromes Specific to Complex Visual Processing: Carbon Monoxide, Mercury Poisoning

Visual form agnosia was described in a number of patients with CO poisoning which damages the white matter tracts extensively but particularly the posterior tracts [135, 136]. Some toxic processes have a particular predilection for the posterior white matter circuitry. In a detailed study of neuropsychological and behavioral study of patient DF with CO poisoning, a visual form agnosia was documented. Quantitative MR analysis revealed that the most profound cortical loss was in the lateral occipital cortex areas with a reduction in the number of white matter tracts from this area to other cortical regions. Detailed study of the patient DF has also contributed significantly to our understanding of the "Two Visual Systems Hypothesis," first proposed by Goodale and Milner, in which the ventral stream is key to constructing visual percepts and the dorsal stream mediating the visual action control and visual guidance of grasping [137, 138]. Mercury poisoning has also been implicated in occipital white matter pathology [139].

Occipital Seizures

Seizural or irritating lesions of the occipital region may cause transient or recurrent complex visual disturbances. A particularly notable report by Agnetti et al. of left-sided occipital pole arachnoid cyst concerns a patient whose episodes of recurrent prosopagnosia were interpreted as epileptic events originating in the left occipital lobe. Successful removal the arachnoid cyst was associated with the disappearance of these prosopagnosic seizures [140]. A similar case by Mesad et al., with a lesion of the right inferolateral temporo-occipital junction, associated with recurrent prosopagnosia, also improved with surgical removal [141]. Severe visual agnosia related to continuous spike and slow-wave activity in the temporo-occipital cortex without a structural lesion has also been documented [142].

Top of the Basilar

The top of the basilar syndrome occurs commonly amongst acute stroke patient admissions and the number of publications does not do justice to the frequency with which this syndrome is encountered. In brief a number of different syndromes are recognized and involve the occipital lobes, brainstem, and thalamus in various combinations often with a bilateral. The presentations may vary from being life-threatening to

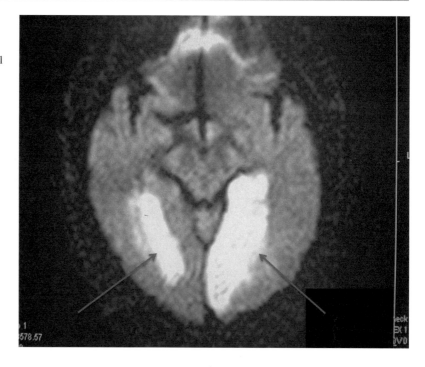

Fig. 4.21 Top of the basilar syndrome, cortical blindness, Anton's syndrome, and bi-occipital infarcts (*red arrows*)

barely noticeable syndromes to frank denial by the patients of any complaint as may be seen with Anton's syndrome and types of anosognosia (Fig. 4.21). An interesting report was that of an elderly artist who suffered a cardioembolic related top of the basilar stroke with predominantly simultanagnosia that improved to near normal, spontaneously over the years [143].

Cerebrovascular Watershed Lesions

Watershed lesions of the brain may be bilateral, unilateral, or internal [144]. The more common mechanisms include hypoxic ischemic mechanisms due to cardiac arrest or bilateral carotid steno-occlusive disease such as may occur with atherosclerotic disease or Moya Moya syndrome. Global hypoperfusion affects certain areas of the brain more than others and includes anterior and posterior watershed areas of the major branches of the circle of Willis, the basal ganglia, the cerebellar folia, the hippocampus, and cortical laminar necrosis with selective involvement of third and fifth layers. Bilateral posterior watershed area damage often manifests with simultanagnosia. In the image (Fig. 4.22), a young man with a global hypoxic event presented with bibrachial paresis more marked proximally with hypotonia, hypertonic crural paresis, frontal disinhibitory behavioral syndrome representing the anterior component of the watershed infarctions, and simultanagnosia of the posterior watershed lesions.

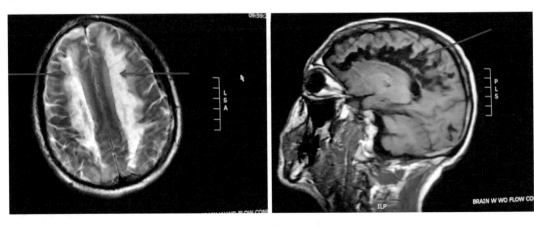

Fig. 4.22 Cerebral watershed lesions (*red arrows*). Frontal behavioral syndromes, simultanagnosia, and "Man in the Barrel Syndrome"

References

1. Zimmerman M. The nervous system in the context of information theory. In: Schmidt RF, Thews G, editors. Human physiology. Berlin: Springer; 1989. p. 166–73.
2. Hassin RR, Uleman JS, Bargh JA. The new unconscious. Oxford: Oxford University Press; 2006.
3. Rowe TB, Macrini TE, Luo ZX. Fossil evidence on origin of the mammalian brain. Science. 2011;332: 955–7.
4. Sereno MI, Allman JM. Cortical visual areas in mammals. In: Levinthal A, editor. The neural basis of visual function. New York, NY: Macmillan; 1991.
5. Jacobs GH, Tootell RB, Fisher SK, Anderson DH. Rod photoreceptors and scotopic vision in ground squirrels. J Comp Neurol. 1980;189: 113–25.
6. Kaas JH. The evolution of brains from early mammals to humans. Wiley Interdiscip Rev Cogn Sci. 2013;4(1):33–45.
7. Kaas JH, Lyon DC. Visual cortex organization in primates: theories of V3 and adjoining visual areas. Prog Brain Res. 2003;134:285–95.
8. Isbell LA. The fruit the tree and the serpent. Cambridge, MA: Harvard University Press; 2009.
9. Lui LL, Dobiecki AE, Bourne JA, Rosa MG. Breaking camouflage: responses of neurons in the middle temporal area to stimuli defined by coherent motion. Eur J Neurosci. 2012;36(1):2063–76. doi:10 .1111/j.1460-9568.2012.08121.
10. Mysore SG, Vogels R, Raiguel SE, Orban GA. Shape selectivity for camouflage-breaking dynamic stimuli in dorsal V4 neurons. Cereb Cortex. 2008;18(6): 1429–43.
11. Ramachandaran VS, Gregory RL. Perceptual filling in of artificially induced scotomas in human vision. Nature. 1991;350:699–702.
12. Weiskrantz L. Blindsight: a case study and its implications. Oxford: Oxford University Press; 1986.
13. Weiskrantz L. Blindsight revisited. Curr Opin Neurobiol. 1996;6:215–20.
14. Heywood CA, Kentridge RW. Affective blindsight? Trends Cogn Sci. 2000;4(4):125–6.
15. Marshall JC, Halligan PW. Blindsight and insight into visuospatial neglect. Nature. 1988;336:766–7.
16. Windmann S, Wehrmann M, Calabrese P, Gunturkun O. Role of the prefrontal cortex in attentional control over bistable vision. J Cogn Neurosci. 2006;18: 456–71.
17. Kandel ER, Schwartz JH, Jessell TM, Siegelbaum SA, Hudspeth AJ, editors. Now updated: the definitive neuroscience resource. 5th ed. New York, NY: McGraw Hill; 2013.
18. Porter PB. Another puzzle-picture. Am J Psychol. 1954;67:550–1.
19. Larsen A, Bundesen C, Kyllingsbaek S, Paulson OB, Law I. Brain activation during mental transformation of size. J Cogn Neurosci. 2000;12:763–74.
20. Schoenfeld MA, Nosselt T, Poggel D, Tempelmann C, Hopf JM, Woldorff MG, et al. Analysis of pathways mediating preserved vision after striate cortex lesions. Ann Neurol. 2002;52:814–24.
21. Rizzo M, Nawrot M, Zihl J. Motion and shape perception in cerebral akinetopsia. Brain. 1995;118: 1105–27.
22. Pitzalis S, Sereno MI, Committeri G, Fattori P, Galati G, Tosoni A, et al. The human homologue of macaque area V6A. Neuroimage. 2013;82: 517–30.
23. Galletti C, Fattori P, Gamberini M, Kutz DF. The cortical visual area V6: brain location and visual topography. Eur J Neurosci. 1999;11:3922–36.
24. Swisher JD, Halko MA, Merabet LB, McMains SA, Somers DC. Visual topography of human intraparietal sulcus. J Neurosci. 2007;27(20):5326–37.

25. Hadjikhani N, Liu AK, Dale AM, Cavanagh P, Tootell RB. Retinotopy and color sensitivity in human visual cortical area V8. Nat Neurosci. 1998;1(3):235–41.

26. Beck DM, Rees G, Frith CD, Lavie N. Neural correlates of change and change blindness. Nat Neurosci. 2001;4:645–50.

27. Rock I, Linnet CM, Grant PI, Mack A. Perception without attention: results of a new method. Cogn Psychol. 1992;24:502–34.

28. Scholte HS, Witteveen SC, Spekreijse H, Lamme VA. The influence of inattention on the neural correlates of scene segmentation. Brain Res. 2006;1076: 106–15.

29. Frith CD. In: Kandel and Schwartz (Eds) Principles of neural science, 5th edition. McGraw Hill, New York, NY; 2013.

30. Guttman SE, Gilroy LA, Blake R. Spatial grouping in human vision: temporal structure trumps temporal synchrony. Vision Res. 2007;47(2):219–30.

31. Singer W, Gray CM. Visual feature integration and the temporal correlation hypothesis. Annu Rev Neurosci. 1995;18:555–86.

32. Overvliet KE, Sayim B. Perceptual grouping determines haptic contextual modulation. Vision Res. 2015;pii:S0042-6989(15)00170-4. doi:10.1016/j.visres.2015.04.016.

33. Tinbergen N. The Herring Gull's world. A study of social behavior of birds. London: Collins; 1953.

34. Ramachandran VS. The tell tale brain. New York, NY: W W Norton & Company; 2011.

35. Carrera E, Tononi G. Diaschisis; past, present, future. Brain. 2014;137:2408–22.

36. Ffytche D, Blom JD, Catani M. Disorders of visual perception. J Neurol Neurosurg Psychiatry. 2010;81:1280–7. doi:10.1136/jnnp.2008.171348.

37. Fugelsang C, Narici L, Picozza P, Sannita WG. Phosphenes in low earth orbit: survey responses from 59 astronauts. Aviat Space Environ Med. 2006;77:449–52.

38. Schott CD. Exploring the visual hallucinations of migraine aura: the tacit contribution of illustration. Brain. 2007;130:1690–703.

39. Landis R, Graves R, Benson F, Hebben N. Visual recognition through kinesthetic mediation. Psychol Med. 1982;12:515–31.

40. Farah M. Visual agnosia. Cambridge, MA: MIT Press; 2004.

41. Kinsbourne M, Warrington EK. The localizing significance of limited simultaneous visual form perception. Brain. 1963;86:697–702.

42. Hartmann JA, Wolz WA, Roeltgen DP, Loverson FL. Denial of visual perception. Brain Cogn. 1991;16:29–40.

43. Zeki S, Ffytche DH. The Riddoch syndrome: insights into the neurobiology of conscious vision. Brain. 1998;121:25–45.

44. Riddoch G. Dissociation of visual perceptions due to occipital injuries, with especial reference to appreciation of movement. Brain. 1917;40:15–57.

45. Wapner W, Judd T, Gardner H. Visual agnosia in an artist. Cortex. 1978;14:343–64.

46. Nishida Y, Hayashi O, Iwami T, Kimura M, Kani K, Ito R, et al. Stereopsis-processing regions in the human parieto-occipital cortex. Neuroreport. 2001;12(10):2259–63.

47. Fortin A, Ptito A, Faubert J, Ptito M. Cortical areas mediating stereopsis in the human brain: a PET study. Neuroreport. 2002;13(6):895–8.

48. Orban GA, Sunaert S, Todd JT, Van Hecke P, Marchal G. Human cortical regions involved in extracting depth from motion. Neuron. 1999;24(4):929–40.

49. Kwee IL, Fujii Y, Matsuzawa H, Nakada T. Perceptual processing of stereopsis in humans: high-field (3.0-tesla) functional MRI study. Neurology. 1999;53(7):1599–601.

50. Jonas J, Frismand S, Vignal JP, Colnat-Coulbois S, Koessler L, Vespignani H, et al. Right hemispheric dominance of visual phenomena evoked by intracerebral stimulation of the human visual cortex. Hum Brain Mapp. 2014;35(7):3360–71.

51. Mittenberg W, Choi EJ, Apple CC. Stereoscopic visual impairment in vascular dementia. Arch Clin Neuropsychol. 2000;15(7):561–9.

52. Tseng CH, Gobell JL, Sperling G. Factors that determine depth perception of trapezoids, windsurfers, runways. Front Hum Neurosci. 2015;9: 182. doi:10.3389/fnhum.2015.00182. eCollection 2015.

53. Heywood CA, Kentridge RW. Achromatopsia, color vision, and cortex. Neurol Clin. 2003;21(2):483–500. Review.

54. Daily MN, Cotrrell GW. Prosopagnosia in modular neural network models. Prog Brain Res. 1999;121: 165–84.

55. Ffytche DH, Howard RJ, Brammer MJ, David A, Woodruff P, Williams S. The anatomy of conscious vision: an fMRI study of visual hallucinations. Nat Neurosci. 1998;1:738–42.

56. Vignal JP, Chauvel P, Halgren E. Localised face processing by the human prefrontal cortex: stimulation-evoked hallucinations of faces. Cogn Neuropsychol. 2000;17(1):281–91.doi:10.1080/026432900380616.

57. Landis T, Cummings JL, Benson DF, Palmer EP. Loss of topographic familiarity. An environmental agnosia. Arch Neurol. 1986;43:132–6.

58. Cavina-Pratesi C, Kentridge RW, Heywood CA, Milner AD. Separate channels for processing form, texture, and color: evidence from FMRI adaptation and visual object agnosia. Cereb Cortex. 2010;20(10): 2319–32.

59. Ffytche DH, Lappin JM, Philpot M. Visual command hallucinations in a patient with pure alexia. J Neurol Neurosurg Psychiatry. 2004;75(1):80–6.

60. Yamagata B, Kobayashi H, Yamamoto H, Mimura M. Visual text hallucinations of thoughts in an alexic woman. J Neurol Sci. 2014;339(1-2):226–8.

61. Cummings JL, Gittinger Jr JW. Central dazzle. A thalamic syndrome? Arch Neurol. 1981;38(6):372–4.

62. Du-Pasquier RA, Genoud D, Safran AB, Landis T. Monocular central dazzle after thalamic infarcts. J Neuroophthalmol. 2000;20(2):97–9.

63. Safran AB, Sanda N. Color synesthesia. Insight into perception, emotion and consciousness. Curr Opin Neurol. 2015;28:36–44.

64. Ptak R, Lazeyras F, Di Pietro M, Schnider A, Simon SR. Visual object agnosia is associated with a breakdown of object selective responses in the lateral occipital cortex. Neuropsychologia. 2014;60:10–20.

65. Montalvo MJ, Kahn MA. Clinicoradiological correlation of macropsia due to acute stroke: a case report and review of the literature. Case Rep Neurol Med. 2014;2014:272084. doi:10.1155/2014/272084.

66. Laudate TM, Nelson AP. "Macropsia", in encyclopedia of clinical neuropsychology. New York, NY: Springer; 2011. p. 1506.

67. Cohen L, Gray F, Meyrignac C, Dehaene S, Degos JD. Selective deficit of visual size perception: two cases of hemimicropsia. J Neurol Neurosurg Psychiatry. 1994;57:73e8.

68. Ffytche DH, Howard RJ. The perceptual consequences of visual loss: positive pathologies of vision. Brain. 1999;122:1247–60.

69. Podoll K, Robinson D. Recurrent Lilliputian hallucinations as visual aura symptom in migraine. Cephalalgia. 2001;21(10):990–2.

70. Uchiyama M, Nishio Y, Yokoi K, Hirayama K, Imamura T, Shimomura T, et al. Pareidolias: complex visual illusions in dementia with Lewy bodies. Brain. 2012;135:2458–69.

71. Uchiyama M, Nishio Y, Yokoi K, Hosokai Y, Takeda A, Mori E. Pareidolia in Parkinson's disease without dementia: a positron emission tomography study. Parkinsonism Relat Disord. 2015;21(6): 603–9.

72. Michel F, Henaff MA. Seeing without the occipito-parietal cortex: simultanagnosia as a shrinkage of the attentional visual field. Behav Neurol. 2004; 15:3–13.

73. Thomas C, Kveraga K, Huberle E, Karnath HO, Bar M. Enabling global processing in simultanagnosia by psychophysical biasing of visual pathways. Brain. 2012;135:1578–85.

74. Hoffmann M, Keiseb J, Moodley J, Corr P. Appropriate neurological evaluation and multimodality magnetic resonance imaging in eclampsia. Acta Neurol Scand. 2002;106(3):159–67.

75. Rizzo M, Robin DA. Simultanagnosia: a defect of sustained attention yields insights on visual information processing. Neurology. 1990;40:447–55.

76. Dalrymple KA, Barton JJ, Kingstone A. A world unglued: simultanagnosia as a spatial restriction of attention. Front Hum Neurosci. 2013;7:145. doi:10.3389/fnhum.2013.00145. eCollection 2013.

77. Naidu K, Moodley J, Corr P, Hoffmann M. Single photon emission and cerebral computerised tomographic scan and transcranial Doppler sonographic findings in eclampsia. Br J Obstet Gynaecol. 1997;104(10):1165–72.

78. Zihl J, Heywood CA. The contribution of LM to the neuroscience of movement vision. Front Integr Neurosci. 2015;9:6. doi:10.3389/fnint.2015.00006. eCollection 2015.

79. Sakurai K, Kurita T, Takeda Y, Shiraishi H, Kusumi I. Akinetopsia as epileptic seizure. Epilepsy Behav Case Rep. 2013;20:74–6. doi:10.1016/j. ebcr.2013.04.002. eCollection 2013.

80. Tsai PH, Mendez MF. Akinetopsia in the posterior cortical variant of Alzheimer disease. Neurology. 2009;73(9):731–2. doi:10.1212/ WNL.0b013e3181b59c07.

81. Ovsiew F. The Zeitraffer phenomenon, akinetopsia, and the visual perception of speed of motion: a case report. Neurocase. 2014;20(3):269–72. doi:10.1080/ 13554794.2013.770877.

82. Wood IC, Tomlinson A. Stereopsis measured by random-dot patterns--a new clinical test. Br J Physiol Opt. 1976;31(4):22–5.

83. Gersztenkorn D, Lee AG. Palinopsia revamped: a systematic review of the literature. Surv Ophthalmol. 2015;60(1):1–35.

84. Critchley M. Types of visual perseveration: 'palinopsia' and 'illusory visual spread.'. Brain. 1951; 74:267–99.

85. Santhouse AM, Howard RJ, Ffytche DH. Visual hallucinatory syndromes and the anatomy of the visual brain. Brain. 2000;123:2055–64.

86. Lopez JR, Adornato BT, Hoyt WF. 'Entomopia': a remarkable case of cerebral polyopia. Neurology. 1993;43:2145.

87. River Y, Ben Hur T, Steiner I. Reversal of vision metamorphopsia: clinical andanatomical characteristics. Arch Neurol. 1998;55:1362–8.

88. Girkin CA, Miller NR. Central disorders of vision in humans. Surv Ophthalmol. 2001;45:379–405.

89. Mattingley JB, Bradshaw JL. Can tactile neglect occur at an intra-limb level? Vibrotactile reaction times in patients with right hemisphere damage. Behav Neurol. 1994;7(2):67–77.

90. Ardila A, Botero M, Gomez J. Palinopsia and visual allesthesia. Int J Neurosci. 1987;32:775–82.

91. Brugger P, Regard M. Illusory reduplication of one's own body: phenomenology and classification of autoscopic phenomena. Cogn Neuropsychol. 1997;2: 19–38.

92. Anzellotti F, Onofrj V, Maruotti V, Ricciardi L, Franciotti R, Bonanni L, et al. Autoscopic phenomena: case report and review of literature. Behav Brain Funct. 2011;7(1):2. doi:10.1186/1744-9081-7-2.

93. Blanke O, Landis T, Spinelli L, Seeck M. Out-of-body experience and autoscopy of neurological origin. Brain. 2004;127:243–58.

94. Whiteside TC, Graybiel A, Niven JI. Visual illusions of movement. Brain. 1965;88:193–210.

95. Kolev OI. Visual hallucinations evoked by caloric vestibular stimulation in normal humans. J Vestib Res. 1995;5:19–23.

96. Uchida H, Suzuki T, Tanaka KF, Watanabe K, Yagi G, Kashima H. Recurrent episodes of perceptual

alteration in patients treated with antipsychotic agents. J Clin Psychopharmacol. 2003;23:496–9.

97. Fftytche D, Blom JD, Catani M. Disorders of visual perception. J Neurol Neurosurg Psychiatry. 2010; 81:1280–7.

98. von Cramon DY, Kerkhoff G. On the cerebral organization of elementary visuospatial perception. In: Gulyas B, Ottoson D, Roland P, editors. Functional organisation of the human visual cortex. Oxford: Pergamon; 1993. p. 211–31.

99. Craver-Lemley C, Bornstein RF, Alexander DN, Barrett AM. Imagery interference diminishes in older adults: age-related differences in the magnitude of the Perky effect. Imagin Cogn Pers. 2009;29(4):307–22.

100. Segal SJ, Gordon PE. The Perky effect revisited: blocking of visual signals by imagery. Percept Mot Skills. 1969;28:791–7.

101. Gray CR, Gummerman K. The enigmatic eidetic image: a critical examination of methods, data, and theories. Psychol Bull. 1975;82:383–407.

102. Brang D, Ramachandran VS. Visual field heterogeneity, laterality, and eidetic imagery in synesthesia. Neurocase. 2010;16(2):169–74. doi:10.1080/13554790903339645.

103. Battelli L, Pascual-Leone A, Cavanagh P. The 'when' pathway of the right parietal lobe. Trends Cogn Sci. 2007;11:204–10.

104. Mauk MD, Buonomano DV. The neural basis of temporal processing. Annu Rev Neurosci. 2004;27: 307–40.

105. Nieder A, Diester I, Tudusciuc O. Temporal and spatial enumeration processes in the primate parietal cortex. Science. 2006;313:1431–5.

106. Bonnet L, Comte A, Tatu L, Millot JL, Moulin T, Medeiros de Bustos E. The role of the amygdala in the perception of positive emotions: an "intensity detector". Front Behav Neurosci. 2015;9:178. doi:10.3389/fnbeh.2015.00178. eCollection 2015.

107. Benarroch EE. The amygdala: functional organization and involvement in neurologic disorders. Neurology. 2015;84(3):313–24.

108. Critchley M. Metamorphopsia of central origin. Trans Ophthalmol Soc U K. 1949;69:111–21.

109. Critchley M. The parietal lobes. New York, NY: Hafner; 1953.

110. Ellis HD, Young AW, Quayle AH, De Pauw KW. Reduced autonomic responses to faces in Capgras delusion. Proc Biol Sci. 1997;264:1085–92.

111. Bischof M, Bassetti CL. Total dream loss: a distinct neuropsychological dysfunction after bilateral PCA stroke. Ann Neurol. 2004;56:583–6.

112. Peña-Casanova J, Roig-Rovira T, Bermudez A, Tolosa-Sarro E. Optic aphasia, optic apraxia, and loss of dreaming. Brain Lang. 1985;26(1):63–71.

113. Santos-Bueso E, Serrador-Garcia M, Porta-Etessam J, Rodríguez-Gómez O, Martínez-de-la-Casa JM, García-Feijoo J, et al. Charles Bonnet syndrome. A 45 case series. Rev Neurol. 2015;60:337–40.

114. Benke T. Peduncular hallucinosis – a syndrome of impaired reality monitoring. J Neurol. 2006;253: 1561–71.

115. Lezak MD, Howieson DB, Bigler ED, Tranel T, editors. Neuropsychological assessment. Oxford: Oxford University Press; 2012.

116. Warrington E, James M. VOSP. Harcourt assessment. London: The Psychological Corporation. The Thames Valley Test Company; 1991.

117. Vidal Y, Hoffmann M. Improvement of Astatikopsia (Riddoch's phenomenon) after correction of vertebral stenoses with angioplasty. Neurol Int. 2012;4(1), e1. doi:10.4081/ni.2012.e1.

118. Marra A, Vargas M, Striano P, Del Guercio L, Buonanno P, Servillo G. Posterior reversible encephalopathy syndrome: the endothelial hypotheses. Med Hypotheses. 2014;82(5):619–22.

119. Call GK, Fleming MC, Sealfon S, Levine H, Kistler JP, Fisher CM. Reversible cerebral segmental vasoconstriction. Stroke. 1988;19(9):1159–70.

120. Fugate JE, Rabinstein AA. Posterior reversible encephalopathy syndrome: clinical and radiological manifestations, pathophysiology, and outstanding questions. Lancet Neurol. 2015;13: S1474–4422.

121. Caroppo P, Belin C, Grabli D, Maillet D, De Septenville A, Migliaccio R, et al. Posterior cortical atrophy as an extreme phenotype of GRN mutations. JAMA Neurol. 2015;72:224–8.

122. Nesse R. The smoke detector principle. Ann N Y Acad Sci. 2006. doi:10.1111/j.1749-6632.2001. tb03472.

123. Koenen KC, Amstadter AB, Nugent NR. Gene-environment interaction in posttraumatic stress disorder: an update. J Trauma Stress. 2009;22(5): 416–26.

124. Raskind MA, Peskind ER, Kanter ED, Petrie EC, Radant A, Thompson CE, et al. Reduction of nightmares and other PTSD symptoms in combat veterans by prazosin: a placebo-controlled study. Am J Psychiatry. 2003;160:371–3.

125. Tang Y-Y, Lu Q, Gen X, Stein EA, Yang Y, Posner MI. Short-term meditation induces white matter changes in the anterior cingulated. Proc Natl Acad Sci U S A. 2010;107:15649–52.

126. Mirzaei S, Knoll P, Koehn H, Bruecke T. Assessment of diffuse Lewy body disease by 2-[18F] fluoro-2-deoxy-D-glucose positron emission tomography (FDG PET). BMC Nucl Med. 2003;3(1):1.

127. De Meyer G, Shapiro F, Vanderstichele H. Diagnosis independent Alzheimer disease biomarker signature in cognitively normal elderly people. Arch Neurol. 2010;67:949–56.

128. Ballanger B, Strafella AP, van Eimeren T. Serotonin 2 a receptors and visual hallucinations in Parkinsons disease. Arch Neurol. 2010;67:416–21.

129. Lim SM, Katsifis A, Villemagne V, Best R, Jones G, Saling M, et al. The 18F FDG PET cingulate island sign and comparison to 123I beta CIT SPECT for

diagnosis of dementia with Lewy bodies. J Nucl Med. 2009;50:1638–45.

130. Loes DJ, Peters C, Krivit W. Globoid cell leukodystrophy: distinguishing early-onset from late-onset disease using a brain MR imaging scoring method. AJNR Am J Neuroradiol. 1999;20(2):316–23.

131. Kim TS, Kim IO, Kim WS, Choi YS, Lee JY, Kim OW, et al. MR of childhood metachromatic leukodystrophy. AJNR Am J Neuroradiol. 1997;18(4):733–8.

132. Lionnet C, Carra C, Ayrignac X, Levade T, Gayraud D, Castelnovo G, et al. Cerebrotendinous xanthomatosis: a multicentric retrospective study of 15 adults, clinical and paraclinical typical and atypical aspects. Rev Neurol (Paris). 2014;170(6–7):445–53.

133. Corlobé A, Taithe F, Clavelou P, Pierre E, Carra-Dallière C, Ayrignac X, et al. A novel autosomal dominant leukodystrophy with specific MRI pattern. J Neurol. 2015;262(4):988–91.

134. Okuda B, Tanaka H, Tachibana H, Iwamoto Y, Takeda M, Kawabata K, et al. Visual form agnosia in multiple sclerosis. Acta Neurol Scand. 1996;94(1):38–44.

135. Benson DF, Greenberg JP. Visual form agnosia. Arch Neurol. 1969;20:82–9.

136. Milner AD, Perrett DI, Johnston RS, Benson PJ, Jordan TR, Heeley DW, et al. Perception and action in "visual form agnosia". Brain. 1991;114:405–28.

137. Whitwell RL, Milner AD, Goodale MA. The two visual systems hypothesis: new challenges and insights from visual form agnosic patient DF. Front Neurol. 2014;5:255.

138. Bridge H, Thomas OM, Minini L, Cavina-Pratesi C, Milner AD, Parker AJ. Structural and functional changes across the visual cortex of a patient with visual form agnosia. J Neurosci. 2013;33(31):12779–91.

139. Landis T, Graves R, Benson F, Hebben N. Visual recognition through kinesthetic mediation. Psychol Med. 1982;12:5150531.

140. Agnetti V, Carreras M, Pinna L, Rosati G. Ictal prosopagnosia and epileptogenic damage of the dominant hemisphere. A case history. Cortex. 1978;14(1):50–7.

141. Mesad S, Laff R, Devinsky O. Transient postoperative prosopagnosia. Epilepsy Behav. 2003;4(5):567–70.

142. Eriksson K, Kylliäinen A, Hirvonen K, Nieminen P, Koivikko M. Visual agnosia in a child with non-lesional occipito-temporal CSWS. Brain Dev. 2003;25(4):262–7.

143. Smith WS, Mindelzun RE, Miller B. Simultanagnosia through the eyes of an artist. Neurology. 2003; 60(11):1832–4.

144. Mangla R, Kolar B, Almast J, Ekholm SE. Border zone infarcts: pathophysiologic and imaging characteristics. Radiographics. 2011;31(5):1201–14.

5

Temporal Lobe Syndromes

Evolution of the Temporal: Some Pertinent Details

This is one of the few cortical regions that are larger in size in humans in comparison studies with apes. Overall there is a relative white matter increase but with a specific gyral white matter increase as opposed to core white matter which is not relatively increased. Gyral white matter increase as opposed to core white matter is thought to mediate much greater interconnectivity, enabled by the short association fibers [1]. In comparative analyses, amongst modern humans, relatively wider orbitofontal cortices, enlarged olfactory bulbs, and larger and more forwardly placed temporal lobe poles are evident, consistent with social brain development [2]. The amygdaloid complex is particularly concerned with the social brain development such as social cognition, coalitions, and emotional regulation. Another unique human development is the relative enlargement of the basolateral nucleus of the amygdaloid group of nuclei (lateral, basal, accessory nuclei). This is also consistent with the general surge in interconnectivity of the temporal lobe association cortices [3, 4].

The Evolutionary Importance of the Social Circuitry and the Social Brain Hypothesis

The importance of social cohesiveness and the challenge of the polyadic relationships are considered to be a major if not key drivers of increasing human brain size (Fig. 5.1) [5]. With the temporal lobe a central component of social processing with the amygdala in particular it is not surprising that some significant human enlargements have been reported in this part of the brain. Human gaze and eye contact are important initial contact modes and ascertaining eye gaze direction of a conspecific or other human and its monitoring are functions of superior temporal lobe [6]. The amygdala also has a key role in interpreting social facial signals from the face. A functional MRI study revealed that during eye contact (and also without eye contact) the direction of gaze activated the left amygdala indicating a general role in monitoring eye gaze. This differed from the right side where only during eye contact was the right amygdala activated [7].

Fig 5.1 The social
brain hypothesis: group
size for primates and
humans and neocortex
ratio. Index of relative
cortex size (neocortex
ratio) is neocortex
volume divided by the
volume of the rest of the
brain. Figure with
permission: Gamble C,
Gowlett J, Dunbar
R. Thinking Big. How
the evolution of social
life shaped the human
mind. Thames and
Hudson, London 2014

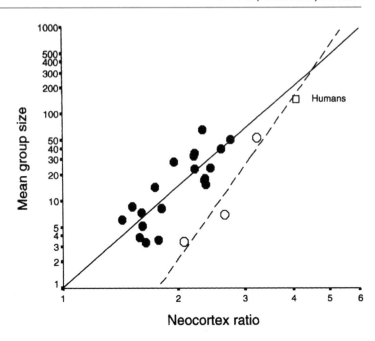

Neuroanatomy and Neurophysiology

The anatomical confines of the temporal lobes
include (Fig. 5.2):

Lateral
Heschl's gyrus, planum polare, planum tempo-
    rale (BA 41, 42, 22)
Superior temporal middle and inferior temporal
    gyrus (BA 22, 21, 20)

Medial
Inferior temporal gyrus (BA 20)
Parahippocampal gyrus (BA 27, 28, 34, 35)
Fusiform gyrus (BA 36)

The posterior temporal lobes are delimited by
an arbitrary line drawn from the parietooccipital
sulcus to the preoccipital notch (indentation in
the inferior temporal gyrus). A horizontal line,
drawn from the midpoint of this particular line to
the lateral sulcus, demarcates the temporal and
parietal lobes [8].

The temporal lobes are intimately tied to all
the other lobes through the association tracts.

Amongst the largest long-range association tracts
include the occipitotemporal and the uncinate
fasciculus. Anterior temporal lobe, inferior frontal lobe lesions, and those of the uncinate fasciculus may be affected together by lesions or disease
states with difficulty in parsing out which is the
most responsible. Accordingly some authors
regard syndromes of the uncinate fasciculus as an
appropriate approach.

Sensory visual and auditory (much less olfactory) inputs mediate evaluation of a conspecific's
or other human's eyes, faces, and body movement. Specialized and separate temporal cortical
areas have been identified for these. Supportive
evidence comes from lesion studies as well as
functional imaging studies with fMRI. The study
by ffytche et al. demonstrated the parts activated
during visual hallucinations for faces, places, and
objects (Fig. 5.3) [9]. These are then subsequently
relayed to superior temporal gyrus and amygdala
for salience evaluation. The mirror neuron circuitry is concerned with theory of mind detection.
In addition the social semantic memory of the
anterior temporal lobe for faces for example forms
part of the social circuitry (Fig. 5.4).

Fig. 5.2 The human temporal lobes and lateral and medial aspects depicted by the Brodmann area map

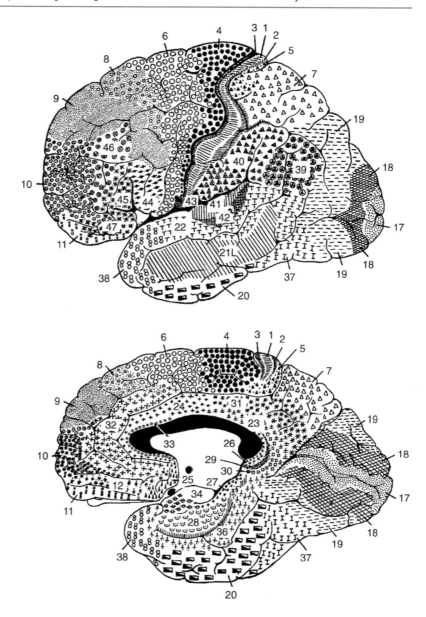

Temporal Lobe Elementary Neurological, Cognitive, and Behavioral Presentations and Syndromes

Right or Left and Bilateral

Elementary

Vertiginous syndromes—vertigo or disequilibrium due to epilepsy or migraine

Olfactory hallucinations due to uncinate lesions or seizures

Gustatory (taste) abnormalities due to medial temporal or insula lesions or seizures

Geschwind-Gastaut syndrome

Kluver–Bucy syndrome (or individual components thereof such as placidity and agnosia)

Neuropsychiatric

Anxiety, agitation, paranoia, aggression

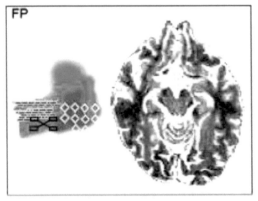

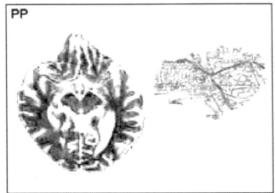

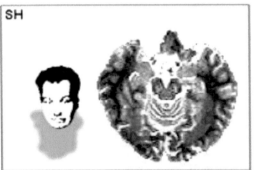

Fig. 5.3 Different infero-occipitotemporal activation associated with different kinds of visual hallucinations in Charles Bonnet syndrome in four different patients. The f-MRI signal elevation is in *red* with the temporal lobe fusiform gyrus in *blue*. The individual hallucinations included *AK* colored, shiny images, cars, Sphinx, *PP* map in *black* and *white*, *SH* colorful cartoon-like face, *FP* brickwork and geometric shapes in colors. Figure with permission: Ffythche DH, Howard RJ, Brammer MJ, David A, Woodruff P, Williams S. The anatomy of conscious vision; an f-MRI study of visual hallucinations. Nature Neuroscience 1998;1:738–742

Medial frontal cortex, anterior cingulate, theory of mind hub

Inferior parietal lobe mirror hub

Inferior frontal gyrus frontal mirror hub

Temporo-occipito-parietal (TOP) region, theory of mind hub

Medial hippocampus CA2 region hub for social memory

Superior temporal sulcus face cells process facial expressions and direction of gaze, head orientation

The right and left amygdala are involved facial social signals. The left is involved in eye gaze direction interpretation and the right is involved with eye gaze directed towards the subject. Also responds especially to aversive social signals

Face cells in inferior temporal cortex for processing facial identity

Fig. 5.4 Schematic social brain circuitry hubs. With permission: Hoffmann M. Brain Beat, Page Publications, New York, 2015

Cognitive

Memory: Korsakoff amnestic state

Cortical deafness

Auditory agnosia (inability to identify sounds despite normal peripheral hearing status)

Auditory paracusias

Auditory hallucinations (simple and complex), illusions (differentiate from peduncular hallucinosis)

Disorders of time perception (time may pass with excessive speed or not at all)

Left

Elementary

Right upper quadrantanopia

Cognitive

Aphasias: Wernicke's, transcortical sensory, and anomic

Memory: Verbal amnesia

Visual agnosia

Amusia: Lexical amusia (impairment in reading music)

Synesthesia

Right

Elementary

Left upper quadrantanopia

Cognitive

Memory: Visuospatial amnesia

Prosopagnosia (occipital–temporal region)

Auditory agnosia—verbal (pure word deafness) and nonverbal (environmental sounds)

Amusias—receptive and expressive

Delusional misidentification syndromes

Theory of mind impairment (semantic dementia) [10–17]

Neuropathological Processes

The more commonly encountered pathologies that involve the temporal lobe in relative isolation include inferior division middle cerebral artery

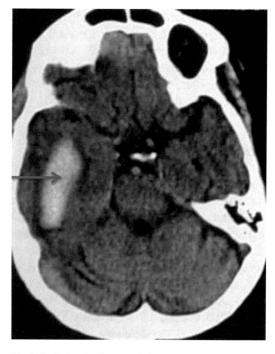

Fig. 5.5 Isolated, discrete, right temporal lobe intracerebral hemorrhage (*arrow*)

territory bland infarction, intracerebral hemorrhage (Fig. 5.5), epilepsy, encephalitis, tumors, and traumatic brain injury. Aside from the aphasic presentations that are typical of left temporal lobe involvement, right temporal lobe syndromes may be more enigmatic or covert. These include Kluver–Bucy syndrome (KBS) and Geschwind-Gastaut syndrome (GGS) presentations or fragments thereof or forme fruste varieties. The KBS, originally described in monkeys, is rare and generally ascribed to bilateral lesions although cases have been reported with unilateral lesions [18]. The presentation includes some or all of the following:

- Visual agnosia
- Hyperorality
- Placidity
- Altered sexual activity both hypersexuality or hyposexuality
- Hypermetamorphosis

Human forms of the KBS are being increasingly described with manifestations such as compulsive social kissing reported in a person with frontotemporal lobe dementia [19], hypersexuality, and hyperphagia post-right temporal lobectomy for seizure management [20]. KBS may be permanent or transient and has been reported in TBI, ICH, FTD, and infectious such as herpes simplex encephalitis TBI and KBS [21, 22].

Geschwind-Gastaut Syndrome

Although the GGS had for many years been described in the context of temporal lobe epilepsy, specifically the interictal phase [23, 24], both isolated bland infarcts and intracerebral hemorrhage and so the right temporal lobe in particular have been correlated with this syndrome. The right temporal lobe had been regarded as one of the so-called silent areas of the brain but in addition to the GGS, delusional misidentification syndromes and a variety of accompanying frontal network syndromes are frequently encountered with lesions of this area, if tested for. Importantly these complex syndromes generally occur without heralding sensorimotor deficits (Fig. 5.6) [25].

This syndrome is comprised of three core features; the diagnostic process is facilitated by the Bear-Fedio Inventory (Table 5.1):

1. Viscous personality
2. Metaphysical preoccupation
3. Altered physiological drives.

The viscous personality may be regarded as the key component of the GG syndrome and may incorporate one or more of the following features:

- Circumstantiality in discourse
- Overinclusive or excessively detailed narrative information
- Excessive detail may present with hypergraphia, painting, drawing
- Undue prolongation of the interpersonal exchange [26]

Metaphysical Preoccupation

1. Incipient and intense intellectual pursuits pertaining to morality, religion, and philosophy.

Other behavioral and physiological deviations such as hyposexuality or hypersexuality, undue fear, or aggression [23, 24, 27].

Uncinate Fasciculus

The UF is a late-maturing (third–fourth decades), major brain long-range association fiber tract connecting the OFC and anterior temporal lobes (Fig. 5.7). Of note is that it is particularly vulnerable to traumatic brain shearing-type injury. Its late maturation makes it a site for neuropsychiatric

Fig. 5.6 Isolated right frontal temporal encephalomalacia (*arrows*) post-hemorrhage and craniotomy in person with classic Geschwind-Gastaut syndrome

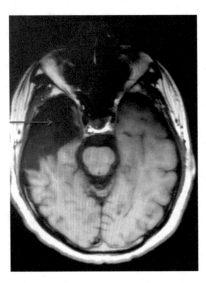 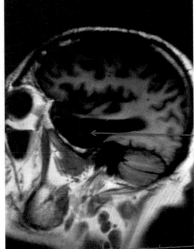

Table 5.1 Bear-Fedio Inventory reproduced from Bear DM, Fedio P. Quantitative analysis of interictal behavior in temporal lobe epilepsy. Arch Neurol 1977;34:454

Bear and Fedio Inventory questions	Yes/no
1. Emotionality—deepening of all emotions, sustained, intense affect	
2. Elation, euphoria—grandiosity, exhilarated mood, manic depressive disease	
3. Sadness—discouragement, tearfulness, self-deprecation, diagnosis of depression, suicide attempts	
4. Anger—increased temper, hostility	
5. Aggression—overt hostility, rage attacks, violent crimes, murder	
6. Altered sexual interest—loss of libido, hyposexualism, fetishism, transvetism, exhibitionism, hypersexual episodes	
7. Guilt—tendency to self-scrutiny and self-recrimination	
8. Hypermoralism—attention to rules with inability to distinguish significant from minor infractions, desire to punish offenders	
9. Obessionalism—ritualism, orderliness, compulsive attention to detail	
10. Circumstantiality—loquacious, pedantic, overly detailed, peripheral	
11. Viscosity—stickiness, tendency to repetition	
12. Sense of personal destiny—events given highly charged personalized significance, divine guidance ascribed to many features of patient's life	
13. Hypergraphia—keeps extensive diaries, notes, writing autobiography, novel	
14. Religiosity—holding deep religious beliefs, often idiosyncratic, multiple conversions, mystical states	
15. Philosophical—nascent metaphysical or moral speculations, cosmological theories	
16. Dependence—cosmic helplessness, "at hands of fate," protestations of helplessness	
17. Humorless—overgeneralized, ponderous concern, no humor, idiosyncratic	
18. Paranoia—suspicious, overinterpretative of motives and events, diagnosis of paranoid schizophrenia	

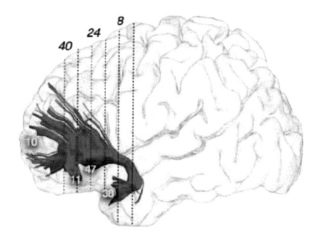

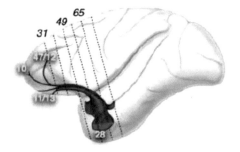

Fig. 5.7 White matter tract expansion: uncinate fasciculus evolution, comparative human and primate analyses. Figure with permission: Thiebaut de Schotten M, Dell'Acqua F, Valabregue R, Catani M. Monkey to human comparative anatomy of the frontal lobe association tracts. Cortex 2012; 48: 82–96

syndromes affecting young adults [28]. Three principal neurophysiological functions of the UF include episodic memory, social emotional, and linguistic processing. Pathological states linked to UF injury include the following:

Social-Emotional Processing Impairment
The anterior temporal pole is the site for storage as well as retrieval of socially related memories and person-related memories and a hub for the theory of mind circuitry.

Uncinate Fits
These are dreamy states, involving olfactory and/or gustatory hallucinations, sexual and other emotional arousals, and involuntary facial or oral activities [29].

Delusional Misidentification Syndromes
The UF is the site of injury at times for the delusional misidentification syndromes such as those with Capgras delusions indicate an impairment in the face-processing cortical regions to limbic areas that convey emotional salience or valence to the faces [30–32].

Cortical Deafness
Due to bilateral lateral temporal lobe lesions involving Heschl's gyrus very similar hearing abnormalities may occur with brainstem stroke due to anterior inferior cerebellar artery occlusion [33].

Auditory Agnosia
An inability to identify sounds that may be either verbal (pure word deafness) or nonverbal, despite normal peripheral hearing status [34].

Auditory Paracusias
These include auditory hallucinations (simple and complex) and illusions. These need to be differentiated from peduncular hallucinosis due to pontine infarcts or other lesions that present with very vivid, colorful, images. These are mostly visual but can be auditory in nature with reports of including people talking or shouting, when not the reality [35–37].

Disorders of Time Perception
The subjective impression that time may pass with excessive speed or not at all [38, 39].

Amusia
This may include subtypes of receptive, expressive, and lexical amusia (impairment in reading music [40].

Other Social Disorders Associated with Temporal Lobe Pathology (Discussed Under Frontal Network Syndromes Chapter)

Frontotemporal lobe disorders.

Right—behavioral FTD syndrome
Left—semantic aphasia

Schizophrenia
Autism spectrum conditions
Involuntary emotional expression disorder (IEED)

Other Social Disorders Associated with Temporal Lobe Pathology (Discussed Under Memory Syndromes Chapter)

Social impairment associated with memory impairments

Urbach–Wiethe disease
Autoimmune encephalitis

Acquired prosopagnosia

Progressive prosopagnosia
Prosopagnosia secondary to stroke

The left temporal pole is implicated in the processing and storage of proper names, important for social interaction. This relationship is supported by both lesion studies and functional imaging activation reports [41].

Traumatic Brain Injury (TBI): Orbitofrontal, Anterior Temporal Lobe and Uncinate Fasciculus Predilection of Injury

Although brain injury may be both focal and diffuse, extensive and widespread processes in both a temporal domain and a spatial domain have been established. In brief vascular, neurotransmitter perturbation, glucose metabolism, and network disturbances have been established. In addition there is a more focal frontotemporal predilection of involvement in TBI. Both blast injury and nonpenetrating head trauma of the anterior temporal lobes, the inferior frontal lobes, and the uncinate fasciculus as a functional unit appear to bear the brunt of injury. This had been established as long ago as 1937 and more recently corroborated by MRI imaging data in 40 patients [42]. Subsequent functional imaging with PET scans has also implicated the inferior frontal and temporal lobes indicating post-acute to chronic injury hypermetabolism [43]. Finally, resting-state network imaging has elucidated the much more diffuse consequences of TBI revealing the extensive cognitive and neuropsychiatric effects of the disconnected hub networks [44].

Williams Syndrome

This syndrome presents during childhood with features of:

Wide mouths
Upturned noses
Small chins
Curious starry eyes
Cardiac abnormalities
Hypercalcemia

Cognitively these children have intellectual prowess in some areas and weakness in others. They are described as being hypermusical, hypernarrative, and hypersocial. Notably they are remarkably social displaying effervescence; readily acquaint strangers are loquacious and seem to delight in story telling. Some of their weaknesses are akin to the autism spectrum people. Certain neuroimaging features have been reported including relatively smaller occipital and parietal cortices and larger temporal lobes. Functional imaging in relation to music has revealed increased activation in the cerebellum, temporal lobes, and amygdala [45].

Emotion Disorders

Although the brain's emotional circuitry is extensive and widespread involving many cortical and subcortical regions, the temporal lobe serves as an important hub. Our understanding of the much more expansive neural circuitry involved has prompted a reappraisal of the critical, necessary, and other involved brain regions in emotional assessment, regulation, and impairments secondary to disease processes (Fig. 5.8).

Evolutionary Insights

The brain has a two-tier system to help cope with survival. Primates and humans inherited vision as the predominant sense which is closely linked to the emotional brain centers. These enable a rapid and nonconscious reaction to environmental stimuli or predators that can be life saving, being a more ancient and "unconscious," system which developed earlier in evolution concerned with more elementary functions and survival, not the least of which is dealing with the vast sensory information that requires processing. This is a standard infrastructure amongst vertebrates in general. The conscious component became a later elaboration of the primate brain (Fig. 5.9) [46]. The occipitotemporal fasciculus is a particularly large fiber tract in humans, part of the inferior longitudinal fasciculus that allows rapid transfer of visual information to the anterior temporal lobe and amygdaloid complex for environmental threat or predator evaluation and human social and emotional salience processing (Fig. 5.10) [47].

As humans we are particularly cooperative as a species and our coordinated behavior is in

Medial view

Lateral view

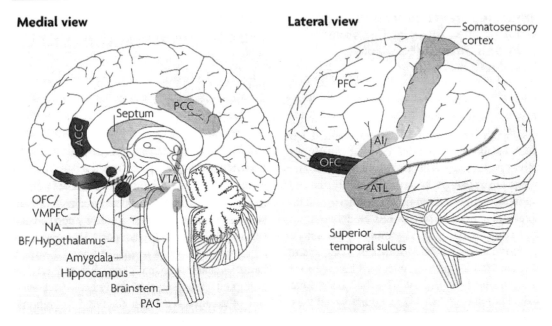

Fig. 5.8 Emotional brain: core (*red* colored) and extended brain (*orange* colored) regions. Core emotional brain: *OFC* orbitofrontal cortex, *VMPFC* ventromedial prefrontal cortex, *ACC* anterior cingulate cortex, *BF* basal forebrain, *NA* nucleus accumbens. Extended emotional brain: *PAG* periaqueductal gray matter, *ATL* anterior temporal lobe, *AI* anterior insula, *PCC* posterior cingulate cortex, *VTA* ventral tegmental area. Figure with permission: Pessoa L. On the relationship between emotion and cognition. Nature Reviews Neuroscience 2008;9:148–158

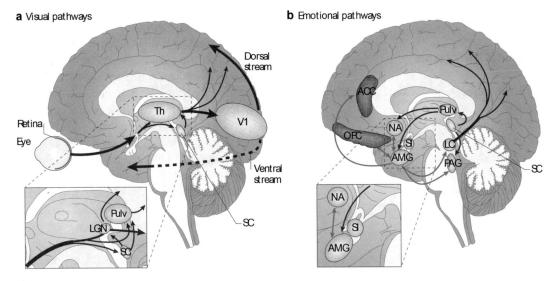

Fig. 5.9 Linking vision and emotion: cortical and subcortical pathways. *Th* thalamus, *V1* primary visual area, *SC* superior colliculus, *LGN* lateral geniculate nucleus, *Pulv* pulvinar, *OFC* orbitofrontal cortex, *ACC* anterior cingulate cortex, *NA* nucleus accumbens, *AMG* amygdala, *SI* substantia innominata, *PAG* periaqueductal gray matter, *LO* locus coeruleus. Figure with permission: Tamietto M, de Gelder B. Neural basis of nonconscious Nature Neuroscience Reviews 2010;11:697–709

Fig. 5.10 Occipito-temporal pathway. Figure with permission: Catani M. Jones DK, Donato R, ffytche DH. Occipito-temporal connections in the human brain. Brain 2003;126:2093–2107

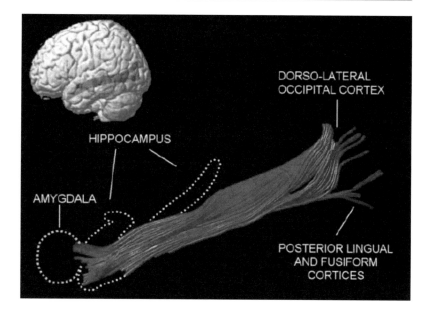

marked contrast to all other animals [48]. Metacognition, a function of the frontopolar cortex (BA 10), represents the apex of higher cortical function ability. This region may be divided into medial lateral and anterior regions within BA 10, mediating self-reflection, introspection, the monitoring and processing of internal states with the externally acquired information, episodic memory subfunctions such as source memory and prospective memory, and contextual and retrieval verification (Fig. 5.11) [49, 50].

Shea et al. have proposed a two-level metacognitive control system, a system 1 and system 2. System 1 (also called cognitively lean system) of metacognition operates nonconsciously or implicitly and is hypothesized to be involved in processing information from the senses for intrapersonal cognition relevant to many animals (Fig. 5.12). System 2 (also termed cognitively rich system) is thought to be unique to humans and is concerned with processing high-level information amongst several conspecifics or people, also termed suprapersonal control (Fig. 5.13). Shea et al. conjecture that the suprapersonal coordination metacognitive system antedated system 1 or that which controlled the intrapersonal cognitive processes that underlie emotional intelligence and inhibitory control [51]. In line

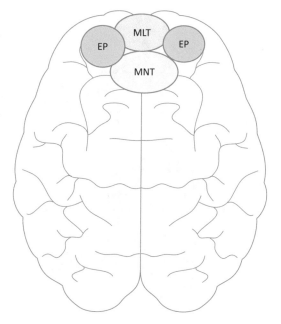

Fig. 5.11 Frontopolar cortex subregion activation patterns (regional cerebral blood flow and blood oxygen level dependent). *MLT* multitasking, *MNT* mentalizing, *EP* episodic memory. Figure adapted from: Gilbert SJ, Spengler S, Simons JS, Steele JD, Lawrie SM, Frith CD, Burgess PW. Functional Specialization within the Rostral Prefrontal Cortex (Area 10): A Meta-analysis. Journal of Cognitive Neuroscience 2006;18:932–948

with these hypotheses is that one of the core frontal functions of disinhibition was a critical development in human evolution.

Fig. 5.12 System one (lean) meta-cognitive system that allows improved oversight of sensory signals and the optimum output in terms of effector organs. Contention scheduling refers to the solving of appropriate output after receiving competing signals. Figure with permission: Shea N, Boldt A, Bang D, Yeung N, Heyes C, Frith CD. Suprapersonal cognitive control and metacognition. Trends in Cognitive Sciences 2014;18:186–193

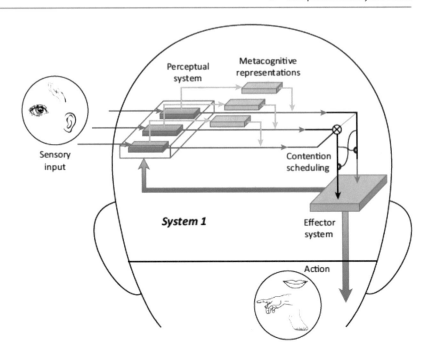

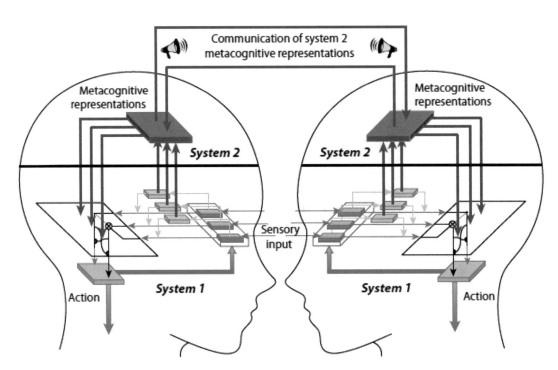

Fig. 5.13 System 2 (rich) meta-cognitive multi-agent cognitive control system. Cooperation between two people. Information from system 1 is available verbal broadcasting translating into increased reliability of the sensory signals. Figure with permission: Shea N, Boldt A, Bang D, Yeung N, Heyes C, Frith CD. Suprapersonal cognitive control and metacognition. Trends in Cognitive Sciences 2014;18:186–193

Clinical

Emotional intelligence (EI) defined differently according to the various clinical brain specialties such as psychology, neurology, and psychiatry. EI has been shown to be important for personal success and career success and in navigating the social complexities in day-to-day living [52]. EI impairment may be apparent in both apparently healthy people and after brain illnesses such as stroke, frontotemporal lobe dementia, Alzheimer's disease, and multiple sclerosis [53, 54]. Baron for example has proposed that EI comprises of the group of abilities that allow one to

1. Understand one's own emotions and be able to express feelings
2. Understand how others that one interacts with feel and how one relates to them
3. Manage and be in control of one's own emotions
4. Use one's emotions in adapting to one's environment
5. Generate positive emotions and use them for self-motivation in facing challenges

The neural circuitry for EI involves orbitofrontal, anterior cingulate, the insula, and amygdaloid complex regions [55, 56]. Processes that injure these areas in particular include traumatic brain injury, stroke, multiple sclerosis, and frontotemporal lobe disorders. A holistic clinical brain injury assessment should therefore involve all the clinical brain specialties, neurological, neuropsychiatric, psychological, and speech and language. EI assessment has already been extensively embraced in the corporate arena in view of its relationship with not only career success but also productivity and institutional health [57, 58] and EI is amenable for behavioral intervention programs [59, 60]. In a study involving stroke patients, many different brain lesions affected EI scores, including with frontal, temporal, subcortical, and subtentorial stroke lesions. The most significant abnormalities were however associated with the frontal and temporal region lesions, supporting Pessoa's extended emotional brain circuitry [61, 62].

Involuntary Emotional Expression Disorder

Involuntary emotional expression disorder (IEED) refers to clinically recognized syndromes that are characterized by intermittent inappropriate laughing and/or crying as well as intermittent emotional lability. Lesion studies have implicated the frontal primary motor (BA 4), premotor (BA 6), supplementary motor (BA 8), dorsal anterior cingulate gyrus, posterior insular and parietal regions, and corticopontine projections to the amygdala, hypothalamus, and periaqueductal gray matter. In addition seizure activity of these circuits may result in laughing (gelastic) or crying episodes (dyscrastic) [63]. Dextromethorphan is a sigma-1 receptor agonist and noncompetitive N-methyl-d-aspartate (NMDA) receptor antagonist. Quinidine is a competitive inhibitor of CYP2D6 which increases plasma levels of dextromethorphan. In placebo, double-blinded studies using validated IEED scales, the Center for Neurologic Study-Lability Scale, in multiple sclerosis and amyotrophic lateral sclerosis patients, the combination drug was superior to placebo [64].

References

1. Schumann C, Amaral DG. Stereological estimation of the number of neurons in the human amygdaloid complex. J Comp Neurol. 2005;491:320–9.
2. Bastir M, Rosas A, Gunz P, Peña-Melian A, Manzi G, Harvati K, et al. Evolution of the base of the brain in highly encephalized human species. Nat Commun. 2011;2:588. doi:10.1038/ncomms1593.
3. Semendeferi K, Barger N, Schenker N. Brain reorganization in humans and apes. The human brain evolving. Gosport, IN: Stone Age Institute Press; 2010.
4. Barton RA, Aggleton JP, Grenyer R. Evolutionary coherence of the mammalian amygdala. Proc Biol Sci. 2003;270:539–43.
5. Dunbar RIM. The social brain: mind, language and society in evolutionary perspective. Ann Rev Anthropol. 2003;32:163–81.
6. Pelphrey KA, Viola RJ, McCarthy G. When strangers pass: processing of mutual and averted social gaze in the superior temporal sulcus. Psychol Sci. 2004;15:598–603.

7. Kawashima R, Sugiura M, Kato T, Nakamura A, Hatano K, Ito K, et al. The human amygdala plays an important role in gaze monitoring. A PET study. Brain. 1999;122:779–83.

8. Kiernan JA. Anatomy of the temporal lobe. Epilepsy Res Treat. 2012;12:176157. doi:10.1155/2012/176157.

9. Ffythche DH, Howard RJ, Brammer MJ, David A, Williams S, Woodruff P. The anatomy of conscious vision; an f-MRI study of visual hallucinations. Nat Neurosci. 1998;1:738–42.

10. Irish M, Hodges JR, Piguet O. Right anterior temporal lobe dysfunction underlies theory of mind impairments in semantic dementia. Brain. 2014;137:1241–53.

11. Caplan LR, Kelly M, Kase CS, Hier DB, White JL, Tatemichi T, et al. Infarcts of the inferior division of the right middle cerebral artery. Mirror image of Wernicke's aphasia. Neurology. 1986;36:1015–20.

12. Earnest MP, Monroe PA, Yarnell PR. Cortical deafness: demonstration of the pathologic anatomy by CT scan. Neurology. 1977;27:1172–5.

13. Devinsky O, Bear D, Volpe BT. Confusional states following posterior cerebral artery infarction. Arch Neurol. 1988;45:160–3.

14. Kluver H, Buc PC. Preliminary analysis of functions of the temporal lobes in monkeys. Arch Neurol Psychiatr. 1939;42:979–1000.

15. Mesulam MM, Waxman SG, Geschwind N, Sabin TD. Acute confusional states with right middle cerebral artery infarctions. J Neurol Neurosurg Psychiatry. 1976;39:84–9.

16. Vignolo LA. Auditory agnosia. Philos Trans Roy Soc Lond. 1982;298:8212.

17. Clark CN, Golden HL, Warren JD. Acquired amusia. Handb Clin Neurol. 2015;129:607–31.

18. Bates GD, Sturman SG. Unilateral temporal lobe damage and the partial Kluver-Bucy syndrome. Behav Neurol. 1995;8(2):103–7. doi:10.3233/BEN-1995-8205.

19. Mendez MF, Shapira JS. Kissing or "osculation" in frontotemporal dementia. J Neuropsychiatry Clin Neurosci. 2014;26(3):258–61.

20. Devinsky J, Sacks O, Devinsky O. Kluver-Bucy syndrome, hypersexuality, and the law. Neurocase. 2010;16(2):140–5.

21. Morcos N, Guirgis H. A case of acute-onset partial Kluver-Bucy syndrome in a patient with a history of traumatic brain injury. J Neuropsychiatry Clin Neurosci. 2014;26(3):E10–11.

22. Gaul C, Jordan B, Wustmann T, Preuss UW. Klüver-Bucy syndrome in humans. Nervenarzt. 2007;78(7):821–3.

23. Trimble M, Mendez MF, Cummings JL. Neuropsychiatric symptoms from the temporolimbic lobes. J Neuropsychiatry Clin Neurosci. 1997;9:429–38.

24. Trimble M, Freeman A. An investigation of religiosity and the Geschwind Gastaut syndrome in patient s with temporal lobe epilepsy. Epilepsy Behav. 2006;9:407–14.

25. Hoffmann M. Isolated right temporal lobe stroke patients present with Geschwind Gastaut syndrome, frontal network syndrome and delusional misidentification syndromes. Behav Neurol. 2008;20:83–9.

26. Bear DM, Fedio P. Quantitative analysis of interictal behavior in temporal lobe epilepsy. Arch Neurol. 1977;34:454.

27. Waxman SG, Geschwind N. Hypergraphia in temporal lobe epilepsy. Neurology. 1974;24:629–36.

28. Johnson CP, Juranek J, Kramer LA, Prasad MR, Swank PR, Ewing-Cobbs L. Predicting behavioral deficits in pediatric traumatic brain injury through uncinate fasciculus integrity. J Int Neuropsychol Soc. 2011;17(4):663–73.

29. Von Der Heide R, Skipper LM, Kobusicky E, Olsen IR. Dissecting the uncinate fasciculus: disorders, controversies and a hypothesis. Brain. 2013;136:1692–707.

30. Hirstein W, Ramachandran VS. Capgras syndrome: a novel probe for understanding the neural representation of the identity and familiarity of persons. Proc Biol Sci. 1997;264(1380):437–44.

31. Feinberg TE, Roane DM. Misidentification syndromes. In: Feinberg TE, Farah MJH, editors. Behavioral neurology and neuropsychology. New York, NY: McGraw Hill; 1997.

32. Forstl H, Almeida OP, Owen AM, Burns A, Howard R. Psychiatric, neurological and medical aspects of misidentification syndromes: a review of 260 cases. Psychol Med. 1991;21(4):905–10.

33. Bamiou DE. Hearing disorders in stroke. Handb Clin Neurol. 2015;129:633–4.

34. Saygin AP, Leech R, Dick F. Nonverbal auditory agnosia with lesion to Wernicke's area. Neuropsychologia. 2010;48(1):107–13.

35. Lo YL, Hameed S, Rumpel H, Chan LL. Auditory hallucinations and migraine of possible brainstem origin. Headache Pain. 2011;12(5):573–5.

36. Cambier J, Decroix JP, Masson C. Auditory hallucinations in lesions of the brain stem. Rev Neurol (Paris). 1987;143(4):255–62.

37. Benke T. Peduncular hallucinosis: a syndrome of impaired reality monitoring. J Neurol. 2006;253(12):1561–71.

38. Perbal-Hatif S. A neuropsychological approach to time estimation. Perbal-Hatif S. dialogues. Clin Neurosci. 2012;14(4):425–32.

39. Oliveri M, Koch G, Caltagirone C. Spatial-temporal interactions in the human brain. Exp Brain Res. 2009;195(4):489–97.

40. Brust JCM. Music and language: musical alexia and agraphia. Brain. 1980;103:367–92.

41. Semenza C. Naming with proper names: the left temporal pole theory. Behav Neurol. 2011;24(4):277–84.

42. Gentry LR, Godersky JC, Thompson B. MR imaging of head trauma: review of the distribution and radiopathologic features of traumatic lesions. Am J Roentgenol. 1988;150(3):663–72.

43. Byrnes KR, Wilson CM, Brabazon F, von Leden R, Jurgens JS, Oakes TR, et al. FDG-PET imaging in

mild traumatic brain injury: a critical review. Front Neuroenergetics. 2014;9:13. doi:10.3389/fnene.2013.00013.

44. Fagerholm ED, Hellyear PJ, Scott G, Leech R, Sharp DJ. Disconnection of network hubs and cognitive impairment after traumatic brain injury. Brain. 2015;138:1696–709.

45. Galaburda AM, Holinger D, Mills D, Reiss A, Korenberg JR, Bellugi U. Williams syndrome. A summary of cognitive, electrophysiological, anatomo-functional, microanatomical and genetic findings. Rev Neurol. 2003;36 Suppl 1:132–7.

46. Kaas J. Convergences in the modular and areal organization of the forebrain of mammals: implications for the reconstruction of forebrain evolution. Brain Behav Evol. 2002;59(5-6):262–72.

47. Catani M, Jones DK, Donato R, Ffytche DH. Occipito-temporal connections in the human brain. Brain. 2003;126:2093–107.

48. Tomasello M. Why we cooperate. Cambridge, MA: MIT Press; 2009.

49. Gilbert SJ, Spengler S, Simons JS, Steele JD, Lawrie SM, Frith CD, et al. Functional specialization within the rostral prefrontal cortex (Area 10): a meta-analysis. J Cogn Neurosci. 2006;18:932–48.

50. Christoff K, Gabrieli JDE. The frontopolar cortex and human cognition: evidence for a rostrocaudal hierarchical organization within the human prefrontal cortex. Psychobiology. 2000;28:168–86.

51. Shea N, Boldt A, Bang D, Yeung N, Heyes C, Frith CD. Suprapersonal cognitive control and metacognition. Trends Cogn Sci. 2014;18:186–93.

52. Goleman DP. Emotional intelligence: why it can matter more than IQ for character, health and lifelong achievement. New York, NY: Bantam Books; 1995.

53. Mendez MF, Lauterbach EC, Sampson SM. An evidence-based review of the psychopathology of frontotemporal dementia: a report of the ANPA committee 16 on research. J Neurospsychiatry Clin Neurosci. 2008;20(2):130–49.

54. Van der Zee J, Sleegers K, Van Broeckhoven C. The Alzheimer disease frontotemporal lobar degeneration spectrum. Neurology. 2008;71:1191–7.

55. Bar-On R, Tranel D, Denburg NL, Bechara A. Exploring the neurological substrate of emotional and social intelligence. Brain. 2003;126:1790–800.

56. Shamay-Tsoory SG, Tomer R, Goldsher D, Berger BD, Aharon-Peretz J. Impairment in cognitive and affective empathy in patients with brain lesions:anatomical and cognitive correlates. J Clin Exp Neuropsychol. 2004;26(8):1113–27.

57. Loehr J, Schwartz T. The making of a corporate athlete. Harv Bus Revn. 2001;79:120–1281l.

58. Gilkey R, Kilts C. Cognitive fitness. Harv Bus Rev. 2007;85:53–4.

59. Cherniss C, Adler M. Promoting emotional intelligence in organizations. Alexandria, VA: American Society for Training and Development; 2000.

60. Goleman D. Working with emotional intelligence. London: Bloomsbury; 1999.

61. Hoffmann M, Benes-Cases L, Hoffmann B, Chen R. Emotional intelligence impairment is common after stroke and associated with diverse lesions. BMC Neurol. 2010;10:103. doi:10.1186/1471-2377-10-103.

62. Pessoa L. On the relationship between emotion and cognition. Nat Rev Neurosci. 2008;9:148–58.

63. Lauterbach EC, Cummings JL, Kuppuswamy PS. Toward a more precise, clinically—informed pathophysiology of pathological laughing and crying. Neurosci Biobehav Rev. 2013;37(8):1893–916.

64. Pioro EP. Review of dextromethorphan 20 mg/quinidine 10 mg (NUEDEXTA®) for pseudobulbar affect. Neurol Ther. 2014;3(1):15–28.

Memory Syndromes

6

Overview

The different types of memory and the related process of learning are all contained within expansive cerebral circuits. For example the recall of a particular event has elements of visual, acoustic, tactile, olfactory, and language-associated details. For memories to be made, there firstly needs to be adequate attention to allow registration of details, after which hippocampal processing occurs, with consolidation of information which is then relegated to storage in a particular area of the cortex. The prefrontal cortex is required for eventual retrieval, the right PFC for autobiographic details, and the left PFC for semantic information [1]. Clinical, cognitive neuroscience and psychological and neuropathological classifications exist for memory and the related disorders. These may be better understood by reviewing what is currently known about memory systems in evolution of animals, mammals, primates, and hominins.

Evolutionary Overview

Episodic Memory Evolution

Episodic memory is that cognitive ability which allows us to travel back in time and forward in time. Specific to episodic memory is an autobiographical association with past experiences and mental time travel into the future. Learning from past experiences and being able to contemplate future scenarios both have key evolutionary survival value. From archeological and biological data there is evidence that the hippocampus was present in early teleost fishes and enabled spatial information processing. Episodic memory is postulated to have developed much later in evolution and allowed information not only about "where" but also "what" and "when" components and is present in mammals and birds. The possible time frame of these evolutionary developments is depicted in Fig. 6.1 [2].

The hippocampus, consisting of the dentate gyrus, subiculum, dentate, and cornu ammonis fields, receives extensive inputs from the entorhinal cortex and major output fibers from CA1 and subiculum connect back to the entorhinal cortex. The parahippocampal region (PHR), that includes the entorhinal, perirhinal, and parahippocampus proper, is an interface structure between the hippocampus and other cortical areas of the brain. The PHR receives both the "what" circuit (perirhinal, lateral entorhinal areas) for processing of object characteristics and the "where" circuit (parahippocampal, medial rhinal, and postrhinal areas) that integrates visuospatial characteristics. From the PHR there is connectivity to all other cortical areas, most notably the prefrontal cortex (the site of delay neurons that mediate working memory) directly from the CA1 region and indirectly from the parahippocampal areas (Fig. 6.2). From a neurobiological functional point of view,

© Springer International Publishing Switzerland 2016
M. Hoffmann, *Cognitive, Conative and Behavioral Neurology*,
DOI 10.1007/978-3-319-33181-2_6

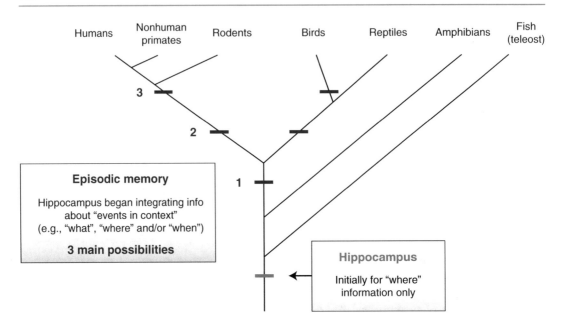

Fig. 6.1 Evolutionary emergence of episodic memory: Hippocampus first processes "where" information and later the what and when information. Figure credit with permission: Allen TA, Fortin NJ. The evolution of episodic memory. PNAS 2013;110:10379–10386

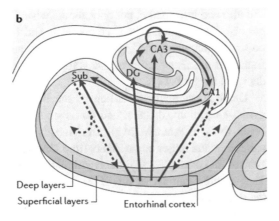

Fig. 6.2 Hippocampus anatomy. The entorhinal cortex connects principally with the dentate gyrus (DG), subiculum, and CA3. The major hippocampal outflow is via CA1 and the subiculum to the entorhinal cortex which in turn connects to other subcortical regions and cortical association areas. Image with permission: Small SA, Schobel SA, Buxton RB, Witter MR, Barnes CA. A pathophysiological framework of hippocampal dysfunction in ageing and disease. Nature Reviews Neuroscience 2011;12:585–601

CA1 is associated with information input integration, the dentate gyrus is thought to be concerned with pattern separation, CA3 with pattern completion, the subiculum with memory retrieval, and the entorhinal cortex is likely implicated in

the short-duration hippocampal memory processing [3] (Fig. 6.3). The basic circuit of episodic memory therefore comprises of the prefrontal, hippocampus, and parahippocampal regions that are present in both mammals and birds or proto-episodic memory system. Information from the primary sensory and secondary sensory association areas is channeled via the PHR and communicates with the hippocampus which separates the where, what, and when types of information. These processes have been termed the:

1. Registration by the sensory systems
2. Encoding and association (sensory association areas)
3. Consolidation (hippocampus)
4. Storage in differing but sensory relevant cortical areas termed engrams
5. Retrieval or ecphory (prefrontal cortex)

The right frontotemporopolar cortical areas enable episodic retrieval deficits and the left frontotemporopolar enables semantic retrieval deficits at least as is supported by lesion studies [4]. Two interacting circuits for the memory and learning include the memory-related Papez (hippocampal) circuit and the emotion-related

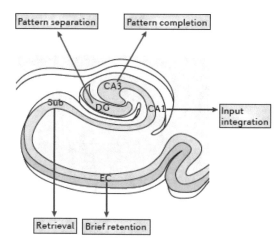

Fig. 6.3 Putative neurobiological functional mapping of the hippocampal corpus ammonis and associated subregions performing distinct cognitive processes; integration of inputs (CA1), pattern separation (dentate gyrus), pattern completion (CA3), memory retrieval (subiculum), and brief retention in memory tasks (entorhinal cortex). Image with permission: Small SA, Schobel SA, Buxton RB, Witter MR, Barnes CA. A pathophysiological framework of hippocampal dysfunction in ageing and disease. Nature Reviews Neuroscience 2011;12:585–601

amygdaloid, basolateral limbic circuit. These subserve the encoding and consolidation phases and allow new information to reverberate in these circuits before ultimate transfer to long-term memory storage in other cortical areas [5].

Episodic recall of events can be seen as a reactivation of circuits that have been cued by the hippocampus and PHR for a particular memory trace. The PFC retrieval process allows evaluation of the retrieved data and allows response planning for action. This demonstrates the overlap of EM and WM for example. In addition the overlap with EM and SM can be understood in terms of the fundamental role of the hippocampus being the assimilation semantic and episodic memory traces. Similar neural circuitry is used for episodic memory retrieval and simulating the future. Episodic memory has been considered to be particularly important in maintaining social relationships and primates, scrub jays, and rodents all of which have both episodic memory and sociality (Fig. 6.4) [6].

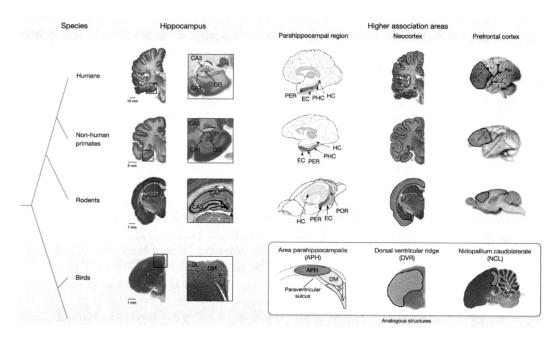

Fig. 6.4 Comparison of the key brain components (hippocampus, parahippocampal regions, cortical association areas, prefrontal cortex) that enable episodic memory in different species. *PER* perirhinal region, *EC* entorhinal cortex, *PHD* parahippocampal cortex, *HC* hippocampal cortex, *DG* dentate gyrus. Figure credit with permission: Allen TA, Fortin NJ. The evolution of episodic memory. PNAS 2013;110:10379–10386 and Furtak SC, Wei SM, Agster KL, Burwell RD. Functional neuroanatomy of the parahippocampal region in the rat: The perirhinal and postrhinal cortices. Hippocampus 2007;17(9):709–722

For example scatter hoarding in birds and mammals is a foraging behavioral technique that promotes social competition as opposed to larder hoarding that is less cognitively challenging. The former is also associated with larger hippocampi in both mammals and birds and allows scatter hoarding mammal or bird to monitor and track conspecific activities in relation to foraging and food. Hence imperative of social intelligence development may have strong ties to the evolution of episodic memory particular in primates and humans [7, 8].

The Next Evolutionary Development: The Dentate Gyrus Evolution

Neurogenesis in humans takes place in the hippocampal dentate gyrus and subventricular zone and this plasticity, mediated through improved network efficiency, by virtue of the foraging activity for example provides a particular advantage to animals in dealing with challenging environments [9, 10].

Although the hippocampus dates back to reptiles, amphibians, and fish, the dentate gyrus appeared very later, only in mammals. Even amongst primates there is a progressive increase in dentate gyrus size when measured between insectivores and monkeys, apes, and humans. One of the proposed neural mechanisms whereby the dentate gyrus enables advantageous and more flexible adaptation may be through the ability of neurons to time stamp information as it is being processed termed pattern separation and not only quantitative neurogenesis being important but also the impact it has on connectivity to ultimately provide a network that is more efficient or advantageous. Motility and changing environments whether due to climate or predators, or simply "moving," are aided by flexibility, being able to draw on past experiences for better predictions. This is one hypothesis why physical activity is linked to neurogenesis. The dentate gyrus neurogenesis also plays a role on affective functions in humans and decreased neurogenesis is seen in schizophrenia and depression and may be as a collateral effect of being an animal with a relatively high rate of neurogenesis (Fig. 6.5) [11].

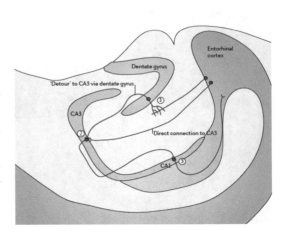

Fig. 6.5 The dentate gyrus evolved later in mammalian evolution. Figure with permission: Kemperman G. New Neurons for survival of the fittest. Nature Neuroscience Reviews 2012;13:727–736

Working Memory Evolution

A definition of working memory is the ability to maintain information online for a short period of time, usually seconds to minutes to process information which could be a sentence, words, numbers, thoughts, and actions. This is usually what is referred to as short-term memory. Whereas working memory is measured in seconds to minutes, episodic memory is measured in several minutes (episodic short term) to years (episodic long term). Working memory may be thought of as a kind of operating system for the brain and underlies the core frontal processes such as disinhibition, attention, and initiation [12].

From an Evolutionary Perspective Working Memory May Be Regarded as a Missing Link in Human Cultural Development and Creativity, Arts, and Language Development

Working memory evolution and more specifically the advent of enhanced working memory between 200 and 40 kya ago are regarded as a key factor in human mind evolution [13]. In Skoyles's view the advent of neural plasticity together with the development of the prefrontal, parietal, and

Working memory hubs

1. Epoptic process: Mid dorsolateral prefrontal cortex monitors information
2. Manipulation of information: Posterior parietal region around the intraparietal sulcus
3. Both regions interact with the superior temporal region for sensory information (auditory, visual verbal, multimodal) resides

Principal fiber tracts sub-serving working memory

A. Superior longitudinal fasciculus
B. Middle longitudinal fasciculus
C. Extreme capsule

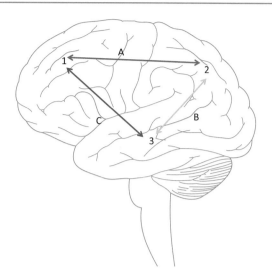

Fig. 6.6 The most fundamental and cardinal function of the frontal lobes is working memory: The extensive evolutionary evolved frontoparietal circuits were exapted for working memory. Adapted from Petrides M. The mid dorsolateral prefrontal-parietal network and the epoptic process. In: Stuss DT, Knight RT. Principles of Frontal Lobe Function 2nd ed. Oxford University Press, Oxford 2013

temporal heteromodal association cortices enabled the development of newly acquired abilities such as recursion in language, visual arts, and sophisticated toolmaking [14].

The working memory circuit is an expansive brain circuit with frontal, parietal, temporal, and subcortical components (Fig. 6.6). BA 46 neurons which keep firing and are active during the delay period after a stimulus and before an action takes place represent the neural substrate for the maintenance of information (auditory, visual, verbal information). Whereas the DLPFC BA areas 46 are key for monitoring of information, termed the epoptic process by Petrides, the posterior parietal cortical regions are concerned with the manipulation of information [15]. For example, activities such as mental arithmetic are examples of how the PPA is concerned with information during working memory activities; it is notable in this regard that the BA 46 target axons connect extensively with levels I–III of sections of the intraparietal sulcus underscoring modulation of the MDLPFC on parietal activity [16]. The inferotemporal regions are involved with visual working memory or object representations and the superior temporal region with auditory working memory representations [17].

The major fiber tracts involved include the extreme capsule, the superior longitudinal fasciculus, and the middle longitudinal fasciculus. The midventrolateral PFC (not DLPFC) is responsible for retrieval of memory traces shown by images studies [18, 19].

A further component of working memory is BA10 which integrates subgoal working memory processes and is one of the areas that is particularly enlarged (allometrically expanded) in humans compared to apes [20].

Extant primate comparative studies have suggested that chimpanzees have a short-term memory level graded 2, perhaps 3, based on a number of studies including nut cracking, object manipulations, and gestures. In general chimpanzees are able to coordinate object manipulation: one at 1 year, two at about 2 years, and more than two 3 years or later, but three object manipulations were recorded as being rare. Other working memory tasks associated with primates include the termite foraging with "fishing sticks" seen in the Congo basin chimpanzees. Developmental research with human infants and evaluating the magnitude of short-term working memory capacity (STWMC) and plotting their growth patterns with chimpanzee brain growth patterns using

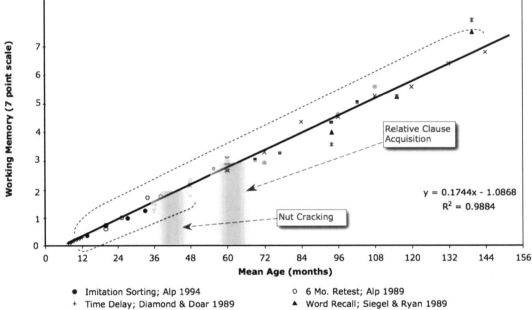

Fig. 6.7 Working memory trajectory in humans, at level 7 compared to Pan Paniscus (Bonobos) at level 2. Figure with permission and modifed from Read D. Working Memory: A Cognitive Limit to Non-Human Primate Recursive. Thinking Prior to Hominid Evolution. Evolutionary Psychology 2008;6(4):676–714

several studies revealed the composite graphs for STWMC over a time period ranging from 7 to 144 months depicted in Figs. 6.7. Taken together, these studies are consistent with the archeological data indicating a WM capacity for the common ancestor of Pan and hominins of 2–3 and 7 in modern humans (Fig. 6.8) [21].

Wynn and Coolidge have assembled a number of archeological findings as evidence of enhanced or modern working memory abilities including the discovery of the following:

Traps and snares

Involve long planning for the future, in both time and space

Reliable weapons

Projectile weapons such as spears, bow, and arrows comprise of several components

Hafting

Involves combining a point or stone tool to a shaft, and is also multi-component assembly, at least three (shaft, stone tool, the haft itself) which is a feature of EWM capacity with the earliest archeological evidence for hafting associating with Neanderthals ~200 kya

Foraging systems

Such as agriculture which not only requires planning ahead, but also being able to practice delayed gratification with respect to a portion of a crop stored for future planting, a function of response inhibition

Hunter-gatherer activities such as planning a hunt, intercepting herds, and meat storage are other functions dependent on working memory

Fig. 6.8 Working memory stages. A possible genetic mutation expanded less developed working memory (WM) into enhanced WM (modern humans have EWM) between 13,000 and 80,000 years BP and shift in short-term WM 6 to short-term WM 7. Figure with permission: Read D, van der Leeuw S. Biology is only part of the story. Phil. Trans. R. Soc. B. 2008;363:1959–1968

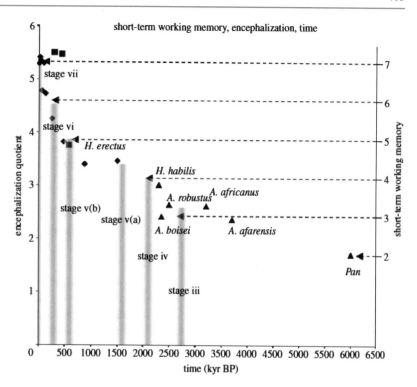

The inherent limitation of restricted working memory capability of the brain was subsequently overcome to storing information outside the brain, thus freeing the episodic buffer from holding information allowing the central executive to process the information. Termed externalization of information, subsequent archeological discoveries have included the following:

1. The Blombos Cave artifacts of engravings on ochre, bone, and marine shells, dated to 77 kya, that presumably provided some externalized information about the wearer
2. The therianthropic figurine, the Hohlenstein-Stadel Lionman, found in Germany dated to 32 kya, representing "externalized abstraction" of a metaphysical or existential challenges of people at the time, the term used by Wynn and Coolidge
3. Engravings found in the Grotte du Tai, Western France, dated to 12 kya

These examples about evolving WM capacity suggest that sometime between 77 and 30 kya (at least according to the artifacts found to date)

our WM expanded from the level 6 attributed to Neanderthals into the modern or enhanced working memory that can be graded at level 7 [22]. More recently these include engraved lines or numbers, written storage in books followed by computer discs and cloud storage.

Procedural Memory Evolution

Stone knapping may have laid the foundations of procedural memory with progressive integration of the frontoparietal cortices and subcortical regions, including the basal ganglia and cerebellum. Procedural related cognitive ability is first evident with the artifacts associated with *Homo erectus* ~1.6 mya with an 880 cc intracranial volume. Being a predominant motor activity it is influenced by both spatial input through the dorsal stream, primarily parietal, and shape appreciation attributed to the ventral pathway, predominantly temporal lobe. It has been proposed that the activity of making a stone tool such as a biface required the combined input of these two and in Wynn and Coolidge's view these were accessed by virtue of the working memory component of visuospatial attention. In essence, these formed the procedural memories of *Homo erectus*. A further leap in the sophistication of visuospatial ability in terms of dorsal and ventral pathway coordination and processing expertise enabled allocentric perception, the ability to imagine other views, not centered on the self-centered viewpoint, regarded as key for three-dimensional hand axe construction, for example. This ability is represented by *Homo heidelbergensis* and the finding of the Shoeningen spears dated at ~400 kya and was also associated with the increase in parietal lobe size. The ability to maintain and also manipulate the visuospatial information is thought to have been aided by the evolution of the visuospatial sketchpad component of working memory during this period. In this manner it has been postulated that spatial cognitive expertise evolved out of technical intelligence that in turn was fostered by stone tool knapping of the evolving hominins. Thereafter, the advent of increased working memory capability of *Homo heidelbergensis*, these technical intelligences and ability, "trapped" within procedural memory allowed allocentric perception, for example, and more advanced coordination of shape and spatial components [23, 24].

Clinical

The different memory systems and the clinically appreciated disorders can be categorized in a number of different ways such as the following:

1. Memory can be categorized temporally such as anterograde and retrograde amnesia
2. Short-term, long-term, and prospective memory
3. Amnesia or hyperthymesia (Savant syndromes)
4. Different neurobiological and clinical categories such as episodic, working, semantic, procedural (Figs. 6.9, 6.10, 6.11, and 6.12)

For example people with hyperthymesia have an inordinately detailed and accurate autobiographic memory. Derived from the Greek hyper (excessive) and thymesis (remembering) it refers to a vastly superior autobiographical memory whereby people can remember almost every day of the life in considerable detail without particular effort that may be ascribed to a subconscious process [25–27].

The different memory systems are all represented in widely distributed but differing circuits that may also have distinct neuropathological signatures (Fig. 6.13). With episodic memory the

Memory subsystem	Information, type / time	Neuroanatomical components
Working (7 digits forward, 5 back, shapes)	seconds to minutes	DLPFC, PPA, TL, subcortex
Episodic (word lists, short)	minutes to years	PFC, FPC, MT ATN, F, MB
Semantic (factual information)	minutes to years	Lateral inferior temporal region
Procedural (biking, walking)	minutes to years	SMA, BG, cerebellum

Legend: PFC-prefrontal cortex, DLPFC – dorsolateral prefrontal cortex, PPA – posterior parietal areas, TL – temporal lobe, PHR – parahippocampal region, SMA – supplementary motor area, BG – basal ganglia

Fig. 6.9 Neurobiology and neurophysiology of the different memory systems

Fig. 6.10 Different kinds of memory impairment. Time-based loss and hypo- and hyperfunction

Retrograde amnesia Anterograde amnesia Amnesias Hyperthymesia

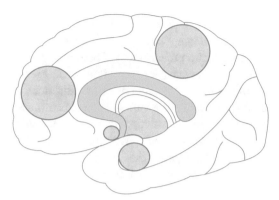

Fig. 6.11 Schematic diagram of the principal episodic memory components: frontal, basal forebrain, amygdala, thalamus, and medial parietal

central theme is autonoesis or the autobiographical perspective of events based in time. The basic episodic memory process involves encoding, consolidation, and retrieval of information and clinical manifestations of impairment include both anterograde and retrograde dysmnesia; a number of different episodic subtypes have been described, including the following.

Memory Classification: A Neurobiological and Clinical System Approach

Episodic memory and subtypes

Item
Source
Associative
Familiarity
Reality monitoring
Retrograde and anterograde

Working memory

Central executive
Episodic buffer
Visuospatial sketchpad
Phonologic loop

Semantic

Category-specific memory loss
Progressive prosopagnosia or facial memory amnesia

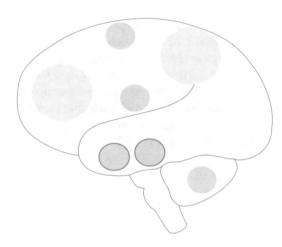

Fig. 6.12 Schematic diagrams of the major hubs of working memory frontal, parietal (*pink*), semantic memory inferotemporal, anterior temporal lobe (*green*), procedural memory; supplementary motor area, basal ganglia, cerebellum (*blue*)

Fig. 6.13 The four memory systems and the principal memory subsystem affected. Other memory systems may also be impaired, but to a lesser degree

Condition	Working	Episodic	Semantic
Procedural			
• Cognitive vascular disorder	√		
• Multiple Sclerosis	√		
• Frontotemporal dementia (B)	√		
• Traumatic brain injury	√		
• Alzheimer's		√	
• Transient global amnesia		√	
• Korsakoff's syndrome		√	
• Urbach Wiethe disease		√	
• FTD/semantic dementia			√
• Parkinson's			√
• Huntington's			√

Procedural
Metacognitive memory
Hyperthymesia (Savant syndromes)

Eidetic memory (vivid art images)
Calendar calculating
Musical genius abilities
Mathematical
Visuospatial/mechanical skills

Episodic Memory

The neuro-anatomical components of the episodic memory circuit include the following (Fig. 6.2):

Medial Temporal, Hippocampus
Cornus ammonis CA1, CA2, CA3, CA4
Dentate gyrus
Postsubiculum

Parahippocampus
Parahippocampal cortex
Entorhinal area
Perirhinal area

As with the different memory systems, the episodic memory subsystems also have particular neuropathological correlations or signatures.

For example, hippocampal lesions are associated with source memory, associative memory, and recollection dysmemory. There is a relative sparing of the subsystems of item memory and familiarity memory. This has been attributed to its role in binding the various elements of material that is learnt. Parahippocampal or extrahippocampal regions are associated with item memory impairment, that is, it is particularly important for storage of individual words or objects [28].

Newly acquired information reverberates in a neural circuit before being transferred to long-term storage. This is for encoding and consolidating information and the region includes the limbic system (hippocampo-entorhinal complex) as the critical component. The role of the frontal lobe in episodic memory has become appreciated with lesion studies and functional imaging reports supporting right fronto-temporopolar damage associated with episodic retrieval difficulty, whereas left frontotemporopolar damage is associated with semantic retrieval difficulty. There are in fact two interacting circuits within the limbic system [29].

Papez circuit (hippocampal)—predominantly subserving memory (Fig. 6.14)
Basolateral limbic circuit (amygdaloid)—predominantly subserving emotion (Fig. 6.15)

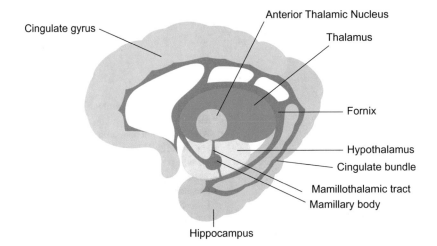

Fig. 6.14 Two interacting circuits for memory: Papez circuit)

Cingulate gyrus

Anterior Thalamic Nucleus

Thalamus

Fornix

Hypothalamus

Cingulate bundle

Mamillothalamic tract

Mamillary body

Hippocampus

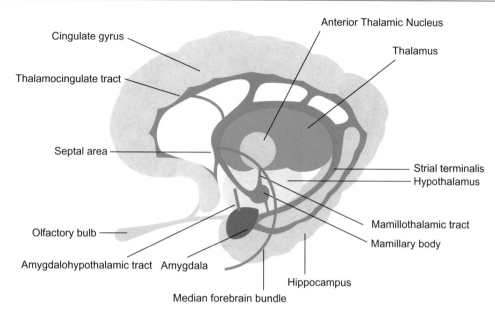

Fig. 6.15 Two interacting circuits for memory: limbic system circuit

The Hippocampus Organizes Memory According to Object, Space, and Time

Particulars or details about objects in the environment are relayed to the perirhinal (PRC) and lateral entorhinal areas (LEA) (the "what" information). Localization-related information ("where" details) is computed by the parahippocampal (PHC) and medial entorhinal areas (MEA). The two-way interaction with these areas may constitute the mechanisms for context or recall via the PHC-MEA and for item recall associations via the PRC-LEA (Fig. 6.16) [28]. Furthermore, in addition to organizing information in terms of spatial and physical dimensions, the hippocampus is also the region where memory is organized in the dimension of time. The appreciation of episodic memory in terms of temporal dimensions has recently been reviewed by Eichenbaum [30].

The specific role of the amygdala in relation to memory is with ascribing valence from an emotional point of view to the encoding of the information as well as in the retrieval process. This process may explain the so-called flash bulb memories in which the memory trace has extensive detail and is very personalized due to the magnitude of importance of a particular event [31].

Neurophysiology and Neurochemistry of Memory Formation

The hippocampus contains three pathways that subserve facilitation to long-term potentiation (LTP) in response to a recent activity:

1. Perforant pathway: connecting entorhinal area to the dentate gyrus
2. Mossy fiber tract: granule cell axons connecting to the CA3 pyramidal cells of the hippocampus
3. Schaffer collateral tract: collateral tract from CA3 to CA1 (excitatory) hippocampal pyramidal cells

The mechanisms of LTP induction within the hippocampal CA1 region include:

1. Postsynaptic depolarization
2. Activation of NMDA receptor activation

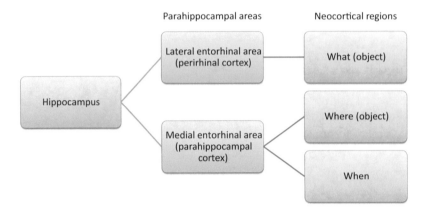

Fig. 6.16 The hippocampus areas compute details about objects, the spatial characteristics of objects, as well as time dimensions. *LEA* lateral entorhinal area, *PRC* perirhinal cortex, *PHC* parahippocampal cortex, *MEA* medial entorhinal area. Figure with permission: Eichenbaum H, Yonelinas AP, Raganath C. The medial temporal lobe and recognition memory. Ann Rev Neurosci 2007;30:123–152 and Eichenbaum H. Time cells in the hippocampus: a new dimension for mapping memories. Nature Reviews Neuroscience 2014;15:732–744

Table 6.1 Episodic memory: summary of frontal lobe and medial temporal lobe impairment profile

Memory subtype	Frontal lobe	Temporal lobe
Free recall	Impaired	Impaired
Item memory	~Normal	Impaired
Source memory	Impaired	Impaired
Recognition memory	~Normal	Impaired
Environmental assistance	Improves	No improvement
False memories	Frequent	Not present

Adapted from Wolk and Budson [32]

3. Influx of calcium
4. Calcium activation by a number of second messengers within postsynaptic cells

These processes enable an overall increase in the sensitivity and total number of postsynaptic AMPA receptors to glutamate and result in increased neurotransmitter release from presynaptic CA3 neurons. LTP has both an early transient and a later consolidation phase of influence (Table 6.1):

Early LTP, duration 1–3 h, without synthesis of new proteins
Late LTP, duration ≥24 h, and involves the synthesis of new proteins [29]

Clinically episodic memory may encompass a number of different subtypes that may sometimes point to the process or pathological state:

Source memory
The recall of the particular context relevant to a particular memory
Associative memory
Joining in memory the different experiences of a particular event inking multiple aspects of an event
Item memory
Recall of a list of items not associated with context
Reality monitoring
The concept of having actually performed a task or act as opposed to having only thought of having performed a task in the past
Familiarity
Maintain they have met an individual but unable to recollect the circumstances or how that happened

Anterograde and Retrograde Amnesia

Retrograde amnesia is usually most evident for information acquired more recently (Ribot's law). This has been argued to be supportive of the standard consolidation model, basically due to time-dependent memory representation. The hippocampus forms bonds with the neocortex pertaining to the particular details of an occurrence and as time passes these cortical representations become relatively independent of the initial critical influence of the hippocampus which subsequently wanes, perhaps to no role at all, and this is the explanation for remote memories being independent of hippocampal impairment.

Metamemory

This refers to the concept of "self-knowledge," or having insight into one's own memory abilities or capabilities as well as the accuracy of both retrospective and prospective performance judgements. This can be tested by tests such as the feeling of knowing (FOK) and judgment of learning (JOL) tests [33, 34].

Frontal Lobe Episodic Memory Impairment

The prefrontal cortex (PFC) episodic memory involvement enhances encoding of memory and is involved in retrieval of information whereas the medial temporal lobe is concerned with information retention. PFC lesions are typically due to lesions such as stroke particularly subcortical white matter abnormalities or leukoaraiosis, traumatic brain injury, multiple sclerosis, frontotemporal lobe degeneration, Lewy body dementia, frontal brain tumors, and depression. In general the associated episodic memory (EM) disorder is milder than that recorded for hippocampal pathologies and the specific subsystems affected tend to involve recollection, source memory, and memory for the temporal sequence of an event with familiarity memory relatively spared. Importantly, free recall is much more impaired than recognition memory and both encoding and retrieval are facilitated by environmental assistance. False memories, memory distortions, and confabulation are also features of frontal lobe episodic memory impairment. Some of these subtypes can be tested by standard memory testing protocols including word lists (item memory), word pairs (associative memory), and learning two word lists and testing for words that came from which particular list. These help differentiate memory impairment from the medial temporal lobe as opposed to the frontal lobe hubs in the episodic memory circuit [35].

Certain Pathological Conditions Pick Out Components of the Episodic Memory (EM) Subsystems

Within the medial temporal lobe, whether the lesions involve the medial hippocampus or the bilateral fornices, anterior thalamic nuclei, mammillary bodies, or posterior cingulate cortex, similar episodic memory disturbances have been reported. With Alzheimer's disease (AD), the EM deficit may be conceived of as a cortico-hippocampal cortex disconnection due to entorhinal layer II neuron fibrillary tangle involvement, disconnecting the perforant tract into the hippocampus. In addition, forebrain pathology due to the AD pathology may also cause a relative cholinergic deficiency [36]. The most common processes affecting episodic memory impairment are cognitive vascular disorders due to small vessel cerebrovascular disease and leukoaraiosis, posterior cerebral artery infarcts with unilateral or bilateral mesial temporal lobe involvement, and AD pathology (Fig. 6.17). In addition there is evidence that EM impairment with aging is predominantly frontal lobe based, probably due to leukoaraiosis affecting the frontal subcortical networks or cortical volume loss or both resulting in relative dopaminergic deficiency (Fig. 6.18) [37].

Neurobiological components	Pathological process
Medial temporal lobe	PCA infarct, AD, LBD, HIE, encephalitis (autoimmune, viral)
Hippocampal?	Transient global amnesia
Hippocampal	Urbach–Wiethe disease
Thalamus, mammillary bodies	Korsakoff's amnesia
Basal forebrain	Anterior communicating aneurysm rupture, Ach deficiency
Parietal lobes	Both midline (retrosplenial) and lateral mostly due stroke

PCA posterior cerebral artery, AD Alzheimer's disease, LBD Lewy body disease, *HIE* hypoxic ischemic injury, *Ach* acetylcholine [38]

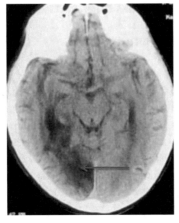
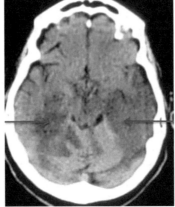
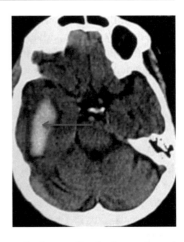

Fig. 6.17 Common causes of episodic memory loss due to different stroke syndromes: unilateral occipitotemporal infarct (*left*), bilateral temporal infarcts (*middle*) usually due to the posterior cerebral artery distribution stroke and right temporal lobe hemorrhage (*right*)

Fig. 6.18 Differing processes affecting memory in ageing: Frontal subcortical, dysexecutive syndrome, hypertension associated and medial temporal, Alzheimer's pathology subtypes. Figure with permission: Buckner RL. Memory and Executive Function in Ageing and AD: Multiple Factors that Cause Decline and Reserve Factors that Compensate. Neuron 2004;44:195–208

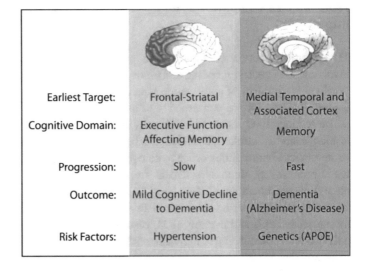

Earliest Target:	Frontal-Striatal	Medial Temporal and Associated Cortex
Cognitive Domain:	Executive Function Affecting Memory	Memory
Progression:	Slow	Fast
Outcome:	Mild Cognitive Decline to Dementia	Dementia (Alzheimer's Disease)
Risk Factors:	Hypertension	Genetics (APOE)

Some Memory-Specific Pathological Conditions

Herpes Simplex Encephalitis, Autoimmune Encephalitis

Both tend to target discrete areas of the medial temporal lobe and sometimes the inferior frontal lobes, rarely the thalamus. Although the presentation may be fulminant, with headache, pyrexia, seizures, encephalopathy, or neuropsychiatric syndromes, the presentation may be covert with subtle forms of memory loss. Seizures, when they occur, alert the clinician, heralding the neurological process. Herpes simplex encephalitis is one such disorder that may easily be missed if MRI neuroimaging is not promptly performed as the neurological deficit may be subtle and neuroimaging is usually pivotal in establishing the diagnosis (Fig. 6.19) [39]. With improved MRI imaging, PCR CSF evaluation, and antiviral treatment the majority of people afflicted survive with good recovery but persistent memory loss may remain [40].

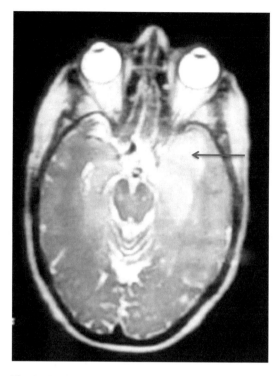

Fig. 6.19 Herpes simplex encephalitis-related left mesial temporal lobe lesion (*arrow*)

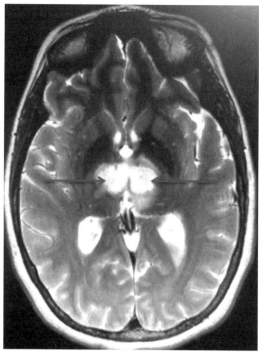

Fig. 6.20 Tegmentothalamic syndrome due to mycoplasma pneumonia encephalitis (*arrows*)

Two cases (unpublished), encountered by the author, demonstrate these presentations:

A 39-year-old woman with relatively sudden onset of initial mutism and encephalopathic state, with subsequent recovery over several weeks to living independently. However in the first year of recovery, the most significant problems were marked episodic memory loss and a Parinaud's syndrome. The dysmemory involved forgetting the day of the week and the date, and frequently misplacing objects such as her keys and wallet. The offending lesions were bilateral tegmentothalamic hyperintensities attributed to mycoplasma pneumonia encephalitis (Fig. 6.20). Other similar cases have been described [41].

A middle-aged man presented to an emergency room accompanied by his wife who was concerned that he suddenly did not recognize their bedroom at home. Examination revealed a mild pyrexia without other general medical or neurological examination abnormalities.

Neuroimaging revealed classic medial temporal hyperintense signal abnormalities consistent with herpes simplex encephalitis, which was confirmed by brain biopsy. The heralding neurological (neuropsychological) presentation was therefore a sudden onset of jamais vu (feeling of unfamiliarity) for his own bedroom.

Transient Global Amnesia

Transient global amnesia (TGA) is a clinical diagnosis, usually by the Oxford criteria (witness with repetitive questioning, duration <24 h); the pathophysiology is not known and currently it is without with no corroborative radiological or laboratory tests [42]. The neurological syndrome is one of evanescent episodic dysmemory lasting a number of hours, without other neurological or neuropsychological deficits, and MRI brain scans are typically normal. However a number of functional imaging studies, single-photon emission computed tomography (SPECT), functional magnetic resonance imaging, magnetic resonance diffusion and perfusion studies, and positron emission computed tomography (PET), have detailed objective thalamic and cortical hypoactivity, and hypoperfusion in the hippocampus and frontotemporal regions. The current pathophysiological mechanisms are thought to involve a primary neuronal hypometabolism in turn due to excitotoxic neurotransmitter perturbation that temporarily impairs memory) [43–48].

Urbach–Wiethe Disease (Ceroid Neuronal or Lipoid Proteinosis)

This autosomal recessive condition is noted by dermatological infiltrate thickening of skin with 50–75 % of patients having medial temporal lobe calcifications, particularly of the amygdala and peri-amygdaloid areas. Common presentations include epilepsy, anxiety, schizophrenia, and other psychoses [49]. The amygdala involvement affects the appropriate interpretation of emotionally important signals and stimuli, in particular with respect to negative emotional facial presentations. This in turn affects their memory processing, as the amygdala is an integral component of memory formation. The amygdala and septal nuclei may have opposing functions with the amygdala increasing emotional response and the septal activation downgrading emotional activity, both being part of the limb memory circuit [50]. Working memory has been reported to be paradoxically increased in three people with Urbach–

Wiethe disease (UWD). As emotional tone is concerned with encoding of episodic memories, the integrity of the amygdala is likely important in this circuitry involving both anterograde and retrograde memory influences. Working memory on the other hand is prefrontal cortex (and posterior parietal) based and the evolving capability of increasing working memory control, ton one of voluntary control, probably allowed persistent emotion inhibition control and with that the emergence of the modern human mind and emotional intelligence. The amygdala, being responsible for continuous salience reconnaissance, exacts a cost on the working (attention) memory and the amygdala is a hub for mediation between salience and attentional, executive circuitry and function. The neurobiological mechanisms whereby this rare disease process probably up-regulated working memory are thought to be paradoxical functional facilitation [51].

The Hippocampal Involvement with Other Common Neurological and Psychiatric Conditions

Extensive cortico-cortical connections to the hippocampus may be an explanation for the appreciation of the "molecular anatomy" within this region that can be considered a type of "melting pot" for clinical neurological and psychiatric disease. A regional vulnerability within the hippocampal formation is evident; for example, the CA1 subfield is especially vulnerable to vascular disease (such as after hypoxic ischemic

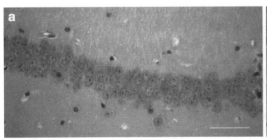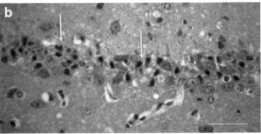

Fig. 6.21 Hippocampal pyramidal cells: normal (**a**), hypoxic damage (**b**), delayed neuronal death, in CA1 Sommer's sector with cytoplasmic eosinophilia, nuclear hyperchromasia and pyknosis or "red dead neurons" (*arrows*). Figure with permission: Quillinan, N, Grewal

H, Deng G et al. Region specific role for GluN2B containing NMDA receptors in injury to Purkinje cells and CA1 neurons following global cerebral ischemia. Neuroscience 2015 January 22;0:555–565. doi:10.1016/j. neuroscience.2014.10.033

injury) (Fig. 6.21) associated with the relatively high expression of NMDA receptors within the CA1 subfield. The dentate gyrus has been implicated pathologically in post-adrenalectomy cases attributed to the high levels of mineralocorticoid receptors represented in this region. Alzheimer's disease, cognitive vascular disorder, and normal ageing show hippocampal hypometabolic states, as assessed by functional and structural MR imaging, whereas schizophrenia, depression, and post-traumatic stress disorder appear as hypermetabolic, in different hippocampal formation components (Fig. 6.22) [3].

Testing

Overall the assessment of episodic (also termed explicit or declarative) memory should include the following formats:

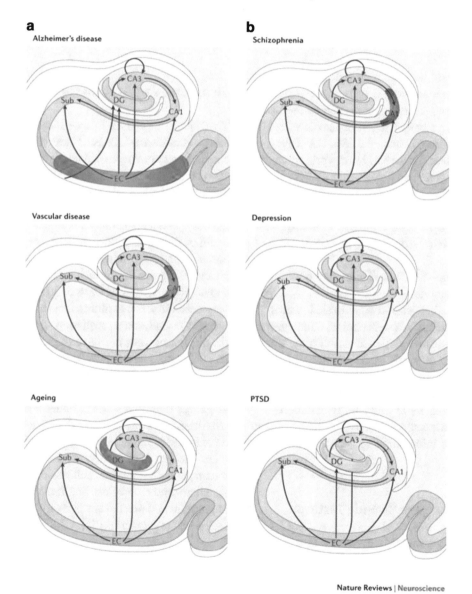

Fig. 6.22 Hippocampal syndromes according to subregions: hypo- or hyperactivity. Figure with permission: Small SA, Schobel SA, Buxton RB, Witter MR, Barnes CA. A pathophysiological framework of hippocampal dysfunction in ageing and disease. Nature Reviews Neuroscience 2011;12:585–601

1. Establish adequate attention by assessing with orientation for five items or serial 7's calculation
2. Short story recall is an ecologically sound method of testing learning and memory
3. Rote learning by means of lists to establish recall for both verbal and visual information with trials; frequently employed are four trials to assess for learning, then a delay to evaluate free recall, and thereafter a recognition or cued recall test
4. Remote memory testing both general information (last five presidents) and autobiographical information (school, college, or other institutional graduation dates, wedding date)

This can be accomplished by more rapid bedside-type assessment or in more quantitative detail. Rapid screening tests such as the MMSE (three-word recall) assess for episodic verbal (not visual) memory and the MOCA (five-word recall) assess for verbal, brief working memory components for example.

Bedside Episodic Memory Testing: Short Term, Long Term, Remote

Orientation—five items; date, day of the week, place of interview

Registration—use five unrelated words (for example, ocean, orange, courage, building, rapid) with immediate repetition to establish registration

Short-term recall—ask for the five words' recall after 1 min

Long-term recall—ask for recall of five words at approximately 15 min after registration

Remote memory—last five presidents or autobiographic information (graduation dates, wedding date)

More Extensive Bedside Testing

The Benton Bedside Memory Test

A semiquantitative bedside memory test developed by Benson is an intermediate type of testing approach. The following words are read with a recall each time with a total of four trials and free recall tested after a 5–10-min delay. In adults, approximate normative values include

being able to learn 7–8 words after the four trials, six words for free recall, and all remaining if supplied cues [52].

Words	Category cue
Cabbage	Vegetable
Table	Furniture
Dog	Animal
Baseball	Sport
Chevrolet	Automobile make
Rose	Flower
Belt	Article of clothing
Blue	Color

Visual and Verbal Memory

The Three Words-Three Shapes Test is a relatively easily administered bedside test that involves copying the three shapes and three words, immediate reproduction which presumably tests working memory and incidental recall, and then after delays of 5, 15, and 30 min testing for episodic long-term memory. The test also has a multiple-choice recognition format at a 30-min delay. Approximate normal values in younger adults <65 years of age recall all six items for the incidental recall measure and approximately four items at 30 min. This is a very versatile test that allows incidental (working memory) recall, study recall trials (determined by the number of trials until five of the six items are recalled), long-term episodic memory (delayed recall), and recognition memory features. In addition the test can be formatted for different languages [53]. In a study of patients with probable Alzheimer's disease, Korsakoff's amnesia, and a control group, the test was able to differentiate these groups when analyzing incidental recall, retention, and recognition abilities. Its value is the simplicity and ease of use and it may be particularly useful as dementia advances from the mild to the moderate categories [54].

Metric/Quantitative

There are varying formats of supraspan testing but a common format is the presentation of word lists, distractor lists, and recognition trials such as with the Rey Auditory Verbal Learning Test. Typically on presentation of a 12-word list, adults (18–41) recall 6, older adults recall 5, and the 66–77 age group 4–5 words [52].

Rey Auditory Verbal Learning Test

A 15-word list category A list followed by a 15-word interference category B list, followed by two post-list B interference recall trials, an immediate and delayed, then a recognition trial with distractor words [55]

Wechsler Memory Scale IV [56]

Probably the most widely used memory test, which is a comprehensive battery that deciphers memory and learning according to the domains of working and episodic into both visual and verbal components as well as recall and recognition scores for both (Fig. 6.23)

California Verbal Learning Test Version II (CVLT-II)

The CVLT II is available as both standard and short forms. The CVLT employs both cued recall and recognition formats so that the following information is obtained:

- Immediate recall
- Short delay free recall
- Short delay cued recall
- Long delay free recall
- Long delay cued recall
- Long delay recognition

The CVLT II SF, which has one list of nine words, three categories, and four learning trials [57].

Hopkins Verbal Learning Test

This tests consists of 12-word lists followed by a 24 recognition list and 6 alternate forms [58].

Repeatable Battery for the Assessment of Neuropsychological Status (RBANS)

This test includes a short story recall which is considered the most important way of testing day-to-day memory usage and has a story and a ten-word list test paradigm [59].

Remote Memory

Famous Faces Test [60]

Visual Recall

Rey-Osterrieth Complex Figure Test [61]

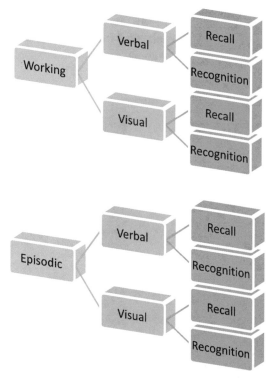

Fig. 6.23 Wechsler Memory Scale IV: working memory and episodic memory assessment

Working Memory

Working memory (WM) may well have been "missing link" of human mind evolution, enabling interconnectivity of the various intelligences such as natural history, social, language, and technical intelligences. Quite appropriately, it has been termed the "new intelligence" [62]. WM allows a task to be accomplished by devoting appropriate attention to allow maintenance of the particular information in active memory and at the same time inhibit both internal and external interference that may prevent accomplishing the task or process at hand. An extensive frontoparietal subcortical circuitry has been proposed by Petrides and the PFC components include not only the DLPFC (episodic buffer) but also the anterior cingulate cortex (central executive), the subcortex (frontal subcortical circuitry), and the parietal (lateral and medial) and superior temporal cortex (Fig. 6.6) [19]. The episodic buffer itself is likely part of a more distributed circuit that includes the left anterior temporal lobe [63]. Research with functional imaging such as PET brain scanning and f-MRI for working memory assessment has corroborated the expansive frontoparietal and fronto-parieto-occipitotemporal circuitry (Figs. 6.24 and 6.25) [64, 65].

The working memory components as proposed by Baddeley and Hitch consist of:

1. Central executive: a pan modal, attention processing unit
2. Episodic buffer: a postulated temporary multimodal memory storage facility
3. Phonological loop: dedicated to auditory modality domain
4. Visuospatial sketch pad: processes visual information domain processor for location and shape of objects (Fig. 6.26)

Working memory as a concept was first proposed by Miller in 1960 and subsequently elaborated and revised several times by Baddeley in 1974, 1983, and 2003 [66–68]. A more recent conceptualization of working memory and its relationship to the other memory subtypes was formulated by Wynn and Coolidge, presented in a particularly lucid diagram (Fig. 6.27) [69].

The central executive is thought to be particularly involved in the setting of novel stimuli or tasks and its functions are presumed to include maintenance of attention during processing, inhibition of distracting stimuli, temporal tagging, planning, and linking to long-term memory storage areas.

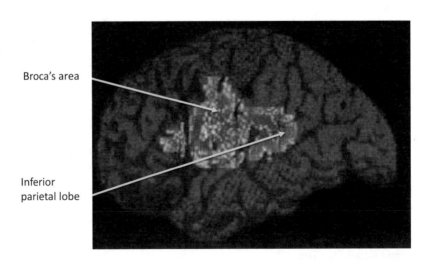

Fig. 6.24 Working memory phonologic loop: activation of Broca's region and inferior parietal region (phonologic store). Statistical parametric mapping of verbal working memory showing significant blood flow increases (*orange*, *yellow*) with verbal memory task compared to non-verbal task. Figure with permission: Paulesu E, Frith CD, Frackowiak RS. The Neural Correlates of Working Memory. Nature. 1993;362(6418):342–5

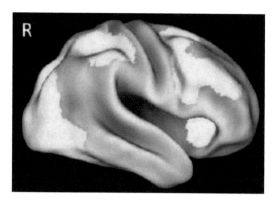

Fig. 6.25 Working memory visuospatial sketchpad: Statistical parametric mapping of visuospatial memory showing significant blood flow increases (in yellow) with visuospatial memory task compared to non-visuospatial task. This shows activation of the right occipital right parietal and right prefrontal cortices with the right angular gyrus representing the visuospatial buffer. Figure with permission: Dumontheil I, Klingberg T. Brain Activity during a Visuospatial Working Memory Task Predicts Arithmetical. Performance 2 Years Later. Cerebral Cortex 2012;22:1078–1085

The episodic buffer component of working memory allows temporary maintenance of information in a multimodal form enabling integration and is thought to have relatively limited capacity. The multimodal code in which information is temporarily held is in turn derived from long-term memory stores and the visual and auditory subsidiary systems [70].

The phonological loop component in turn is considered to have both a short-term phonological store subserving speech for example and what is termed an articulatory loop for the maintenance of either vocal or subvocal information. It has been postulated that through the expansion of working memory to enhanced working memory or level 7, this allowed language evolution together with the episodic memory expansion with more complex memory storage and access.

The visuospatial sketchpad component is proposed as a temporary store of visuospatial information [71].

Fig. 6.26 Schematic diagram of the working memory (working attention) model according to Baddeley A and Hitch G

- Central executive (pan modal processor)
- Episodic buffer (temporary memory storage)
- Phonological loop (auditory processing)
- Visuospatial store (visual processing location, shapes)

Fig. 6.27 Revised working memory model. Figure adapted from memory working concepts of Wynn T and Coolidge FL. The rise of homo sapiens. The evolution of modern thinking. Wiley Blackwell 2009 Oxford

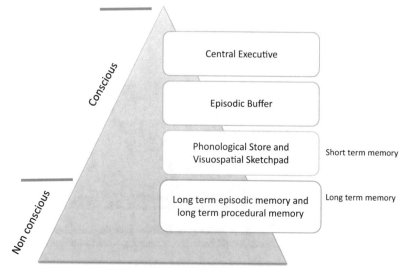

There is a complex and as yet poorly understood relationship between working memory and
short-term episodic memory. For example,
Baddeley considered the episodic buffer WM
component to be closely related to the episodic
working memory as understood in Tulving's episodic memory model which has both a short-term
and long-term memory storage system. Working
memory, episodic short-term, and episodic long-
term memory and attention overlap interact with
each other (Fig. 6.28). From research to date it
may be deduced that the most important factor
for short-term memory processing is in the control of attention [72].

Clinical manifestations of working memory
impairment are protean. Cerebrovascular disease

due to many different mechanisms, but small
vessel disease in particular, as well as schizophrenia and depression are associated with a significant working memory disorder [73, 74].

The more extensive the leukoaraiosis, which
may present in manifold ways, the more working
memory impairment is evident, but other cognitive processes are affected as well, such as visuospatial dysfunction, episodic memory, and
language function (Fig. 6.29). In a study of 83
cognitive impairment patient group with cognitive vascular disease or Alzheimer's associated
with neuroimaging evidence of leukoaraiosis that
was quantified by MRI scanning, leukoaraiosis
ranged from less than 1 to 23 % and associated
with a significant impact on working memory

Fig. 6.28 Schematic
diagram of relationships
between different
memory systems. The
red central sphere
represents the attentional
domain, integral to all
memory and learning
systems, in addition to
the working memory
domain

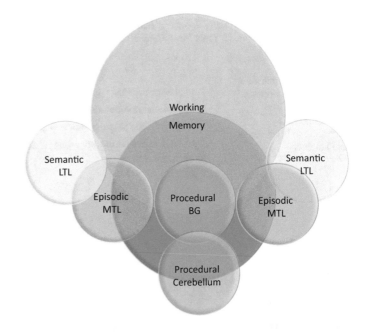

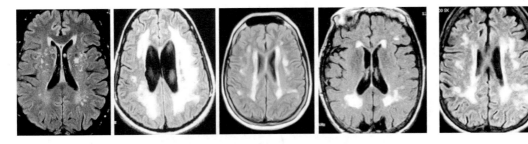

Fig. 6.29 Leukoaraiosis: different neuroimaging features including discrete subcortical hyperintensities, extensive confluent subcortical hyperintensities, periventricular rimming, ventricular capping, and mixed patterns

function when the degree exceeded 3 % [75]. In vascular cognitive impairment studied after transient ischemic attack presentations, one or more cognitive domains were affected in over one-third of patients. In a study of 107 TIA patients without previous dementia or stroke, tested 3 months after the event, impairments in working memory occurred most frequently (25 %) followed by attention (22 %) and speed of information processing (16 %) [76]. Working memory is impaired also in other cognitive vascular disease mechanisms such as internal carotid artery (ICA) stenosis. The ICA stenosis group of patients, studied by f-MRI, revealed relatively greater bilateral middle frontal gyrus activations, left more than right which decreased with increasing stenosis. Both frontal lobe dysfunction correlated with degree of ICA stenosis and working memory in particular was associated with left ICA stenosis more than right ICA stenosis [77].

Unsurprisingly the mainstay NIH stroke score scale underestimates cognitive function in the stroke setting. Neuropsychological assessment in 223 patients with first stroke was assessed in the subacute stroke period and again at 6 months postictus. Although all patients had cognitive deficits if the NIHSS was ≥4, importantly one or more cognitive deficits were present in 41 % of patients with normal NIHSS scores [78].

Testing Working Memory

Various methods can be used with the aim of holding information in active state for further manipulation or computation. Both bedside testing, a number of different neuropsychological tests, and computerized testing can be employed:

Bedside Testing
Serial 7 subtractions

Digit span testing: 7 digits forward and 5 backwards or not differing by >2, is regarded as average for adults

The MMSE subtest of spelling WORLD backwards

Recitation of months of the year or days of the week in reverse order

Neuropsychological Tests
Wechsler Memory Scale IV (also WAIS IV) working memory subtests [79]

Trail Tests such as the Comprehensive Trail Making Test (5 subtests) [80]

Paced Auditory Serial Addition Test (PASAT) [81].

Delis Kaplan Executive Function System (DKEFS) consisting of 9 subtests including a trail-making test that tests working memory [82]

Computerized Testing
CNS-Vital Signs [https://www.cnsvitalsigns.com] CANTAB [83]

Semantic Memory

This refers to the knowledge we possess about the world and people such as both general and more specific information about concerning historical events, facts about objects, animals, and people. It is different from episodic memory insofar that the retrieval of this knowledge or the facts is not associated with an autobiographical "tag" or autonoesis, which refers to the linking to either in a time-based (temporal) or spatial contextual capacity. Semantic memory impairments usually present with naming difficulty and specifically cueing does not assist in the recall of information [84].

The neurobiology of semantic memory involves distributed circuitry with a hub in the inferolateral cortices bilaterally. Whereas the left DLPFC is implicated with the encoding of episodic memories, the right DLPFC is concerned with the retrieval of episodic memories. Semantic memory retrieval is a function of the left hemisphere, more specifically the left inferolateral temporal lobe. However semantic memory related to faces is a right anterior temporal lobe-associated syndrome that can be tested with the Famous Faces Test [60]. Semantic dysmemory fractionation for category-specific deficits such as fruits, vegetable, animals, and tools have been reported with selective impairment in animate versus non-animate items for example. There appears to be a category-specificity gradient within the occipito-temporal cortices. Using f-MRI activation pattern analysis to study object-specific and categorical semantic representations, increasing specificity of the information obtained is seen moving along the ventral stream from the visual cortex (image based, categorical based) to the object-specific semantic particulars in the perirhinal cortex and anterior temporal cortex that allow more fine-grained analysis amongst objects [85].

There is some overlap between episodic memory and semantic memory in that the long-term or more remote autobiographic memories merge into the semantic knowledge repositories. During our development from childhood onwards, there is a progressive formation of episodic memories, many of which become eventual semantic memory traces. An initial learning process of learning how to use a tool, implement, or device begins as an episodic memory experience, later to become long-term memory and semantic memory or factual information. In the latter the exact associations of time, place, and personal involvement or autobiographic association is lost. Hence, semantic memory is initially derived from episodic memory experiences. With time there is a gradual loss of associations with the particulars and events associated with the initial episodic memory experience and many then become semantic memories or facts about the world and people [86].

Neuropathological States Associated with Semantic Memory Disorders

- Semantic aphasia subtype of the frontotemporal lobe degenerations and dementias
- The primary progressive aphasia syndrome with Alzheimer's pathology (Fig. 6.30)
- Traumatic brain injury syndrome
- Stroke in the posterior cerebral medial branch syndrome, amyloid angiopathy-related temporal lobe hemorrhage
- Herpes simplex encephalitis

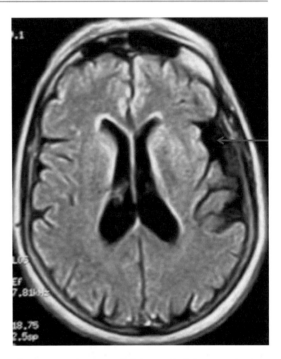

Fig. 6.30 Left temporal lobe, perisylvian atrophy in a person with primary progressive aphasia

Testing

- Boston Naming Test including the cueing option
- Category fluency for animals is impaired with semantic dysmemory.
- Pyramids and Palms Test
- Cambridge Semantic Memory Test Battery [87–91]

The differentiation between semantic memory disorders and aphasic anomia is important. In people with anomia, there may be an inability to name an animal such as sheep but there will be knowledge that it provides wool and is commonly reared in Australia, for example. However, with semantic memory impairment there is no help from cueing, or providing multiple choices. In addition, performance in category fluency tasks is worse than with letter fluency and facial recognition may also be affected with semantic memory disorders in a condition known as progressive prosopagnosia. There is a difficulty with recognizing faces but with retained recognition of other categories such as flowers and buildings in the case reported by Evans et al. Neuroimaging revealed a right anterior temporal lobe hypoperfusion and atrophy akin to the progressive non-fluent aphasia syndromes seen with left temporal lobe degeneration [92].

The second version of the Boston Naming Test has a cueing option, which is well suited for testing this and differentiating it from pure anomia. When semantic dysmemory is present, then non-verbal ability tested by matching pictures is also abnormal as well as impairment in describing or defining objects when given their names. In general, the tests best suited for assessing semantic memory include category fluency tests, picture to word matching tests, and the Pyramids and Palm Trees Test [89].

Procedural Memory

This refers to knowing how to perform or do something, without there being specific awareness for the task at hand. Procedural memory examples and the results of technological activity with stone tool production are perhaps the most evident in archeological research and records of artifacts. Advanced working memory was not required to a great extent and the stone tool making was enabled by motor action sequences that are the basis of long-term memory and procedural memory programs acquired over many years of practice and relatively independent of language [93]. Aside from the vast number of stone tools, particularly biface hand axes that are testimony of this acquired skill over 1–2 million years, nowadays we learn skills such as playing a musical instrument, driving a car, bicycle riding, tying shoelaces, as well as cognitive skills such as reading. These are abilities that are automatic, operate at the non-conscious level, and are subtype of long-term memory, an implicit memory type and learned through repetitive activity to finally produce an activity more or less automatically and, more specifically, a type of implicit memory [94]. Sometimes referred to as muscle memory, with protracted training in a sport for example, there is a gradual shift of the circuitry

involvement from prefrontal cortical areas to the subcortical components, the basal ganglia, and cerebellum. Paying specific attention to the details of a learned motor act may disrupt or downgrade the performance. For example drawing attention to your opponent's skill or mastery in a racquet ball game may have the effect of inducing errors in their performance.

Functional Neuroanatomy of Procedural Memory

Damage to basal ganglia (BG), supplementary motor cortex (SMA) and cerebellum have been correlated impairment in learning procedural skills. In addition, these areas are activated during functional imaging when a new procedural memory task is being processed or performed (Fig. 6.12) [95].

Neurophysiology

During training for a motor task, activation is seen initially in the cortical regions and there is a progressive transfer to the striatum with continuation of the training procedure. This may be reflective of the encoding and consolidation of the motor skills and gradual transfer to a basal ganglia automatic or semiautomatic type of program. One hypothesis of Fitts is that skill acquisition posits that this may be acquired through a number of distinct neurophysiological stages including:

- Cognitive phase
- Associative phase
- Autonomous phase

The final autonomous phase reflects procedural memory or the procedural phase [96].

Pathophysiology

People with early symptoms and signs of Parkinson's disease show impairment of procedural memory in the context of normal episodic memory function. Such presentations are also sent in people with toxicity associated with the basal ganglia (CO poisoning, HIE) Huntington's disease, strokes or tumors of these procedural memory hubs (BG, SMA, cerebellum), depression, and olivopontocerebellar atrophy. Contrariwise, the basal ganglia, cerebellum, and supplementary motor area are spared in early AD who have impaired episodic memory function [97].

Tests

Bedside
Inquiry usually towards the spouse, family, or caregiver will illuminate behavior of inability to perform certain mundane or routine tasks that they had been able to perform flawlessly throughout their lives.

Metric Tests
More formal evaluation may include the following:

- Serial reaction time task
- Pursuit rotor task—visual motor task that can be accomplished by following a mark on a computer screen in a circular fashion by operating a computer mouse
- Mirror tracing task
- Weather prediction task [98–101]

Savant Conditions and Hyperthymesia

In his review of savant syndromes, Treffert maintained that all were associated with a prodigious memory that may be primarily a extraordinary working memory capacity. Examples include the memorizing over 6000 books [102] and a surgeon struck by lightening and then becoming an accomplished musician [103]. From an epidemiological point of view about 1/10 people with autism spectrum condition have savant abilities and alternatively about 50% of people with savant abilities are autistic. The condition may be congenital or acquired, males predominate in an approximate 6:1 ratio, and IQ may range from superior to very low. Savant skills are invariably associated with right hemisphere hyperfunction or "release" after left hemisphere brain damage. The particular savant skills may be singular but sometimes several co-occur [104]. The type of extra ordinary abilities include:

- Complex calendar calculating
- Superlative musical abilities or musical genius, invariably associated with mental impairment and blindness
- Art genius associated with eidetic memory with vivid recollection of images
- Mathematical genius with remarkably rapid calculations and other number skills such as the calculations of square roots and prime numbers
- Visuospatial and mechanical genius skills [105]

Less commonly encountered savant abilities include

- Polyglots with exceptional language abilities
- Extrasensory-type perception ability and "paranormal abilities"
- Motor abilities such as described with extraordinary golfing or pinball accuracy

In his multidecadal research on people with savant syndromes, Treffert suggested categorization of savant people into the following categories:

1. Sudden savant: Acquired savant are normal people or neurotypicals who develop the trait after injury, dementia, or stroke. Acquired savants have been reported after frontotemporal lobe degeneration by Miller who developed very superior visual art ability [106]
2. Splinter savant—characterized by obsessive memorization or preoccupation with music, historical facts, sports, or numbers
3. Talented savant—refers to people that are cognitively impaired that have musical or artistic abilities
4. Prodigious savant (know things but never learnt them) [107]

Some Proposed Neurobiological Mechanisms

People with savant abilities tend to have domain-specific extraordinary abilities with superior memory, increased low-level perception processing, inadequate social cognition, impaired executive function, and compulsions, frequently in the context of overall cognitive or mental disabilities. There may be a type of rewiring process with recruitment or release of areas of "dormant brain capacity" that are released after left brain insult or injury. All savant types described are characteristic of one or more right brain abilities such as mathematics, art, and music. Some theories have been proposed including:

1. Paradoxical functional facilitation [108]
2. A release phenomenon or compensatory function [109]
3. Autistic savant model that is based on the weak central coherence theory proposed by of Frith et al. [110]
4. Underconnectivity theory that emerged from autism spectrum condition neuroimaging findings by Just et al. as a consequence of the impaired long-range connectivity tracts but relatively intact or superlative local connectivity [111]
5. Epigenetics-based theory: Knowing things that cannot be explained, akin to genetic memory, epigenes may explain remembering things that a person never learned [112]

Memory Treatment Considerations

Before engaging in specific memory treatment programs, the mitigation of confounding influences of anticholinergic medications, sedative medications, and substance abuse needs attention. Further promotion of adequacy (generally 8 h per night) is important for memory, learning, and the rehabilitation of memory. Philosophically the approaches employ mnemonics with the optimization of encoding as well as retrieval abilities, usage of external electronic memory aids (smartphone applications), and usage of spared neural circuits such as other sensory systems to help memory retrieval. Some examples of memory rehabilitation approaches include the following:

Episodic Memory Rehabilitation

1. PQRST
   The PQRST method includes the steps of: Preview (P), question (Q), read (R), state (S), and test (T) steps have been shown to be effective in improving long-term memory in people with mild dysmemory due to prefrontal cortical lesions such as traumatic brain injury [113]
2. Spaced retrieval training
   Using the USMART (Ubiquitous Spaced Retrieval-based Memory Advancement and Rehabilitation Training) system efficacy and good tolerance were demonstrated in a mild cognitive impairment group of patients [114]
3. Vanishing cues (VC) and errorless learning (EL) techniques
   VC and EL in combination were associated with better results that one or the other in a controlled trial involving probable Alzheimer's patients [115]
4. Ecologically orientated neuro-rehabilitation of memory (EON-MEM)
   In a study by Stringer and Small, patients with stroke, traumatic brain injury, and other neurologically based dysmemories all showed statistically significant improvements in their memory with this technique [116].

Working Memory Improvement and Rehabilitation

Working memory allows retention of information over short periods of time, within seconds to minutes, and is a central function and core function for all other cognitive activities. WM impairment is a core feature of traumatic brain injury, multiple sclerosis, stroke, attention-deficit hyperactivity disorder, schizophrenia, and many other neurological and neuropsychiatric syndromes. Working memory training is associated with up-regulation of dopamine receptors and increased cerebral activities in the frontoparietal cortex and basal ganglia [117]. The neural basis for both WM and attention share the same neural circuitry for sustained neural activation with a top-down excitatory effect on frontoparietal networks and might be difficult or impossible to separate even at a neuronal level. Computerized WM training programs have been shown to improve performance in other tasks for which there was no specific rehabilitation training. This supports the induction of plasticity that is due to the WM training which may transfer to improved reasoning, disinhibition, and attentive focus.

For example, Klingberg developed a computerized training of WM developed for children with ADHD whereby 30–40-min training per day, 5 days per week, for 5 weeks total 15 h. In a randomized controlled trial with 53 children with ADHD, there was an improvement in working memory, response inhibition, reasoning, and inattention symptoms [118, 119].

Brain functional imaging may help monitor WM ability and improvement or deterioration. WM capacity in people has been correlated with intra-parietal cortex and DLPFC activity and a decline in WM capacity associated with a decrease in this activity [12, 120].

The Importance of Sleep

Sleep promotes new memory formation including episodic memory and the integration of new knowledge that may be linked to slow-wave sleep and creativity that may be linked to REM sleep. Procedural memory also benefits from sleep with respect to the motor memory reorganization that may be associated with modulatory effects from emotions. Sleep disruption has also been associated with disordered memory formation and the development of neuropsychiatric syndromes [121].

Working Memory Improvement Can Be Facilitated by Physical Exercise

Physical exercise may be a powerful WM booster. A number of studies support this premise and Sibley and Beilock conducted a 30-min treadmill study on health adults and working memory assessments [122]. In a community-based study of 2747 people enrolled in the CARDIA study, there was improved verbal memory and psychomotor speed were correlated with better cardiorespiratory fitness, 25 years earlier, in people aged 43–55 years [123]

Neuropharmacological Treatment, Transcranial Magnetic Stimulation, and Transcranial Direct Current Stimulation of Memory Disorders

Discussed in the chapter on frontal systems.

References

1. Tulving E. Memory and consciousness. Can Pyschol. 1985;26:1–12.
2. Allen TA, Fortin NJ. The evolution of episodic memory. Proc Natl Acad Sci U S A. 2013;110: 10379–86.
3. Small SA, Schobel SA, Buxton RB, Witter MR, Barnes CA. A pathophysiological framework of hippocampal dysfunction in ageing and disease. Nat Rev Neurosci. 2011;12:585–601.
4. Aggleton JP, Brown MW. Interleaving brain systems for episodic and recognition memory. Trends Cogn Sci. 2006;10:455–63.
5. Kopelman MD. Disorders of memory. Brain. 2002;125:2152–90.
6. Clayton NS, Bussey TJ, Dickinson A. Can animals recall the past and plan for the future? Nat Rev Neurosci. 2004;4(8):685–91.
7. Jacobs LF. The role of social selection in the evolution of hippocampal specializations. In: Tommasi L, Petereson MA, Nadel L, editors. Cognitive biology. London: MIT Press; 2009.
8. Van Strien NM. Nat Rev Neurosci. 2009; 10:272–82.
9. Bateson P, Gluckman P. Plasticity, robustness, development, and evolution. Cambridge: Cambridge Universit Press; 2011.
10. Kemperman G, Chesler EJ, Lu L, Williams RW, Gage FH. Natural vcariation and genetic covariance

in adult hippocampal neurogenesis. Proc Natl Acad Sci U S A. 2006;103:780–5.

11. Kemperman G. New neurons for survival of the fittest. Nat Neurosci Rev. 2012;13:727–36.

12. Rajah MN, D'Esposito M. Region-specific changes in prefrontal function with age: a review of PET and fMRI studies on working and episodic memory. Brain. 2005;128:1964–83.

13. Wynn T, Balter M. Did working memory spark creative culture? Science. 2010;328:160–3.

14. Skoyles J. Med Hypotheses. 1997;48:499–501.

15. Petrides M. The mid-dorsolateral prefrontal-parietal network and the epoptic process. In: Stuss DT, Knight RT, editors. Principles of frontal lobe function. Oxford: Oxford University Press; 2012.

16. Medalla M, Barbas H. Diversity of laminar connections linking periarcuate and lateral intraparietal areas depends on cortical structure. Eur J Neurosci. 2006;23(1):161–79.

17. Petrides M. Dissociable roles of the mid dorsolateral prefrontal cortex and anterior inferotemporal cortex in visual working memory. J Neurosci. 2000;20:7496–503.

18. Champod AS, Petrides M. Dissociation within the frontoparietal network in verbal working memory: a parametric functional magnetic resonance imaging study. J Neurosci. 2010;30:3849–56.

19. Champod AS, Petrides M. Dissociable roles of the posterior parietal and the prefrontal cortex in manipulation and monitoring process. Proc Natl Acad Sci U S A. 2007;104:14837–42.

20. Semendeferi K, Barger N, Schenker N. Brain reorganization in humans and apes. The human brain evolving. Gosport, IN: Stone Age Institute Press; 2010.

21. Read D, van der Leeuw S. Biology is only part of the story. Philos Trans R Soc B. 2008;363:1959–68.

22. Wynn T, Coolidge FI. The implications of the working memory model for the evolution of modern cognition. Int J Evol Biol. 2011;2011:741357.

23. Wynn T, Coolidge FL. The rise of Homo Sapiens. Oxford: Wiley Blackwell; 2009.

24. Kosslyn SM, Alpert NM, Thompson WL, Chabris CF, Rauch SL, Anderson AK. Identifying objects seen from different viewpoints. A PET investigation. Brain. 1994;117:1055–71.

25. LePort A, Mattfeld A, Dickinson-Anson H, Fallon J, Stark C, Kruggel F, et al. Behavioral and neuroanatomical investigation of highly superior autobiographical memory (HSAM). Neurobiol Learn Mem. 2012;98:78–92.

26. Parker ES, Cahill L, McGaugh JL. A case of unusual autobiographical remembering. Neurocase. 2006;12:35–49.

27. Ally B, Hussey E, Donahue M. A case of hyperthymesia: rethinking the role of the amygdala in autobiographical memory. Neurocase. 2013;19(2):1–16.

28. Eichenbaum H, Yonelinas AP, Raganath C. The medial temporal lobe and recognition memory. Ann Rev Neurosci. 2007;30:123–52.

29. Kandel ER, Schwartz JH, Jessell TM, Siegelbaum SA, Hudspeth AJ, editors. Principles of neural science. 5th ed. New York, NY: McGraw Hill; 2013.

30. Eichenbaum H. Time cells in the hippocampus: a new dimension for mapping memories. Nat Rev Neurosci. 2014;15:732–44.

31. Conway MA, Anderson SJ, Larsen SF, Donnelly CM, McDaniel MA, McClelland AG, et al. The formation of flash bulb memories. Mem Cognit. 1994;22:326–43.

32. Wolk DA, Budson AE. Memory systems. Continuum. 2010;16:15–28.

33. Fleming SM, Dolan RJ. The neural basis of metacognitive ability. Philos Trans R Soc B. 2012;367:1338–49. doi:10.1098/rstb.2011.0417.

34. Pronin E. Perception and misperception of bias in human judgment. Trends Cogn Sci. 2007;11:37–43.

35. Wheeler MA, Stuss DT, Tulving E. Frontal lobe damage produces episodic memory impairment. J Int Neuropsychol Soc. 1995;1:525–36.

36. Grothe MJ, Schuster C, Bauer F, Heinsen H, Prudlo J, Teipel SJ. Atrophy of the cholinergic basal forebrain in dementia with Lewy bodies and Alzheimer's disease dementia. J Neurol. 2014;261(10):1939–48.

37. Buckner RL. Memory and executive function imaging and AD: multiple factors that cause decline and reserve factors that compensate. Neuron. 2004;44:195–208.

38. Wagner AD, Shannon BJ, Kahn I, Buckner RL. Parietal lobe contributions to episodic memory retrieval. Trends Cogn Neurosci. 2005;9:445–53.

39. Sili U, Kaya A, Mert A. HSV Encephalitis Study Group. Herpes simplex virus encephalitis: clinical manifestations, diagnosis and outcome in 106 adult patients. J Clin Virol. 2014;60(2):112–8.

40. Grydeland H, Walhovd KB, Westlye LT, Due-Tønnessen P, Ormaasen V, Sundseth Ø, et al. Amnesia following herpes simplex encephalitis: diffusion-tensor imaging uncovers reduced integrity of normal-appearing white matter. Radiology. 2010;257(3):774–81.

41. Gerwig M, Kastrup O, Wanke I, Diener HC. Adult post-infectious thalamic encephalitis: acute onset and benign course. Eur J Neurol. 2004;11(2):135–9.

42. Hodges JR. Transient global amnesia: clinical and neuropsychological aspects. London: WB Saunders; 1991.

43. Strupp M, Bruning R, Wu RH, Deimling M, Reiser M, Brandt T. Diffusion weighted MRI in transient global amnesia: elevated signal intensity in the left mesial temporal lobe in 7 of 10 patients. Ann Neurol. 1998;43:164–70.

44. Matsuda H, Higashi S, Tsuji S, Sumiya H, Miyauchi T, Hisada K, et al. High resolution TC-99 m HMPAO SPECT in a patient with transient global amnesia. Clin Nucl Med. 1993;18:46–9.

45. Heiss WD, Pawlik G, Holthoff V, Kessler J, Szelies B. PET correlates of normal and impaired memory functions [Review]. Cerebrovasc Brain Metab Rev. 1992;4:1–27.

46. Baron JC, Petit-Taboue MC, Le Doze F, Desgranges B, Ravenel N, Marchal G. Right frontal cortex hypometabolism in transient global amnesia A PET study. Brain. 1994;117:545–52.

47. Gass A, Gaa J, Hirsch J, Schwartz A, Hennerici M. Lack of evidence of acute ischemic tissue changes in transient global amnesia on single shot echo planar diffusion weighted MRI. Stroke. 1999;30:2070–2.

48. Budson AE, Schlaug G, Briemberg HR. Perfusion and diffusion weighted magnetic resonance imaging in transient global amnesia. Neurology. 1999;53: 239–40.

49. Thornton HB, Nel D, Thornton D, van Honk J, Baker GA, Stein DJ. The neuropsychiatry and neuropsychology of lipoid proteinosis. J Neuropsychiatry Clin Neurosci. 2008;20:86–92.

50. Hurlemann R, Wagner M, Hawellek B, Reich H, Pieperhoff P, Amunts K, et al. Amygdala control of emotion-induced forgetting and remembering: evidence from Urbach-Wiethe disease. Neuropsychologia. 2007;45(5):877–84.

51. Morgan B, Terburg D, Thornton HB, Stein DJ, van Honk J. Paradoxical facilitation of working memory after basolateral amygdala damage. PLoS One. 2012;7(6), e38116.

52. Lezak MD, Howieson DB, Bigler ED, Tranel D. Neuropsychological assessment. 5th ed. Oxford: Oxford University Press; 2012.

53. Mesulam M-M. Principles of behavioral and cognitive neurology. Oxford: Oxford University Press; 2000.

54. Weintraub S, Peavy GM, O'Connor M, Johnson NA, Acar D, Sweeney J, et al. Three words three shapes: a clinical test of memory. J Clin Exp Neuropsychol. 2000;22(2):267–78.

55. Rey A. L'examen clinique en psychologie. Paris: Presses Universitaires de France; 1964.

56. Wechsler D. Wechsler Memory Scale and Wechsler Adult Intelligence Scale IV technical manual. San Antonio, TX: Psychological Corporation; 2009.

57. Delis DC, Kramer JH, Kaplan E, Ober BA. California verbal learning test. 2nd ed. San Antonio, TX: Texas Psychological Corporation; 2000.

58. Brandt J. The Hopkins Verbal Learning Test: development of a new memory test with six equivalent forms. Clin Neuropsychol. 1991;5:125–42.

59. Randolph C. Manual: repeatable battery for the assessment of neuropsychological status. San Antonio, TX: Psychological Corporation; 1998.

60. Hodges JR, Salmon DP, Butters N. Recognition and naming of famous faces in Alzheimer's disease: a cognitive analysis. Neuropsychologia. 1993;31:775–88.

61. Osterrieth PA. Le test de copie d'une figure complexe. Arch Psychol. 1944;30:206–356. and [Rey A. L'examen psychologique dans les cas d'encephalopathie traumatique. Arch Psychol 1941;28:286–340].

62. Alloway TP, Cowan N, Balota D. Working memory: the new intelligence. London: Psychology Press; 2011.

63. Berlingeri M, Bottini G, Basilico S, Silani G, Zanardi G, Sberna M, et al. Anatomy of the episodic buffer: a voxel-based morphometry study in patients with dementia. Behav Neurol. 2008;19:29–34.

64. Paulesu E, Frith CD, Frackowiak RS. The neural correlates of the verbal component of working memory. Nature. 1993;362(6418):342–5.

65. Paulesu E, Connelly A, Frith CD, Friston KJ, Heather J, Myers R, et al. Functional MR imaging correlations with positron emission tomography. Initial experience using a cognitive activation paradigm on verbal working memory. Neuroimaging Clin N Am. 1995;5:207–25.

66. Miller G, Galanter E, Pribam K. Plans and the structure of behavior. New York, NY: Holt Rinehart and Winston; 1960.

67. Baddley A. Working memory: looking back and looking forward. Nat Rev Neurosi. 2003;4:829–39.

68. Baddeley AD, Hitch G. Working memory. In: Gordon HB, editor. Psychology of learning and motivation: advances in research and theory. New York, NY: Academic; 1974.

69. Wynn T, Coolidge FL. The rise of homo sapiens. The evolution of modern thinking. Oxford: Wiley Blackwell; 2009.

70. Baddeley A. The episodic buffer: a new component of working memory? Trends Cogn Sci. 2000;4(11): 417–23.

71. Shah P, Miyake A. The separability of working memory resources for spatial thinking and language processing: an individual differences approach. J Exp Psychol Gen. 1996;125:4–27.

72. LaRocque JJ, Lewis-Peacock JA, Postle BR. Multiple neural states of representation in short term memory? It's a matter of attention. Front Hum Neurosci. 2014;8:5. doi:10.3389/fnhum.2014.00005.

73. Galecki P, Talarowska M, Moczulski D, Bobinska K, Opuchlik K, Galecka E, et al. Working memory impairment as a common component in recurrent depressive disorder and certain somatic diseases. Neuro Endocrinol Lett. 2013;34(5):436–45.

74. Roussel M, Dujardin K, Hénon H, Godefroy O. Is the frontal dysexecutive syndrome due to a working memory deficit? Evidence from patients with stroke. Brain. 2012;135:2192–201.

75. Price CC, Mitchell SM, Brumback B, Tanner JJ, Schmalfuss I, Lamar M, et al. MRI-leukoaraiosis thresholds and the phenotypic expression of dementia. Neurology. 2012;79(8):734–40.

76. van Rooij FG, Schaapsmeerders P, Maaijwee NA, van Duijnhoven DA, de Leeuw FE, Kessels RP, et al. Persistent cognitive impairment after transient ischemic attack. Stroke. 2014;45(8):2270–4.

77. Zheng S, Zhang M, Wang X, Ma Q, Shu H, Lu J, et al. Functional MRI study of working memory impairment in patients with symptomatic carotid artery disease. Biomed Res Int. 2014;2014:Article ID 327270. doi:10.1155/2014/327270.

78. Kauranen T, Laari S, Turunen K, Mustanoja S, Baumann P, Poutiainen E. The cognitive burden of

stroke emerges even with an intact NIH Stroke Scale Score: a cohort study. J Neurol Neurosurg Psychiatry. 2014;85(3):295–9.

79. Wechsler D. Wechsler Memory Scale IV. San Antonio, TX: Pearson; 2009.

80. Reynolds CR. Comprehensive trail making test. Austin, TX: Pro-ed; 2002.

81. Gronwall DM. Paced auditory serial-addition task: a measure of recovery from concussion. Percept Mot Skills. 1977;44(2):367–73.

82. Delis DC, Kaplan E, Kramer JH. Delis Kaplan executive function system. San Antonio, TX: Pearson; 2001.

83. Cambridge Cognition Ltd, Bottisham, Cambridge, UK; 2015.

84. Tulving E. Episodic memory: from mind to brain. Annu Rev Psychol. 2002;53:1–25.

85. Clarke A, Tyler LK. Object-specific semantic coding in human perirhinal cortex. J Neurosci. 2014; 34(14):4766–75.

86. Zannino GD, Caltagirone C, Carlesimo GA. The contribution of neurodegenerative diseases to the modelling of semantic memory: a new proposal and a review of the literature. Neuropsychologia. 2015;75:274–90.

87. Kaplan E, Goodglass H, Weintraub S. Boston naming test. 2nd ed. Boston, MA: Lippincott Williams & Wilkins; 2001.

88. Howard D, Patterson K. Pyramids and palm trees: a test of semantic access from pictures and words. Bury St Edmunds: Thames Valley Test Company; 1992.

89. Adlam A-LR, Patterson K, Bozeat S, Hodges JR. The Cambridge Semantic Memory Test Battery: detection of semantic deficits in semantic dementia and Alzheimer's disease. Neurocase. 2010. doi:10.1080/13554790903405693.

90. Robbins TW, James M, Owen AM, Sahakian BJ, McInnes L, Rabbitt P. Cambridge Neuropsychological Test Automated Battery (CANTAB): a factor analytic study of a large sample of normal elderly volunteers. Dementia. 1994;5:266–81.

91. Gualtieri CT, Johnson LG. Reliability and validity of a computerized neurocognitive test battery, CNS vital signs. Arch Clin Neuropsychol. 2006;21:623–43.

92. Evans JJ, Heggs AJ, Antoun N, Hodges JR. Progressive prosopagnosia associated with selective right temporal lobe atrophy. A new syndrome? Brain. 1995;118:1–13.

93. Stout D, Toth N, Schick K, Chaminade T. Neural correlates of Early Stone Age toolmaking: technology, language and cognition in human evolution. Philos Trans R Soc Lond B Biol Sci. 2008;363(1499): 1939–49.

94. Squire LR. Memory systems of the brain: a brief history and current perspective. Neurobiol Learn Mem. 2004;82:171–7.

95. Brashers-Krug T, Shadmehr R, Bizzi E. Consolidation in human motor memory. Nature. 1996;382:252–5.

96. Fitts PM. The information capacity of the human motor system in controlling the amplitude of movement. J Exp Psychol. 1954;47:381–91.

97. Foerde K, Shohamy D. The role of the basal ganglia in learning and memory: insight from Parkinson's disease. Neurobiol Learn Mem. 2011;96:624–36.

98. Balota DA, Connor LT, Ferraro FR. Implicit memory and the formation of new associations in nondemented parkinson's disease individuals and individuals with senile dementia of the Alzheimer type: a serial reaction time (SRT) investigation. Brain Cogn. 1993;21:163–80.

99. Bullemer P, Nissen MJ, Willingham DB. On the development of procedural knowledge. J Exp Psychol Learn Mem Cogn. 1989;15:1047–60.

100. Allen JS, Anderson SW, Castro-Caldas A, Cavaco S, Damasio H. The scope of preserved procedural memory in amnesia. Brain. 2004;127:1853–67.

101. Beilock SL, Carr TH, MacMahon C, Starkes JL. When paying attention becomes counterproductive: impact of divided versus skill-focused attention on novice and experienced performance of sensorimotor skills. J Exp Psychol. 2002;8(1):6–16.

102. Treffert DA. The savant syndrome: an extraordinary condition. A synopsis: past, present, future. Philos Trans R Soc Lond B Biol Sci. 2009;364(1522): 1351–7.

103. Sacks O. A neurologist's notebook: a bolt from the blue: where do sudden passions come from? New Yorker. 2007;23:38–42.

104. Treffert DA. Savant syndrome: realities, myths and misconceptions. J Autism Dev Disord. 2014;44(3):564–71.

105. Takahata K, Saito F, Muramatsu T, Yamada M, Shirahase J, Tabuchi H, et al. Emergence of realism: enhanced visual artistry and high accuracy of visual numerosity representation after left prefrontal damage. Neuropsychologia. 2014;57:38–49.

106. Miller BL, Boone K, Cummings JL, Read SL, Mishkin F. Functional correlates of musical and visual ability in frontotemporal dementia. Br J Psychiatry. 2000;176:458–63.

107. Treffert D. Islands of genius. London: Jessica Kingsley Publishers; 2010. and [Treffert D. Stephen Wiltshire. Prodigious drawing ability and visual memory. Wisconsin Medical Society. Retrieved 7 Nov 2007].

108. Kapur N. Paradoxical functional facilitation in brain behavior research. Brain. 1996;119:175–1790.

109. Geschwind N, Galaburda AM. Cerebral lateralization: biological mechanisms, associations and pathology. Cambridge, MA: MIT Press; 1987.

110. Happé F, Briskman J, Frith U. Exploring the cognitive phenotype of autism: weak "central coherence" in parents and siblings of children with autism: I. Experimental tests. J Child Psychol Psychiatry. 2001;42(3):299–307.

111. Just MA, Keller TA, Malave VL, Kana RK, Varma S. Autism as a neural systems disorder: a theory of

frontal-posterior underconnectivity. Neurosci Biobehav Rev. 2012;36(4):1292–313.

112. Takahata K, Kato M. Neural mechanism underlying autistic savant and acquired savant syndrome]. Brain Nerve. 2008;60(7):861–9.

113. Ciaramelli E, Neri F, Marini L, Braghittoni D. Improving memory following prefrontal cortex damage with the PQRST method. Front Behav Neurosci. 2015;9:211. doi:10.3389/fnbeh.2015.00211.

114. Han JW, Oh K, Yoo S, et al. Development of the ubiquitous spaced retrieval-based memory advancement and rehabilitation training program. Psychiatry Investig. 2014;11(1):52–8.

115. Haslam C, Moss Z, Hodder E. Are two methods better than one? Evaluating the effectiveness of combining errorless learning with vanishing cues. J Clin Exp Neuropsychol. 2010;32(9):973–85.

116. Stringer AY, Small SK. Ecologically-oriented neuro-rehabilitation of memory: robustness of outcome across diagnosis and severity. Brain Inj. 2011;25(2):169–78.

117. Klingberg T. Training and plasticity of working memory. Trends Cogn Sci. 2010;14:317–24.

118. Klingberg T, Forssberg H, Westerberg H. Training of working memory in children with ADHD. J Clin Exp Neuropsychol. 2002;24:781–91.

119. Klingberg T, Fernell E, Olesen PJ, Johnson M, Gustafsson P, Dahlström K, et al. Computerized training of working memory in children with ADHD – a randomized controlled trial. J Am Acad Child Adolesc Psychiatry. 2005;44:177–86.

120. Persson J, Nyberg L. Altered brain activity in healthy seniors: what does it mean ? Prog Brain Res. 2006;157:45–56.

121. Landmann N, Kuhn M, Piosczyk H, Feige B, Baglioni C, Spiegelhalder K, et al. The reorganisation of memory during sleep. Sleep Med Rev. 2014;18(6):531–41.

122. Sibley BA, Beilock SL. Exercise and working memory: an individual differences investigation. J Sport Exerc Psychol. 2007;29:783–91.

123. Zhu N, Jacobs DR, Schreiner PJ, Yaffe K, Bryan N, Launer LJ, et al. Cardiorespiratory fitness and cognitive function in middle age. The CARDIA study. Neurology. 2014;82:1339–46.

Left Hemisphere Syndromes: Apraxias

7

Definition

Apraxia refers to an impairment in the performance of skilled motor acts. This is in the context of normal or near-normal elementary neurological function such as motor power and sensation. In addition a long list of specific exclusions and conditions that may masquerade as apraxia need to be excluded. These include the domains of attention, cognition, motor deficits, and movement disorders:

Exclusions

Cognitive: aphasia, abulia, inattention, visuospatial impairment, neglect, agnosias

Movement disorders: tremor, dystonia, myoclonus, chorea, athetosis

Motor: significant weakness for example less than −4/5 (MRC grading) of the limb

Cerebellar syndromes: ataxia, intention tremor.

Evolutionary Insights

The praxis and language largely overlap with critical nodes in the left inferior frontal, inferior parietal, and superior temporal cortices. How might this have occurred? Hominin evidence of toolmaking dates to over 3 million years (Dikika Ethiopia 3.3 mya) where stone flakes or so-called mode 1 tool (Oldowan techonology) evidence has been found that relates mainly to the fronto-parietal motor-perceptual circuitry rather than cognitive circuitry, inferred by neuroimaging evidence. Subsequent mode 2 (Acheulean) blades or handaxe, biface technology appeared more recently. Mode 2 bifaces were outside the range of ape abilities and were suitable for a variety of tasks including butchery. They had bilateral symmetry and required attention to shape. This activity required combined spatial information and visual processing spatial information (ventral and dorsal stream, respectively) [1].

Social learning by observation, imitation, and repetition, provided by the mirror neuron circuitry, in particular the left hemisphere inferior frontal lobe and anterior IPS sulcus, likely subserved the biface technology [2]. A PET brain scan study revealed the differences between the familiar and novel motor imitation circuit and how they may have laid the foundations of the praxis circuits in hominins. The inferior temporal lobe is regarded as the site for categorizing tools and objects and differentiating them from animate objects and the superior temporal lobe is a visual data coding region for actions, which then project to the posterior parietal (PP) mirror neuron hub. Here PP mirror neuron region integrates kinesthetic details of actions, which in turn project to the inferior frontal mirror neuron region. This basic scheme enables a visuomotor integration of an action (Fig. 7.1). Novel motor activity is perceived to use additional neural substrates.

M. Hoffmann, *Cognitive, Conative and Behavioral Neurology*,
DOI 10.1007/978-3-319-33181-2_7

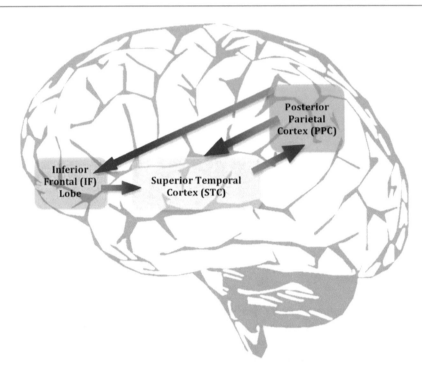

Fig. 7.1 Familiar motor imitation circuit. Figure with permission : Subiaul F. Dissecting the imitation faculty: The multiple imitation mechanisms (MIM) hypothesis. Behavioural Processes 2010;83:222–234 and Chaminade T, Meltzoff AN, Decety J. Does the end justify the means? A PET exploration of mechanisms involved in human imitation. Neuroimage 2002;15:318–328. Carr L, Iacoboni M, Dubeau MC, Mazziotta JC, Lenzi GL. Neural mechanisms of empathy in humans: a relay from neural systems for imitation to limbic areas. Proceedings of the National Academy of Sciences 2003;100:5497–5502

These include the left posterior cerebellum and dorsal as well as ventral prefrontal cortex, both key areas subserving the intention to imitate as well as observational learning (Fig. 7.2) [3, 4].

Bifaces probably represented symbols that endured over time and were widespread geographically. They were symbols that antedated language by 2–3 million years [5]. Bifaces are particularly noteworthy in that they represent an important period of time in evolution where hominin technology diverged significantly from ape technology [6].

Converging clinical, neuroimaging, archeological, and extant primate studies suggests an overlapping of the networks for tool use, praxis, and language. This suggests hypothetically that praxis and language evolved from tool use and that gestural communication served as an intervening system between stone tool production and language evolution. The circuitry of both praxis and language has hub areas in ventral premotor, inferior parietal, and superior temporal cortices [7]. The syntax, semantic, and lexicon architecture of language may have their origins in the praxis circuits that in turn were developed by stone knapping, With stone knapping, the object (stone in right hand), verb (action of striking one stone with another), and subject (the stone being knapped may underlie sentence structure). The cross-modal linking of the visual and auditory information and motor maps was enabled by inferior parietal lobes and synkinesis of the inferior frontal lobe for the motor output. This is the basis of the synesthetic bootstrapping hypothesis, proposed by Ramachandran [8].

Why Test for Apraxias? Pathophysiological Associations

Apraxias are important to test for. Firstly they are more common than generally appreciated as judged by the reported literature. In general, just over half (57%) of patients with aphasia have apraxia [9] and occur in approximately 35% of

Fig. 7.2 Novel motor imitation circuit. Figure with permission: Subiaul F. Dissecting the imitation faculty: The multiple imitation mechanisms (MIM) hypothesis. Behavioural Processes 2010;83: 222–234 and Chaminade T, Meltzoff AN, Decety J. Does the end justify the means? A PET exploration of mechanisms involved in human imitation. Neuroimage 2002;15:318–328. Petrosini L. "Do what I do and do how I do": different components of imitative learning are mediated by different neural structures. Neuroscientist 2007;13:335–348

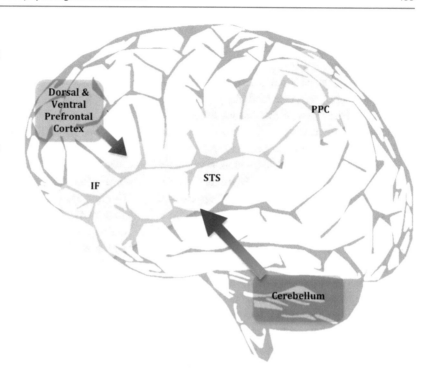

Alzheimer's patients and in 35.9% of stroke patients [10, 11]. They constitute one of several neurological syndromes that may not be volunteered by patients or even denied [12]. Furthermore they have an important outcome if they constitute part of the neurological syndrome with respect to the rehabilitation success.

Apraxia is mostly due to left hemisphere pathology, typically stroke, and often co-occurs with aphasia. However, two-way dissociations have been recorded clinically, with either one or the other occurring in isolation. Despite these reports, some consider that apraxia and aphasia both constitute impairments in symbolization, with aphasia representing the verbal symbolization deficit and apraxia the non-verbal symbolization deficit with tools, pantomime, and emblems [13]. Critical hubs in the neural praxis circuitry include the left inferior parietal lobe, superior parietal lobule, supplementary motor area, premotor cortex, and corpus callosum. In addition subcortical components, the basal ganglia, thalamus, and white matter connecting these structures with the cortex may also cause ideomotor apraxia (Fig. 7.3). Motor engrams or praxicon circuitry

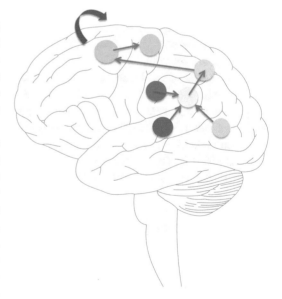

Fig. 7.3 Schematic apraxia diagram of the left hemisphere. Sensory information is transferred from the parietal (*dark blue*), temporal (*purple*), and occipital regions (*pink*) to the critical hub areas, which include the left inferior parietal lobe (*yellow*), superior parietal lobe (*light blue*), supplementary motor area (*green*), and primary motor area (*orange*). After information arrives at the left SMA it crosses via the corpus callosum to the homologous right SMA and innervates the right motor areas (*red arrow*)

hubs are located posteriorly, where visuokinetic data of visual, auditory, and tactile information is integrated within the supramarginal and angular gyri of the parietal lobe and anteriorly in the supplementary motor area (the medially situated part of the premotor area) of the frontal lobe where movement is effectuated.

The majority of our concepts concerning apraxias stem from lesion studies, often single-case reports and limited case series. These have implicated a number of left hemisphere regions and point to a widely distributed left hemisphere frontoparietal network subserving praxis. At its simplest level this comprises of a frontoparietal circuit for grasping and a frontal-subcortical circuit responsible for sequencing of motor acts. When specifically tested for, apraxia occurs in ~50 % of patients with left hemisphere lesions and less than 10 % with right hemisphere pathology implying some degree of bilateral praxis circuitry [14].

In addition to frequent association with stroke, movement disorders also prominently feature apraxia syndromes. These include Huntington's disease, progressive supranuclear palsy, Parkinson's, and corticobasal degeneration [15]. Corticobasal degeneration is a complex syndrome defined by the constellation of akinesia, rigidity, alien limb syndrome, and long tract signs that may include motor weakness and sensory loss. Apraxia is a prominent feature among movement disorders in general, and is most frequently associated with corticobasal degeneration, occurring in up to 70 % of cases [16, 17]. The various apraxias and their hypothetical circuitry are summarized in Fig. 7.3.

Classification of Apraxias

Limb
  Ideomotor
  Melokinetic
  Sympathetic
  Callosal
  Conduction
  Dissociation
  Conceptual
  Verbal motor dissociation
  Visuomotor
  Tactile motor dissociation apraxia

Pantomime agnosia
Axial
Parkinson's with axial apraxia
Progressive supranuclear palsy associated
Buccofacial
Oral apraxia
Complex (frontal lobe based)
Ideational
Ideomotor prosodic
Ocular apraxia
Component of Balint's syndrome
Task specific
Gait apraxia
Apraxia of eyelid opening
Ocular apraxia
Dressing apraxia
Constructional apraxia
Apraxic agraphia

Ideomotor Apraxia

Associated with left hemisphere lesions with estimates of approximately 50 % of patients with LH injury having ideomotor apraxia (IMA), mostly of the inferior parietal lobe and rarely described with right hemisphere lesions. In right-handed people movement representations (praxicons), containing spatial and temporal variables for action, are stored in a circuit with a hub in the left supramarginal gyrus (Fig. 7.4).

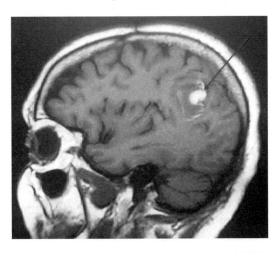

Fig. 7.4 Ideomotor apraxia in a patient with a small left inferior parietal hemorrhage (*red arrow*)

Destruction of these praxicons causes deficits in pantomiming to command as well as imitating and use of the actual object. Another hub area is the supplementary motor area, involved with storage of praxicons. Lesions here tend to cause melokinetic apraxia, rather than IMA [18]. Lesions of the putamen, which have robust connections with the premotor areas and the thalamus, can cause IMA but tend to be milder. The types of errors that may occur in people with IMA include:

Postural errors—internal configuration at fault

Egocentric movement error—the incorrect joint employment or coordination of joints

Allocentric movement errors—incorrect targeting of the intended object

Body part as tool error—with transitive act pantomiming, the hand or fingers are used as the tool [19]

Limb Kinetic Apraxia or Melokinetic Dyspraxia

A relatively benign or mild form of apraxia defined as an impairment of the deftness, nimbleness, or swiftness of finger and hand movements. A number of rapid bedside-type tests have been devised including the following:

1. Thumb to finger opposition test: The rapid opposition of the thumb to each of the fingers in sequence and then in the reverse direction with comparison to the good hand (mostly the left)
2. Coin rotation test: With American nickel in palms rotate between thumb index and middle finger as rapidly as possible for 20 rotations [20]
3. Pegboard tests [21]

The cerebral impairment has been found in the region of the supplementary motor area (SMA) and premotor cortex. With right handedness, a left hemisphere (not so with right hemisphere) injury can cause both ipsilateral and contralesional melokinetic dyspraxia (not so with RH lesions). The left hemisphere has apparently stronger ipsilateral control of spinal motor neurons compared

with the right hemisphere. The SMA and premotor regions have been correlated with the programming of deft movements by functional magnetic resonance imaging [22].

Sympathetic and Callosal Apraxias

A corpus callosal (CC) lesion may disconnect movement formulas embedded within the left hemisphere from the right hemisphere motor areas. The left hemisphere in right-handed people may contain spatial temporal movement representations (movement formula) and a CC lesion disconnects these from the right hemisphere motor areas inducing a unilateral left IMA. Neurological processes such as multiple sclerosis in particular involve the corpus callosum and adjacent white matter tracts. Lunardelli et al. described a young woman with both callosal apraxia and alien hand syndrome, both syndromes dependent on the interhemispheric disconnection due to anterior corpus callosum lesions. In particular this raised the contemporary concept of lesions having both inhibitory and excitatory effects, in this case apraxia and alien hand syndrome, respectively [23].

Sympathetic apraxia may occur with lesions in the white matter adjacent to Broca's area in the left hemisphere that also disconnects the two motor association areas of the two hemispheres. Fiber tracts from the left hemisphere therefore do not reach the right hemisphere motor association area and motor engrams of the left hemisphere are not transmitted to the otherwise intact left arm with normal strength [24].

Conduction (Gesture Imitation) Apraxia

People with IMA generally imitate transitive gestures better than pantomiming transitive gestures to verbal instruction. However case reports of imitation of learned transitive movements that were more impaired than pantomiming the same movements were reported by both Ochipa et al. [25] and Politis et al. [26].

Dissociation Apraxia

Although the embedded praxicons are intact, white matter lesions, usually, impair access to these movement representations from a stimulus in a particular modality such as visual or verbal [27].

Conceptual Apraxia

This type of apraxia refers to the loss of the specific mechanical knowledge of a tool or an implement and can be subdivided into the following deficiencies:

Tool selection deficit—unable to recognize which tool is associated with which object—such as a screw and a screwdriver.
Tool action association deficit—unable to recall the specific actions associated with a particular tool.
Impaired mechanical knowledge of a particular tool such as the case of a partially sawn piece of wood would require a saw to complete the job, or an unshaven face requiring a razor.
The specific test used for such impairments is the 45-item line drawings of the FLART (Florida Action Recall Test) of items that suggest the institution of a particular action [28].

Verbal Motor Dissociation Apraxia

A specific defect is present that when asked to gesture there is no response but at all. Speech comprehension however is intact in such individuals who are unable to activate the correct praxicon in response to verbal instruction (Fig. 7.1) [29].

Visuomotor Apraxia

In this form of apraxia, there is a failure to execute gestures in response to visual input. However, gestures are executed correctly to verbal instruction [30].

Tactile Motor Dissociation Apraxia

Deficient tactile object recognition refers to an impairment of recognizing object shape in the context of normal sensorimotor functions. Tactile recognition requires integrated exploratory hand movements. The case of a young woman with a large right hemisphere stroke was described by Valenza et al. with tactile apraxia due to deficient integration of sensorimotor and stored tactile shape [31].

Pantomime Agnosia

The inability to comprehend visually presented gestures but with retained ability to perform gestures to command as well as being able to imitate normally. The cases reported by Rothi had lesions in the left temporo-occipital region [32].

Axial Apraxia

First described in Parkinson's patients referring to the difficulties of abnormal postures turning over while recumbent, arising from supine or sitting position, particularly in the syndrome of progressive supranuclear palsy representing an extreme form of axial dyspraxia. In addition flexing, extending the head, turning the head to the right and left, shrugging shoulders and stepping forward, and stepping backwards are tests deployed by Poeck et al., for example, in their evaluations of axial apraxia [33–35]. In general the available literature points to an axial sparing of praxis in the setting of aphasia and associated limb apraxia [36, 37]. Studies in rats point to damage of the lateral hypothalamus and labyrinths are associated with axial rotation and righting reflexes [38].

Buccofacial Apraxia (Oral Apraxia)

There is difficulty with performing skilled movements of the lips, tongue face, pharynx, and larynx. Errors reported in performing these movements included movement substitutions, verbal descriptions instead of the movement, perseverations, content errors, and spatial errors [39].

Complex Apraxia Forms

Ideational Apraxia

An impairment in performing multistep (three or more) sequential actions to command. An often used test is that used in the MiniMental State Examination involving the folding of a piece of paper and placing on the table and the more intricate challenge of the correct sequences required for meal preparation. Responsible lesions have been reported in the left parieto-occipital region and prefrontal cortex.

Ideomotor Prosodic Apraxia

This form of apraxia was reported by Zakzanis who described a patient with Alzheimer's disease who had normal prosody during spontaneous speech but unable to produce specific acoustical variations to verbal command. Hence this was termed ideomotor prosodic apraxia and was tested by a 10-item prosodic apraxia scale incorporating phrases selected from the patient's spontaneous speech. Presumptive pathology is considered to be the right prefrontal cortex [40].

Ocular Apraxia

May present as a component of Balint's syndrome with simultanagnosia, ocular ataxia, and ocular apraxia (see chapter on complex visual processing).

Task-Specific Apraxias

Not all researchers agree that the following entities should be subsumed under the apraxias, but historically the following bear the apraxia label.

Whereas gait apraxia perhaps satisfies most of the criteria for apraxia, there is much less certainty about the inclusion of constructional apraxia and dressing apraxia for example. However understanding the mechanisms of these disorders can be very insightful with respect to treatment options such as the visual cues and musical cadence facilitation in gait apraxia for example.

Gait Apraxia

Perhaps the most at-risk subtype is gait apraxia, because of the propensity to fall causing disabling and costly conditions such as intracerebral hemorrhage, subdural hematoma, and hip fractures. Gait ignition failure is one of the better categorized examples of gait apraxia and patients present with hesitation in starting to walk and in turning, unaccompanied by other elementary neurological impairment such as motor, sensory, or cerebellar impairment. Current understanding of the neurobiology and neurophysiology of normal gait includes an SMA and BG interaction with preparatory movement within the SMA, which fires just prior to the initiation of gait, and is then transmitted to the motor area M1 area. The basal ganglia provide timing clues and phasic influence which switches off the SMA. External sensory cues (auditory, visual, proprioceptive information) influence this process and may also bypass it [41, 42].

The proposed classification of Liston et al. delineates gait ignition apraxia (GIA), equilibrium gait apraxia (EGA), and a mixed gait apraxia (MGA) entities. GIA is associated with lesions of the SMA and BG, EGA with PMA and white matter connections, and MGA with SMA, BG, PMA, and white matter connections between these centers. Both GIA and MGA may be influenced by visual cues and auditory cadence [43]. Causative lesions have been described in the supplementary motor cortex, parasagittal premotor, the DLPFC, and white matter lesions. Disease processes have included multiple sclerosis, Binswanger's, Parkinson's, stroke, and hydrocephalus. Marsden has postulated a disconnection pathophysiology to the SMA. Common processes such as small vessel cerebrovascular disease-related leukoaraiosis and normal-pressure hydrocephalus interrupt the white matter fiber tracts to and from the premotor and supplementary motor areas [44].

In normal-pressure hydrocephalus the triad of symptoms, sometimes referred to as the Hakimov triad (gait apraxia, urinary incontinence, dementia) [45], has a prominent component as gait disturbance that is best described as a gait apraxia by the abovementioned criteria. The gait apraxia consists of movement initiation difficulty, truncal mobility impairment, initiating movement, problems with raising the foot off the floor (so-called magnetic gait), imbalance, falls, postural reflex impairment resulting in imbalance, and falls and short steps. Also referred to as Bruns ataxia or magnetic gait, it may be improved considerably in correctly diagnosed patients by relieving CSF pressure, presumably by relieving pressure off the frontopontocerebellar tract or Arnold's bundle linking frontal lobe to the cerebellum via pontine relays. Leukoaraiosis is particularly associated with gait disturbance [46, 47]. Leukoaraiosis has been) reported to be an independent predictor of people with gait impairment gait [48]. The corticopontothalamocortical network has only recently been appreciated and is a relatively recent acquired fiber tract associated with bipedality in hominin evolution [49].

Apraxia of Eyelid Opening

Patients remain with eyes closed, unable to open them to command, usually after right hemisphere brain infarction for example. However they may at times, in the recovery period, spontaneously open their eyes, but when they close again, there is an inability to open them to verbal command, often but not always associated with a number of movement disorders such as blepharospasm and Parkinsonism, typically due to right frontal infarction, basal ganglia, or putaminal infarction or hemorrhage [50–53]. Interestingly, wearing goggles has been reported to improve the ALO in Parkinsonian patients [54].

Ocular Apraxia

Usually seen in conjunction with Balint's syndrome (optic apraxia optic ataxia, simultanagnosia—see visual processing chapter) and Cogan's ocular motor apraxia also known as saccadic initiation failure. The latter congenital or acquired condition was first described by Cogan in 1952 and characterized by the absence of voluntary gaze, absence of the fast-phase component of the horizontal optokinetic response, jerky-type head movements in an attempt to gaze left or right, and normal vertical eye movements [55]. The pathobiology is usually caused by bilateral lesions and is one of saccade initiation failure (voluntary production of saccades from caudal middle frontal gyrus) or associated with parietal eye field lesions in the region of the IPS that subserve reflexive production of saccades or vestibular ocular reflex failure (posterior). Other causes include bifrontal infarctions (affecting the frontal eye fields) and posterior cortical atrophy or Benson's syndrome [56, 57].

Dressing Apraxia

Dressing requires intact visuospatial functioning and most cases reported are thought to be due to an impairment in this faculty. Typical examples include putting a shirt or trousers on back to front or inside-out and has been reported in patients with Alzheimer's disease, corticobasal degeneration, and right hemisphere stroke [58, 59].

Constructional Apraxia

Frequently referred to also as visuospatial dysfunction and much depends on the precise wording of these disorders. It involves several cognitive processes that at the very least include attention, visuospatial perception, planning, and motor output. It was first defined by Kleist in 1934 as "a disturbance in the activities of drawing, assembling and building in which the spatial form of the product proves to be unsuccessful as assembling, building and drawing, in which the spatial form of the product proves to be unsuccessful, without there being an apraxia for single movements." [60]. It differs from IMA in that it is not a motor planning impairment but is conceived as a deficiency between visual perceptual and the linking of motor activity. Bedside assessment involves the copying of two-dimensional (example a flower) and three -dimensional (example a 3-cube) figures and a clock face and by quantitative, metric tests such as the Rey Osterrieth Complex Figure test, the WAIS block design subtest, and Judgment of Line Orientation tests [61]. Most studies support a right parietal predominance in about two-thirds of patients, but both left hemisphere lesions and subcortical lesions may also cause such deficits [62].

Apraxic Agraphia

In this type of apraxia there is slow, effortful, imprecise letter formation with distortions that lead to illegible handwriting. This is in the context of otherwise normal sensory motor function, language function, and cerebellar function and without Parkinsonian syndromes or dystonia. It may be associated with IMA or appear without it. Causative lesions are usually within the left parietal lobe, in particular the superior parietal lobule, the intraparietal sulcus, and anterior intraparietal sulcus. Sakurai has delineated a number of apraxic agraphias [63]. Apraxic agraphia (pure agraphia) without associated IMA has been reported with lesions of either parietal lobe as well as frontal lobe or thalamic lesions [64].

Testing

As the condition may not be apparent to the patient or family, it is worth performing a brief screening test in most patients, especially in those presenting with mild cognitive complaints or aphasia and right-side paresis. Frequently, family members report a clumsiness while eating. In addition it is worth recalling that many of the tests employed may be culturally specific or

weighted. Overall the analysis of errors made when apraxic includes deficits of the spatial course of action, the temporal course of action, or the correct sequencing of the movement components. The suggested order of testing that appears below can also double as a simple bedside, severity grading of apraxias that can be employed:

1. Mild—when pantomime by verbal command impaired
2. Moderate—when ability to imitate examiner's movements is impaired
3. Severe—using a tool or object (transitive movements) is impaired

The instructed, verbal actions will improve when imitating the examiner and typically improve further when using the actual tool because of visual and tactile cues. The overall testing strategy can be further divided into the following:

The movement system such as limb (arm, leg) versus axial (truncal, neck)
Input modality, whether verbal (command), visual (imitation), or tactile
Type of limb gesture, whether transitive (using an object/tool) or intransitive (abstract)
Complexity of action, whether single gestures or a sequence of gestures

Imitation relies on the dorsal stream that allows visuomotor conversion and pantomiming requires both the ventrodorsal and ventral streams that enable access to the action engrams representing tool-action associations [65].

Ultra-short (screening) tests (1–2 min)
The author uses a combination of:

Ideomotor Limb Intransitive

Arm
Wave goodbye
Beckon, come here
Leg
Place one foot in front of the other as if tandem walking
Describe a circle with the foot

Ideomotor Limb Transitive

Arm
   Pretend to use a pair of scissors cutting a piece of paper
   Pretend to use a toothbrush, brushing your teeth
   Leg
   Kick a ball
   Pretend to accelerate by pushing on a gas pedal

Buccofacial

Whistle
   Pretend to blow out a candle

Axial

Stand like a boxer
   Stand like a golfer getting ready to swing
   Bend the head forward and back
   Shrug or lift both shoulders

Melokinetic

Rapid opposition of the thumb to each of the four fingers forward and back
   Compare right to left hand

Ideational

Write your address on a piece of paper, fold it in half, and place inside a book or folder.

Short Tests
Heilman and Rothi Use the Following Short Test Battery
Intransitive limb gestures
   Wave goodbye
   Hitch-hike
   Salute
   Become "come here"

Stop
Go
Transitive limb gestures
Open a door with a key
Flip a coin
Open a catsup bottle
Use a screwdriver
Use a hammer
Use scissors
Intransitive buccofacial
Stick out tongue
Blow a kiss
Transitive buccofacial gesturers
Blow out a match
Suck on a straw
Serial acts
Show how you would make a sandwich
Show how you would write and mail a letter

In the more elaborate Brief Apraxia Screening test version by Efros et al., a 24-item version of pantomime commands also has accompanying error description and error attribute components to allow tabulation of a score [66].

The STIMA test of Tessari et al. addressed the direct and semantic imitation route of assessment and provides a semiquantitative assessment of IMA. This test employed a total of 36 gestures divided into 18 known gestures (nine proximal, nine distal) and 18 new (nine proximal, nine distal) with specific scoring rules. Overall the test group had improved scores with known gestures and proximal gestures. Some examples of the known gestures included the following: I am tired, I am hungry, military salute, hitch-hiking, and crazy. Some examples of new gestures included hand over opposite shoulder, back of hand on contralateral cheek, hand over opposite shoulder, and horizontal back of hand under chin fist with the thumb extended toward the body midline. Scoring was graded 2 point for correct imitation, 1 point for correct imitation after a second trial presentation, and 0 if failed on both attempts, allowing a total score range of 0–72 points, and generally testing time was 2–3 min. Age-adjusted scores and percentile normative data were generated for the subcomponents and total scores [67].

The Van Heugten Test for Apraxia reported by Smits et al. evaluated mild cognitive impairment patients and also included validity measurements such as interrater and intrarater reliability. In the populations evaluated apraxia was present in 10 % of MCI patients and 35 % of Alzheimer's and absent in controls. Scored from 0 (unable), 1, and 2 (degrees of impairment) to 3 (fully compliant) they evaluated three object use procedures (example, key, hammer, toothbrush); two trials allowed for ideational apraxia and six gestures (stick tongue out, blowing out a candle, close your eyes, wave goodbye, saluting, making a fist). A total score of 90 represents no mistakes and ≤86 was the cutoff for apraxia [10].

Other recently reported brief tests with controls and validity data include the Short Apraxia Screening Test of Leiguarda et al. who reported high sensitivity and specificity for their test. They evaluated one of the largest groups ($n = 70$) to date with limb apraxia, including 12 testable items, intransitive and transitive gestures by visual, verbal, and tactile modality instruction and imitation of both meaningful and meaningless movements and postures [68].

Comprehensive Tests

The much more detailed, experimental edition, the Florida Apraxia Battery of Heilman et al. deemed for research purposes, comprises of:

1. Florida Action Recall Test
2. Postural Knowledge Test
3. Brief Apraxia Screening Test
4. Coin Rotation Task
5. Praxis Imagery Test
6. Florida Apraxia Battery
   (a) Gesture to Command Subtest
   (b) Gesture to Visual Tool and Gesture to Tactile Tool Subtest
   (c) Gesture Imitation Subtest
   (d) Tool to Object Matching Subtest
   (e) Gesture to Named Object Movement Verification Subtest
   (f) Gesture to Conceptual Tool Verification [20, 69–71]

Another comprehensive test, TULIA, with validity data (internal consistency, inter- and intra-rater, test-retest reliability, criterion, construct validity) of Vanbellingen et al., used 48 items, including both imitation and pantomime for intransitive, transitive, and nonsymbolic gestures, comprising of six subtests, six-point scoring system, and total score ranging from 0 to 240. Reliability and validity were confirmed for upper limb apraxia [72].

Treatment/Management

Bearing in mind that some degree of anosognosia is usual in people with apraxia, once elicited by one of the tests a behavioral training program such as one devised by Smania et al. seems worthwhile. The behavioral training program consisted of gesture-production exercises either with or without symbolic value. Significant improvements in error production, in both ideational and ideomotor apraxic patients, were reported by the group [73]. There is a paucity of pharmacotherapeutic trials with pharmacotherapy, but cognotropic therapy with devices appears promising. Bolognini et al. evaluated anodal transcranial direct current stimulation (t-DCS) in six left brain-damaged patients with IMA and six control patients, tested with an IMA testing of 24 intransitive gestures (12 symbolic and 12 nonsymbolic), as well as six of the seven items in the Jebsen Hand Function Test. Improvement was shown in skilled motor performance in the left hand of both patients and controls when stimulating the left, posterior parietal cortex. Furthermore it was discovered that stimulation of the left PPC rescued planning times and stimulation of the right M1 motor cortex reduced execution times [74]. Similarly the use of t-DCS improved gesture processing in the patients with corticobasal ganglionic syndrome. Apraxia is particularly frequent and disabling symptom in this condition. Anodal t-DCS improved the treatment group significantly as judged by the De Renzi Ideomotor Test scores [75].

Gait apraxia is generally not improved by Levodopa but may be improved by dopamine agonists such as ropinirole, with several case reports attesting to dopamine agonists [76]. In addition many case reports support music therapy and visual cueing as being beneficial [Sacks O. Musicophilia. Vintage Books, Random House Inc, New York 2008]. Dopamine release in striatum occurs during a music listening and may be one of the mechanisms whereby this therapy is effective [77, 78]. Anodal tDCS together with regular physiotherapy has been shown to improve patients with leukoaraiosis-associated gait apraxia [79].

References

1. Stout D, Chaminade T. The evolutionary neuroscience of tool making. Neuropsychologia. 2007;45(5): 1091–100.
2. Stout D, Passingham R, Frith C, Apel J, Chaminade T. Technology, expertise and social cognition in human evolution. Eur J Neurosci. 2011;33:1328–38.
3. Chaminade T, Meltzoff AN, Decety J. Does the end justify the means? A PET explotration of mechanisms involved in human imitation. Neuroimage. 2002;15:318–28.
4. Petrosini L. "Do what I do and do how I do": different components of imitative learning are mediated by different neural structures. Neuroscientist. 2007;13: 335–48.
5. Donald M. The origins of the modern mind: three stages in the evolution of culture and cognition. Cambridge: Harvard University Press; 1991.
6. Coolidge FL, Wynn T, Overmann KA, Hicks JM. Cognitive archeology and the cognitive sciences. In: Bruner E, editor. Human paleoneurology. New York, NY: Springer; 2015.
7. Roby-Brami A, Hermsdörfer J, Roy AC, Jacobs S. A neuropsychological perspective on the link between language and praxis in modern humans. Philos Trans R Soc B. 2012;367:144–60.
8. Ramachandran VS. The tell tale brain. New York, NY: WW Norton and Company; 2011.
9. Heilman KM. A tapping test in apraxia. Cortex. 1975;11:259–63.
10. Smits L, Flapper M, Sistermans N, Pijnenburg YA, Scheltens P, van der Flier WM. Apraxia in mild cognitive impairment and Alzheimer's disease: validity and reliability of the Van Heugten Test for apraxia. Dement Geriatr Cogn Disord. 2014;38:55–64.
11. Civelek GM, Atalay A, Turhan N. Association of ideomotor apraxia with lesion site, etiology, neglect, and functional independence in patients with first ever stroke. Top Stroke Rehabil. 2015;22(2):94–101.
12. Rothi LJG, Mack L, Heilman KM. Unwareness of apraxic errors. Neurology. 1990;40 Suppl 1:202.
13. Goldstein K. Language and language disorders. New York, NY: Grüne and Stratton; 1948.
14. Barbieri C, De Renzi E. Patterns of neglect dissociation. Behav Neurol. 1989;2(1):13–24.
15. Zadikoff C, Lang AE. Apraxia in movement disorders. Brain. 2005;128:1480–97.
16. Leiguarda R, Lees AJ, Merello M, Starkstein S, Marsden CD. The nature of apraxia in corticobasal degeneration. J Neurol Neurosurg Psychiatry. 1994;57:455–9.
17. Litvan I, Bhatia KP, Burn DJ, Goetz CG, Lang AE, McKeith I, et al. SIC task force appraisal of clinical diagnostic criteria for parkinsonian disorders. Mov Disord. 2003;18:467–86.
18. Poizner H, Merians AS, Clark MA, Macauley B, Rothi LJ, Heilman KM. Left hemispheric specialization for learned, skilled, and purposeful action. Neuropsychology. 1998;12(2):163–82.
19. Poizner H, Mack L, Verfaellie M, Rothi LJG, Heilman KM. Three dimensional computer graphic analysis of apraxia. Brain. 1990;113:85–101.
20. Hanna-Pladdy B, Mendoza JE, Apostolos GT, Heilman KM. Lateralised motor control: hemispheric damage and the loss of deftness. J Neurol Neurosurg Psychiatry. 2002;73:574–7.
21. Matthews CG, Klove H. Instruction manual for the adult neuropsychology test battery. Madison, WI: University of Wisconsin Medical School; 1964.
22. Nirkko AC et al. Different ipsilateral representations for distal and proximal movements in the sensorimotor cortex: activation and deactivation patterns. Neuroimage. 2001;13:825–35.
23. Lunardelli A, Sartori A, Mengotti P, Rumiati RI, Pesavento V. Intermittent alien hand syndrome and callosal apraxia in multiple sclerosis: implications for interhemispheric communication. Behav Neurol. 2014;2014:873541. doi:10.1155/2014/873541.
24. Geschwind N. Sympathetic dyspraxia. Trans Am Neurol Assoc. 1963;88:219–20.
25. Ochipa C, Rothi LJG, Heilman KM. Conduction apraxia. J Clin Exp Psychol. 1990;12:89.
26. Politis DG. Alterations in the imitation of gestures (conduction apraxia). Rev Neurol (Paris). 2004; 38:741–5.
27. Merians AS, Clark M, Poizner H, Macauley B, Gonzalez Rothi LJ, Heilman KM. Visual imitative dissociation apraxia. Neuropsychologia. 1997;35: 1483–90.
28. Schwartz RL, Adair JC, Raymer AM, Williamson DJG, Crosson B, Rothi LJG, et al. Conceptual apraxia in probable Alzheimer's disease as demonstrated by the Florida Action Recall Test. J Int Neuropsychol Soc. 2000;6:265–70.
29. Heilman KM. Ideational apraxia. A re-definition. Brain. 1973;96:861–4.
30. De Renzi F, Faglioni P, Sorgato P. Modalith specific and supramodal mechanisms of apraxia. Brain. 1982;105:301–12.
31. Valenza N, Ptak R, Zimine I, Badan M, Lazeyras F, Schnider A. Dissociated active and passive tactile

shape recognition: a case study of pure tactile apraxia. Brain. 2001;124:2287–98.

32. Rothi LJG, Mack L, Heilman KM. Pantomine agnosia. J Neurol Neurosurg Psychiatry. 1986;49:451–4.

33. Poeck K, Lehmkuhl G, Willmes K. Axial movements in ideomotor apraxia. J Neurol Neurosurg Psychiatry. 1982;45(12):1125–9.

34. Lakke JP. Axial apraxia in Parkinson's disease. J Neurol Sci. 1985;69:37–46.

35. Lakke JP, van Weerden TW, Staal-Schreinemachers A. Axial apraxia, a distinct phenomenon. Clin Neurol Neurosurg. 1984;86(4):291–4.

36. Howes DH. Ideomotor apraxia: evidence for the preservation of axial commands. J Neurol Neurosurg Psychiatry. 1988;51:593–8.

37. Hanlon RE, Mattson D, Demery JA, Dromerick AW. Axial movements are relatively preserved with respect to limb movements in aphasic patients. Cortex. 1998;34(5):731–41.

38. Pellis SM, Pellis VC, Chen YC, Barzci S, Teitelbaum P. Recovery from axial apraxia in the lateral hypothalamic labyrinthectomized rat reveals three elements of contact-righting: cephalocaudal dominance, axial rotation, and distal limb action. Behav Brain Res. 1989;35(3):241–51.

39. Raade AS, Rothi LJG, Heilman KM. the relationship between buccofacial and limb apraxia. Brain Cogn. 1991;16:130–46.

40. Zakzanis KK. Ideomotor prosodic apraxia. J Neurol Neurosurg Psychiatry. 1999;67:694–5.

41. Georgiou N, Iansek R, Bradshaw JL, Phillips JG, Mattingley JB, Bradshaw JA. An evaluation of the role of internal cues in the pathogenesis of parkinsonian hypokinesia. Brain. 1993;116:1575–87.

42. Marsden CD. Slowness of movement in Parkinson's disease. In: Fahn S, Marsden CD, editors. Movement disorders 4 (Suppl 1). New York, NY: Raven; 1989. p. 26–37.

43. Liston R, Mickelborough J, Bene J, Tallis R. A new classification of higher level gait disorders in patients with cerebral multi-infarct states. Age Ageing. 2003;32:252–8.

44. Abou Zeid NE, Weinshenker BG, Keegan BM. Gait apraxia in multiple sclerosis. Can J Neurol Sci. 2009;36(5):562–5. and [Nadeau SE. Gait apraxia: further clues to localization. Eur Neurol. 2007;58(3):142–5].

45. Mihalj M, Titlic M, Marovic A, Bulovic B, Srdelic-Mihalj S. Gait apraxia. Bratisl Lek Listy. 2010;111(2):101–2.

46. Poggesi A, Gouw A, van der Flier W, Pracucci G, Chabriat H, Erkinjuntti T, et al. Cerebral white matter changes are associated with abnormalities on neurological examination in non-disabled elderly: the LADIS study. J Neurol. 2013;260(4):1014–21. doi:10.1007/s00415-012-6748-3.

47. Masdeu JC, Wolfson L. White matter lesions predispose to falls in older people. Stroke. 2009;40(9), e546. doi:10.1161/STROKEAHA.109.558122.

48. Briley DP, Wasay M, Sergent S, Thomas S. Cerebral white matter changes (leukoaraiosis), stroke, and gait disturbance. J Am Geriatr Soc. 1997;45(12):1434–8.

49. Ramnani N. The primate cortico-cerebellar system: anatomy and function. Nat Rev Neurosci. 2006;7:511–22.

50. Lin YH, Liou LM, Lai CL, Chang YP. Right putamen hemorrhage manifesting as apraxia of eyelid opening. Neuropsychiatr Dis Treat. 2013;9:1495–7. doi:10.2147/NDT.S50974.

51. Hirose M, Mochizuki H, Honma M, Kobayashi T, Nishizawa M, Ugawa Y. Apraxia of lid opening due to a small lesion in basal ganglia: two case reports. J Neurol Neurosurg Psychiatry. 2010;81(12):1406–7.

52. Algoed L, Janssens J, Vanhooren G. Apraxia of eyelid opening secondary to right frontal infarction. Acta Neurol Belg. 1992;92(4):228–33.

53. Johnston JC, Rosenbaum DM, Picone CM, Grotta JC. Apraxia of eyelid opening secondary to right hemisphere infarction. Ann Neurol. 1989;25(6):622–4.

54. Hirayama M, Kumano T, Aita T, Nakagawa H, Kuriyama M. Improvement of apraxia of eyelid opening by wearing goggles. Lancet. 2000;356(9239):1413.

55. Cogan DG. A type of congenital ocular motor apraxia presenting jerky head movements. Trans Am Acad Ophthalmol Otolaryngol. 1952;56(6):853–62.

56. Chen J, Thurtell M. Acquired ocular motor apraxia due to bifrontal hemorrages. Neurol Neurosurg Psychiatry Pract Neurol. 2012;83(6):1117–9.

57. Kirshner HS, Lavin PJ. Posterior cortical atrophy: a brief review (Review). Curr Neurol Neurosci Rep. 2006;6(6):477–80. and [Greene JD. Apraxia, agnosias, and higher visual function abnormalities. J Neurol Neurosurg Psychiatry. 2005;76(Suppl 5):25–34].

58. Takayama T, Sugishia M, Hirose S, Akiguchi I. Anosodiaphoria for dressing apraxia: contributory factor to dressing apraxia. Clin Neurol Neurosurg. 1994;96:254–6.

59. Okuda B, Tanaka H, Kawabata K, Kodam N, Tachibana H. Dressing apraxia in corticobasal degeneration. Geriatr Gerontol Int. 2003;3:64–7.

60. Kleist K. Gehirnpathologie. Leipzig: Barth; 1934.

61. Hamsher K, Carpruson DX, Benton A. Visuospatial judgment and right hemisphere disease. Cortex. 1992;28:493–5. and [Osterreith P. Le test de copie d'unie figure complexe. Arch Psychol (Frankfurt) 1944;30:206–356].

62. Arrigoni G, De Renzi E. Constructional apraxia and hemisphere locus of lesions. Cortex. 1964;1:170–94. and [Black FW, Strub RL. Constructional apraxia in patients with discrete missile wounds of the brain. Cortex 1976;12:212–20].

63. Sakurai Y, Onuma Y, Nakazawa G, Ugawa Y, Momose T, Tsuji S, et al. Characterization of abnormal writing stroke sequences, character formation and character recall. Behav Neurol. 2007;18:99–114.

64. Carey MA, Heilman KM. Letter imagery deficits in a case of pure apraxic agraphia. Brain Lang. 1988;34:147–56.

65. Hoeren M, Kummerer K, Bormann T, Beume L, Ludwig VM, Vry MS, et al. Neural basis of imitation and pantomine in acute stroke patients: distinct streams for praxis. Brain. 2014;137:2796–810.
66. Efros DB, Cimino-Knight AM, Morelli CA. A comparison of two assessment methods for ideomotor limb apraxia. San Diego, CA: American Speech Language Hearing Association Annual Convention; 2005.
67. Tessari A, Toraldo A, Lunardelli A, Zadini A, Rumiati RI. STIMA: a short screening test for ideo-motor apraxia, selective for action meaning and bodily district. Neurol Sci. 2015;36:977–84.
68. Leiguarda R, Clarens F, Amengual A, Drucaroff L, Hallett M. Short apraxia screening test. J Clin Exp Neuropsychol. 2014;36(8):867–74. doi:10.1080/1380 3395.2014.951315.
69. Rothi LJR, Raymer AM, Heilman KM. Limb praxis assessment and glossary. The neuropsychology of action. London: Psychology Press; 1997.
70. Rothi LJG, Raymer AM, Ochipa C, Maher LM, Greenwald ML, Heilman KM. Florida apraxia battery. Experimental Ed, 1992.
71. Mozaz M, Gonzalez-Rothi LJ, Anderson JM, Crucian GP, Heilman KM. Postural knowledge of transitive pantomines and intransitive gestures. J Int Neuropsychol Soc. 2002;8:958–62.
72. Vanbellingen T, Kersten B, Van Hemelrijk B, Van de Winckel A, Bertschi M, Müri R, et al. Comprehensive assessment of gesture production: a new test of upper limb apraxia (TULIA). Eur J Neurol. 2010;17(1):59–66.
73. Smania N, Girardi F, Domenicali C, Lora E, Aglioti S. The rehabilitation of limb apraxia: a study in left-brain-damaged patients. Arch Phys Med Rehabil. 2000;81(4):379–88.
74. Bolognini N, Convento S, Banco E, Mattioli F, Tesio L, Vallar G. Improving ideomotor limb apraxia by electrical stimulation of the left parietal cortex. Brain. 2015;138:428–39.
75. Bianchi M, Cosseddu M, Cotelli M, Manenti R, Brambilla M, Rizzetti MC, et al. Left parietal cortex transcranial direct current stimulation enhances gesture processing in corticobasal syndrome. Eur J Neurol. 2015. doi:10.1111/ene.12748.
76. Taskapilioglu O, Karli N, Erer S, Zarifoglu M, Bakar M, Turan F. Primary gait ignition disorder: report of three cases. Neurol Sci. 2009;30:333–7.
77. George EM, Coch D. Music training and working memory: an ERP study. Neuropsychologica. 2011;49:1083–94.
78. Burunat I, Alluri V, Toiviainen P, Numminen J, Brattico E. Dynamics of brain activity underlying working memory for music in a naturalistic condition. Cortex. 2014;57:254–69.
79. Kaski D, Dominguez RO, Allum JH, Bronstein AM. Improving gait and balance in patients with leuko-araiosis using transcranial direct current stimulation and physical training: an exploratory study. Neurorehabil Neural Repair. 2013;27(9):864–71. doi:10.1177/1545968313496328.

8

Parietal Lobe Syndromes

Evolution

A convenient time to pick up the parietal lobe evolution is perhaps with *Homo erectus* ~1.8 mya after the brain size had doubled to about 880 cc from the average of 450 cc in the Australopithecines. The basic stone knapping techniques used by these early hominins presumably promoted the elaboration of the parietal spatial network and shape recognition by temporal lobe ventral pathways. This was due to learning by observation, prolonged practice, and sequences of actions that were eventually incorporated into procedural memory circuits. Much later in our evolution, *Homo heidelbergensis* artifact evidence points to improved visuospatial sketchpad components of working memory model, exemplified by the finding of the Schoningen spears dated to ~400 kya. This suggested an overall improvement in both their working memory and technical ability when compared to *Homo erectus*, also supporting the emergence of symbolic thinking and allocentric perception (points of view not pertaining to one's own viewpoint). With the emergence of Neanderthals around 300–200 kya, with the invention of hafting they are sometimes referred to as the technical experts but are thought to have lacked enhanced working memory, extended theory of mind, and shared attention. In Wynn and Coolidge's viewpoint they lacked enhanced working memory which is the basis of advanced episodic memory, intelligence, language, and

theory of mind and therefore probably also lacked modern, recursive-type language [1, 2]. Finally, a more dramatic parietal lobe enlargement occurred relatively recently with parietal globularization (150–100 kya) with vertical expansion and anterior widening (klinorhynchy), unique to modern humans [3]. This may have facilitated neural intra-connectivity, cortico-cortico hyperconnectivity, further visuospatial integration, multimodal processing, as well as improved social communication.

The progressive frontoparietal integration has been considered to be the major "pacemaker," thus increasing complexity of higher cortical functions in developing humans. In the Australopithecines there was a progressive posterior displacement of the lunate sulcus in response to the gradually enlarging posterior parietal cortex implying an early reorganization of the brain components as far back as ~5 mya, antedating frontal lobe enlargement. The intraparietal sulcus (IPS) in particular is associated with object discrimination, toolmaking, analysis of number, space, time, hand–eye coordination, 3-dimensional visuospatial functions, and motor planning (praxis). This gradually led to representation of both internal and external space representation, and cross-modal sensory functioning which is required for metaphor and language comprehension [4, 5].

Whereas the inferior parietal lobule is mainly connected to DLPFC, the upper parietal lobule is connected mainly to the dorsomedial PFC.

The upper parietal lobule is involved in the visuo-spatial sketchpad of the working memory model [2]. The superior parietal lobe also includes the IPS and precuneus, both responsible for numerosity and prospective memory. From functional imaging studies it has been determined that the precuneus has the highest metabolic activity of all brain regions especially during the default mode resting state. The neurobiological function of this area is considered to be integration of different neural networks responsible for both self-consciousness [6]. The medial parietal and posterior cingulated regions are the most prominent hypometabolic areas in Alzheimer's disease. This may reflect the recent enlargement of the parietal lobes and vascular metabolic requirements. This has led Bruner to hypothesize that vascular and metabolic constraints consistent with the hub failure hypothesis may have led to the concept of AD being a metabolic syndrome and the parietal lobes presenting a type of Achilles heel of the hominin brain enlargement [7].

Neuro-Anatomical and Neurophysiological Aspects of Parietal Lobes (Fig. 8.1)

Primary Somatosensory Cortex (BA 1,2,3)

Tactile, proprioception, and pressure sensations.

Association Somatosensory (I) (BA 5)

The superior parietal lobule: cortical sensory abilities such as two-point discrimination, tactile localization, determination of shape, weight and size of object and graphesthesia, sensory motor integration for movements, 3-dimensional orientation.

Association Somatosensory Cortex (BA 7)

Posterior parietal lobule and precuneus: Involved in motor imagery, visuospatial imagery,

self-reflection episodic memory retrieval, and consciousness. The precuneus may also include BA 31 [6].

Multimodal Association Area (BA 39, 40)

The supramarginal gyrus (SMG) and angular (AG) enable multimodal tactile, visual, and auditory associations that were critical for modern language development. In general the left inferior parietal lobe (IPL) is key for praxis and communication whereas the right hemisphere IPL is considered dominant for visuospatial function and attention.

Intraparietal Sulcus

Up to 20 mm in depth in some areas is key for numerosity [8], motor perceptual coordination such as reaching, grasping, voluntary eye movement saccades and object manipulation, visuospatial working memory, and theory of mind. The IPS may be divided into ventral, caudal anterior, medial, and lateral functional components:

- Ventral and lateral—eye movement saccades, visual attention
- Ventral and medial—visuomotor, pointing, reaching
- Caudal—stereopsis
- Anterior—grasping, hand movement control [9–11]

Parietal Networks

In addition to the extensive frontoparietal networks (discussed in frontal systems chapter), the visual ventral and dorsal streams involve the occipito-temporal and occipito-parietal regions. The dorsal stream may be divided into three pathways that target the prefrontal, premotor areas and medial temporal lobe cortices. The dorsal parietal to medial temporal lobe tract traverses the retrosplenial and posterior cingulate cortices (Fig. 8.2).

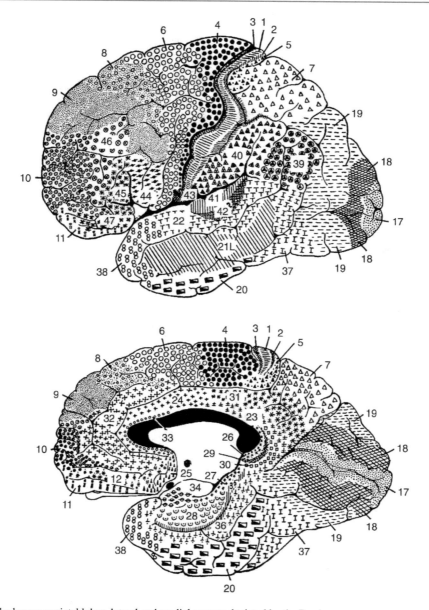

Fig. 8.1 The human parietal lobes, lateral and medial aspects depicted by the Brodmann areas map

These three principal dorsal stream pathways are involved in conscious and non-conscious activity with respect to:

- Parietoprefrontal—spatial working memory
- Parieto-premotor—grasping and other visually guided tasks (predominantly non-conscious processing)
- Parieto-medial temporal—navigation and representation of space references for both egocentric and allocentric vantage points [12]

Microanatomy and receptor architectural studies by Caspar and Zilles have delineated seven different inferior parietal regions with connectivity characteristic suggesting a three-region (rostral, middle, caudal entities) model. This receptor fingerprint model supports IPL connectivity motor, visual, auditory (language) cortices. Broca's region showed extensive connectivity to all three areas whereas the middle IPL group connected more with the superior parietal area and the caudal group mostly with the extrastriate visual cortical areas (Fig. 8.3).

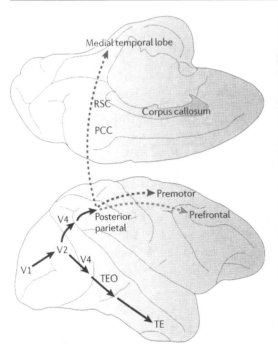

Fig. 8.2 Dorsal stream subpathways to the prefrontal, premotor areas and medial temporal lobe directly via the retrosplenial and posterior cingulate cortices. *PCC* posterior cingulate cortex, *RSC* retrosplenial cortex, *TE* rostral inferior temporal cortex, *TEO* posterior inferior temporal cortex, *V1* primary visual area 1. Figure with permission: Kravitz DJ, Saleem KS, Baker CI, Mishkin M. A new neural framework for visuospatial processing. Dwight J. Nature Reviews Neuroscience 2011;12:217–230

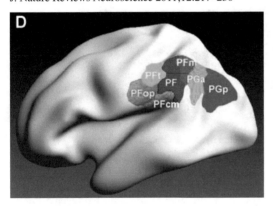

Fig. 8.3 The three-region inferior parietal model (and seven subdivisions) of Caspers and Zilles according to cytoarchitecture and receptor fingerprinting studies. Legend per Caspers et al.: rostral cluster (areas PFt, PFop, and PFcm): shades of *green*; middle cluster (areas PF and PFm): shades of *red*; caudal cluster (areas PGa and PGp): shades of *blue*. Figure with permission: Caspers S, Schleicher A, Bacha-Trams M, Palomero-Gallagher N, Amunts K, Zilles K. Organization of the Human Inferior Parietal Lobule Based on Receptor Architectonics. Cerebral Cortex March 2013;23:615–628

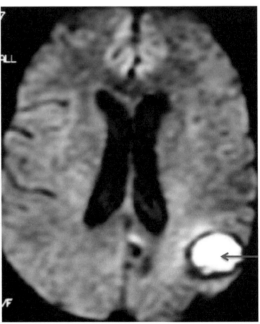

Fig. 8.4 Left parietal, discrete, amyloid angiopathy-related hemorrhage with the sole clinical deficit; Gerstmann's syndrome

Clinically Appreciated Parietal Lobe Syndromes

The parietal lobes may be involved in isolation (Fig. 8.4), even subcomponent involvement, such as the precuneus (Fig. 8.5), by relatively discrete lesions such as stroke, cerebral hemorrhage due to amyloid angiopathy, tumors, and more rarely parietal lobe seizures.

Right or Left Parietal Lesion Syndromes

Cortical Sensory Impairment

- Astereognosis—loss of depth perception
- Two-point discrimination impairment
- Agraphesthesia—palm number tracing
- Abaragnosia—loss of weight perception discrimination
- Tactile agnosia
- Ahylognosia
- Parietal pain syndromes

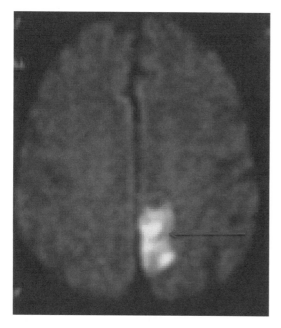

Fig. 8.5 Isolated precuneus infarct with obtundation, due to distal thrombosis of anterior cerebral artery callosmarginal branch occlusion

Visuospatial Dysfunctions

- Balint's syndrome—simultanagnosia, visual apraxia, visual ataxia
- Optokinetic reflex and optokinetic nystagmus (OKN), smooth pursuit/saccadic eye movement disorders

Attention and Consciousness Hubs (Precuneus and Claustrum)

- Minimally conscious states
- Anesthesia
- Transcendental states

TOP Junction Functions and Other Cross-Modal Sensory Abnormalities

- Autoscopy—seeing one's double often from a vantage point also referred to as an out-of-body experience
- Acquired synesthesia—cross-modal sensory experiences whereby a sensory stimulus of one modality triggers sensations in another modality. Color grapheme is most common
- Cotard's syndrome—denial of existence, a perceptual emotional derangement [13]
- Migraine phenomena—autokinesis (apparent movement of objects in the dark), inversion of two- and three-dimensional vision, body weight, and size perception disturbance [14]

Miscellaneous Conditions

- Parietal arm levitation—outstretched hands, spontaneous contralateral arm elevation [15]
- Poikilotonia—erratic and variable tone alteration from hypotonia to hypertonia [15]
- Parietal avoidance syndrome [16]
- Parietal wasting syndrome—rare occurrence of wasting of contralateral arm or leg attributed to parietal lobe lesions [16–18].
- Pain processing in parietal lobes and phantom limb pain
- Algodiaphoria—indifference to painful stimuli [17]
- Parietal seizures
- Conversion disorders

Left

- Anomias—word production, word selection, semantic
- Apraxias—ideomotor, melokinetic
- Gerstmann's syndrome—acalculia, agraphia, R/L orientation, finger anomia
- Angular gyrus syndrome—features of Gerstmann's plus alexia and anomia
- Autotopagnosia—inability to name or localize body parts (inferior parietal)

Right

- Visuoconstructive impairment
- Visuospatial impairment
- Aprosodias
- Neglect syndromes
- Anosognosias and related syndromes
- Geographic disorientation
- Allesthesias

- Aphonognosia—unable to recognize familiar voices
- Three-dimensional sense appreciation impairment
- Dressing apraxia
- Event timing
- Perception of sound movements

Right or Left Parietal Syndromes

Cortical Sensory Function and Deficits (Right or Left)

In addition to the self-evident agraphesthesia, abaragnosia, and astereognosis, less commonly appreciated disorders may include ahylognosia, an inability to distinguish texture of objects and a pseudoradicular parietal sensory impairment may be appreciated. Tactile agnosic patients are able to describe accurately the size, shape, contour, and position of an object but are however unable to recognize the object as an entity nor state what the function might be or represent. With loss of two-point discrimination testing, there is a relative sensitivity of the hand to distance of two points as determined by the "compass test." Most accurate are the finger tips and least in the palm of the hand, with finger-tip separation ability of two points (2–3 mm) and palm of hand up to 20 mm [16, 19].

Visuospatial Function

The optokinetic reflex is a normal feature that develops within the first few months of life consisting of a combination of a slow phase tracking the visual stimulus or smooth pursuits and fast, corrective saccades termed optokinetic nystagmus. It is tested by a rotating drum or by moving a strip with alternating white and colored areas. The reflex allows visual pursuit of objects in motion with the head stationary. The normal nystagmus may be impaired in disease states of the dorsal stream visual pathway or with cerebellar lesions [20].

Pain and the Parietal Lobes

Whereas BA 1/2/3 is referred to as the primary somatosensory cortex (SI), BA 43 or the parietal operculum is also referred to as the secondary somatosensory cortex (SII). This area is lateral to SI, occupies the dorsal part of Sylvian fissure, and is concerned with nociceptive stimuli processing. Investigative and case report studies have found that electrical stimulation of the parietal lobe in people that have phantom limb pain resulted in spontaneous pain [21]. A case report of spontaneous, intermittent pain of right arm associated with a left parietal subcortical glioma, more specifically the parietal operculum, resolved with surgical excision with resolution of pain [22]. Episodic pain associated with a tumor in the parietal operculum: a case report and literature review. SII and the posterior insular region are essential for normal pain and tactile perception. In people with parietal lobe epilepsy there may be an associated pain syndrome, described as a burning-type sensation in the contralateral hemibody or abdomen or in up to 23.6 % [23].

Attention and Consciousness Hubs (Precuneus and Claustrum)

Consciousness is viewed as an emergent property of the brain due to expression of but multiple constituent components, something that is not possible by only some of these components. An alteration of the conscious circuitry may present with coma, persistent vegetative state, minimally conscious states, or transcendental states. Key neurobiological or electrochemical systems for consciousness include:

- Claustrum
- The 40 Hz hypothesis; cortical neurons fire rhythmically every 20–30 ms
- Layer five pyramidal cells
- Precuneus [24]

The claustrum is thought to allow integration of diverse modalities of information that are regarded as key to perception, action, cognition, and cognition. Situated below the insula and just above the putamen, its neurons are particularly sensitive to timing of the inputs and hence being able to bind events dispersed in time. Most cerebral cortical areas project to the claustrum and it is likely that the claustrum is involved in cross-modal processing and enables experiences of objects and events in an integrated fashion. The neurochemical/anatomical basis may involve the gap junctions characteristic of the claustral interneurons (connexin proteins) that have a bidirectional, low-resistance communication between these neurons. These interneurons fire rhythmically and in synchrony in the 30–70 Hz range that may be the basis of integrating widespread cortical neurons. Hence a global integration of information in a rapid time frame is made possible by the claustrum and its proposed role in consciousness [25].

The precuneus has also been implicated to play an important role in the integration of tasks involving episodic memory, visuospatial functions, and self-consciousness. It is noteworthy that the posteromedial parietal area including the precuneus has the highest metabolism as measured by PET brain scanning during rest or when the default mode network is engaged. Contrariwise, precuneus hypometabolism has been documented in altered consciousness states such as anesthesia, persistent vegetative states, and minimally consciousness states. Rarely involved in stroke, an isolated distal anterior cerebral callosomarginal territory infarct of the precuneus was associated with somnolent state, resolving after several days (Fig. 8.5) [6].

The Temporo-Occipito-Parietal (TOP) Junction Functions

Integrates external tactile, visual, and proprioceptive sensory input with interoceptive information that allows an appreciation of spatial orientation

in three dimensions and balance. When affected by hypoxic ischemic encephalopathy during cardiac arrest, during migraine or bilateral stroke this may lead to autoscopic phenomena. These may also be induced by various prolonged sensory stimulation leading to a trance and transcendental state.

Synesthesia

Synesthesia occurs in 2–4% of the normal population, more common in artists and poets, and provides a neurological model that aids in the understanding of language, abstract thought, and metaphor. Furthermore, inferior parietal lobe damage is associated with loss of metaphor conceptualization. Synesthetes may arbitrarily associate a particular color with a particular letter or number (Fig. 8.6). Hupé's "palimpsest hypothesis" regards

Fig. 8.6 Synesthete number-color associations. Figure credit: Brang D, Ramachandran VS. Survival of the Synesthesia Gene: Why do people hear colors and taste words ? PLoS Biology 2011 Nov;9(11):e1001205. doi: 10.1371/journal.pbio.1001205

this extraneuronal circuitry that some have as a variant of Dehaene's neuronal recycling hypothesis, proposed for reading. This refers to the "recycling" of certain brain regions and circuits originally subserving other functions [26]. Sometimes regarded as an evolutionary spandrel, synesthetic experiences may amplify or boost detection and attention to critical stimuli in the environment. Neuroimaging data have supported inferior temporal lobe MRI brain scan DTI-FA value increases that represent increased white matter and connectivity in this region [27]. Synesthesia is associated with augmented sensory processing and multimodality sensory integration. Color-grapheme synesthesia (64.4%) is the most commonly encountered form but about 60 different synesthesia combinations have been described. The next most common form is time unit association such as month or day of the week with color synesthesia (22.4% with third most common being musical to color synesthesia (18.5%)) [28]. Testing can be performed by the validated synesthesia battery [29].

Parietal Seizures

Localization-related phenomena related to parietal onset may present with a subjective levitation prior to seizure onset and may be key to the parietal onset [30]. In addition to basic pins and needle type of presentations more complex ones may occur that may include a creeping type of the face or limb that slowly progresses to involve contiguous areas termed fourmillement.

Body image disturbances with a feeling of the distal part of a limb being shorter than the more proximal part movement, termed paraschematia [16].

Algodiaphoria and Asymbolia for Pain

These are likely within the same spectrum of disorders as algodiaphoria due to a channelopathy with the causative mutation in the *CLTCL1* gene which codes for CHC22 clathrin heavy-chain protein [31]. Described mostly with inferior parietal lobe lesions, there is sensibility for painful stimuli but an altered response or indifference occurs. Other terms have included agnosia for pain or pain apraxia, with the most appropriate being algodiaphoria.

Conversion Disorders

In the last 100 years a number of current-day neurological diagnoses were previously thought to be hysterical or conversion-type disorders. These have included amongst others:

Task-specific dystonia such as writer's cramp and
    musician's dystonia
Epilepsia partialis continua
Frontal lobe seizures
Phantom limb pain
Sensory trick in dystonia
Alexia without agraphia
Pure word deafness
Various atypical tremors
Geschwind-Gastaut syndrome
Foreign accent syndrome

In a study of conversion paresis by fMRI, identification of abnormal parietal supramarginal gyrus, precuneus, and prefrontal cortex underactivation was found, which was specific for conversion paresis [32].

Left Parietal Syndromes

In general there is a left parietal specialization for motor attention reflective of the left hemisphere dominance for temporal processing and item-based attention. This contrasts with right parietal R parietal dominance for both left- and right-sided spatial processing. Language, praxis, angular gyrus syndrome, Gerstmann's syndrome, reading, calculations of various types with consequent alexias, and acalculias with damage to the areas are all syndromes that have hub areas in the left parietal cortices.

Fig. 8.7 Right parietal lobe function—event timing. Adapted from Battelli L, Pascual-Leone A, Cavanagh P. The 'when' pathway of the right parietal lobe. Trends in Cognitive Sciences 2007;11:204–210

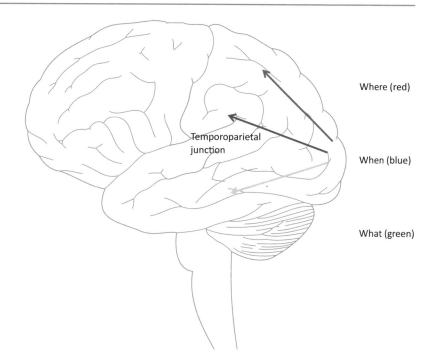

Where (red)

Temporoparietal junction

When (blue)

What (green)

Right Parietal Syndromes

For neglect syndromes, aprosodias, impairments, anosognosias, allesthesias, visuoconstructive and visuospatial dysfunctions, and xenomelia, see right hemisphere chapter.

Event Timing

The right inferior parietal lobe is involved in the evaluation of "when" information or the temporal sequencing or order, whether events occur simultaneously (Fig. 8.7) [33]. This has been demonstrated for example by transcranial magnetic stimulation focused to the right parietal region [33].

Sound Movement Perception

The evaluation of alteration in amplitude and delay of a sound are the informational cues for cerebral computation of movement of a sound. Functional brain imaging with fMRI and PET brain scanning has supported this process to be critically dependent on right parietal lobe [34].

Aphonagnosia

Aphonagnosia or phonagnosia refers to the ability to recognize familiar voices and is impaired in the people with right parietal lesions. Although usually associated with right parietal damage, it may also occur with left hemisphere parietotemporal lesions [35]. This is akin to the right hemisphere predominance for facial recognition and loss (prosopagnosia) [36, 37].

Parietal Lobe Function Testing: Basic Sensory, Secondary Sensory Association, Supramodal and Multimodality Testing

In addition to the language- and praxis-related tests of the left hemisphere and the visuoconstructive and visuospatial related to the right hemisphere tests, more specific tests related to the basic and multimodality processing capabilities of the parietal lobes include the following:

Establish basic sensory abilities (BA 1, 2, 3)
Establish cortical sensory abilities (BA 5)

1. Interlocking of both index fingers and thumbs forming a rings

2. Little fingers hooked together

3. Interlacing of the fingers of both hands

4. Index and middle finger of left hand inserted between middle and ring fingers of right hand

Fig. 8.8 Using simple bedside tests to test for multimodal parietal function: The interlocking finger test. Sequences after: Moo LR, Slotnick SD, Tesoro MA, Zee DS, Hart J. Interlocking finger test: a bedside screen for parietal lobe dysfunction. Journal of Neurology, Neurosurgery & Psychiatry 2003; 74(4): 530–532. Image credit: Shutterstock.com

Fig. 8.9 Bouba-kiki effect. Figure adapted from Köhler W. Gestalt Psychology, 2nd ed. Liveright, New York, 1947 and Ramachandran VS, Hubbard EM. Synaesthesia: A window into perception, thought and language. Journal of Consciousness Studies 2001; 8: 3–34

Optokinetic nystagmus (rotating drum or strip) BA 7

Interlocking finger test (screening test for parietal function) (Fig. 8.8) [38]

Inferior Parietal Lobe Function Test

1. Boston Naming Test (BNT) for assessment for anomias—words are multimodality (parietal) symbols for ideas, objects

2. Using the cueing option of the BNT), a differentiation between anomias and semantic memory loss (anterior temporal lobe) may be made

3. Metaphor testing: This can be tested for example by the proverb sub-test of the DKEFS which provides a scoring system and normative data. DKEFS proverbs include the following:
   - You can't judge a book by its cover.
   - Don't count your chickens before they are hatched.
   - Rome wasn't built in a day.
   - Too many cooks spoil the soup.
   - People who live in glass houses shouldn't throw stones.
   - An old ox plows a straight row.
   - A small leak will sink a large ship.
   - No bread is without a crust [39].

4. Cross-modal testing by the bouba kiki phenomenon

This refers to the mapping of vocal sounds to the visual features of a particular object that is considered to be related, that is, non-arbitrary. First described by Köhler in 1929 [40], he noted that there was a strong preference for association of the jagged shape with "takete" and the rounded shape to "baluba" as it was first called. Ramachandran later devised an experiment amongst college students and Tamil people in India and revealed that between 95 and 98 % chose the wavy shape association with "bouba" and the jagged shape with "kiki" (Fig. 8.9) [41].

Superior Parietal Lobe Function Test

1. Visuospatial—testing for Balint's, optokinetic reflex
2. Synesthesia—multimodal processing
3. Praxis (see Chap. 7)

References

1. Wynn T, Coolidge FL. The implications of the working memory model for the evolution of modern cognition. Int J Evol Biol. 2011;2011:741357. doi:10.4061/2011/741357.
2. Bruner E. Morphological differences in the parietal lobes within the human genus. Curr Anthropol. 2010;51:77–88.
3. Bruner E. Geometric morphometrics and paleneurology: brain shape evolution in the genus Homo. J Hum Evol. 2004;47:279–303.
4. Orban GA, Caruana F. The neural basis of human tool use. Front Psychol. 2014;5:310. doi:10.3389/fpsyg.2014.00310.
5. Bruner E. The evolution of the parietal cortical areas in the human genus: between structure and cognition. In: Broadfield D, Yuan M, Schick K, Toth N, editors. The human brain evolving. Gosport, IN: Stone Age Instiute Press; 2010.
6. Cavanna AE, Trimble MR. The precuneus: a review of its functional anatomy and behavioural correlates. Brain. 2006;129:564–83.
7. Bruner E, Jacobs HI. Alzheimer's disease: the downside of a highly evolved parietal lobe? J Alzheimers Dis. 2013;35(2):227–40. hub failure hypothesis.
8. Dehaene S. Number sense. Oxford: Oxford University Press; 2011.
9. Andersen RA. Visual and eye movement functions of the posterior parietal cortex. Annu Rev Neurosci. 1989;12:377–403.
10. Culham JC, Kanwisher NG. Neuroimaging of cognitive functions in human parietal cortex. Curr Opin Neurobiol. 2001;11:157–63.
11. Todd JJ, Marois R. Capacity limit of visual short-term memory in human posterior parietal cortex. Nature. 2004;428(6984):751–4.
12. Kravitz DJ, Saleem KS, Baker CI, Mishkin M. Visuospatial processing – new neural framework. Nat Rev Neurosci. 2011;12:217–30.
13. Ramirez-Bermudez J, Aguilar-Venegas LC, Crail-Melendez D, Espinola-Nadurille M, Nente F, Mendez MF. Cotard syndrome in neurological and psychiatric patients. J Neuropsychiatry Clin Neurosci. 2010; 22(4):409–16.
14. Jürgens TP, Schulte LH, May A. Migraine trait symptoms in migraine with and without aura. Neurology. 2014;82(16):1416–24.
15. Ghika J, Ghika-Schmid F, Bogousslasvky J. Parietal motor syndrome: a clinical description in 32 patients in the acute phase of pure parietal strokes studied prospectively. Clin Neurol Neurosurg. 1998;100(4): 271–82.
16. Critchley M. The parietal lobes. London: Hafner Press; 1953.
17. Sandyk R. Spontaneous pain, hyperpathia and wasting of the hand due to parietal lobe haemorrhage. Eur Neurol. 1985;24:1–3.
18. Maramattom BV. Parietal wasting and dystonia secondary to a parasagittal mass lesion. Neurol India. 2007;55(2):185.
19. Luria AR. Higher cortical functions in man. London: Tavistock; 1966.
20. Dix MR. The mechanism and clinical significance of optokinetic nystagmus. Review. J Laryngol Otol. 1980;94(8):845–64.
21. Lewin W, Phillips CG. Observations on partial removal of the post-central gyrus for pain. J Neurol Neurosurg Psychiatry. 1952;15(3):143–7.
22. Potagas C, Avdelidis D, Singounas E, Missir O, Aessopos A. Episodic pain associated with a tumor in the parietal operculum: a case report and literature review. Pain. 1997;72(1–2):201–8.
23. Duncan GH, Albanese MC. Is there a role for the parietal lobes in the perception of pain? Adv Neurol. 2003;93:69–86.
24. Koch C. Consciousness. Cambridge, MA: MIT Press; 2012.
25. Crick FC, Koch C. Phil Trans R Soc B. 2005; 360:1271–9.
26. Hupé JM. Synesthesia as a neuronal palimpsest. Med Sci (Paris). 2012;28(8-9):765–71.
27. Brang D, Ramachandran VS. Survival of the synesthesia gene: why do people hear colors and taste words? PLoS Biol. 2011;9(11), e1001205. doi:10.1371/journal.pbio.1001205.
28. Safran AB, Sanda N. Color synesthesia. Insight into perception, emotion and consciousness. Curr Opin Neurol. 2015;28:36–44.
29. Carmichael DA, Down MP, Shillcock RC, Eagleman DM, Simner J. Validating a standardised test battery for synesthesia: does the synesthesia battery reliably detect synesthesia? Conscious Cogn. 2015;33:375–85.
30. Gibbs SA, Figorilli M, Casaceli G, Proserpio P, Nobili L. Sleep related hypermotor seizures with a right parietal onset. J Clin Sleep Med. 2015;15:953–5.
31. Nahorski MS, Al-Gazali L, Hertecant J, Owen DJ, Borner GH, Chen YC, et al. A novel disorder reveals clathrin heavy chain-22 is essential for human pain and touch development. Brain. 2015;138(8):2147–60.
32. Van Beilen M, de Jong BM, Gieteling EW, Renken R, Leenders KL. Abnormal parietal function in conversion paresis. PLoS One. 2011;6(10), e25918. doi:10.1371/journal.pone.0025918.

33. Battelli L, Pascual-Leone A, Cavanagh P. The 'when' pathway of the right parietal lobe. Trends Cogn Sci. 2007;11:204–10.

34. Griffiths TD, Rees G, Tees A, Green GRA, Witton C, Rowe D, et al. Right parietal lobe is involved in the perception of sound movements in humans. Nat Neurosci. 1998;1:74–9.

35. Van Lancker DR, Kreiman J, Cummings J. Voice perception deficits: neuroanatomical correlates of phonagnosia. J Clin Exp Neuropsychol. 1989; 11(5):665–74.

36. Biederman I, Herald S, Xu X, Amir O, Shilowich B. Phonagnosia, a voice homologue to prosopagnosia. J Vis. 2015;15(12):1206. doi:10.1167/ 15.12.1206.

37. Van Lancker DR, Cummings JL, Kreiman J, Dobkin BH. Phonagnosia: a dissociation between familiar and unfamiliar voices. Cortex. 1988;24(2):195–209.

38. Moo LR, Slotnick SD, Tesoro MA, Zee DS, Hart J. Interlocking finger test: a bedside screen for parietal lobe dysfunction. J Neurol Neurosurg Psychiatry. 2003;74(4):530–2.

39. Delis DC, Kaplan E, Kramer JH. D. Delis Kaplan executive function system. San Antonio, TX: Pearson; 2001.

40. Köhler W. Gestalt psychology. New York, NY: Liveright; 1929.

41. Ramachandran VS, Hubbard EM. Synaesthesia: a window into perception, thought and language. J Conscious Stud. 2001;8:3–34.

With almost universally preserved language, right hemisphere (RH) lesions reveal a conundrum of cognitive disorders. However, assessment beyond sensorimotor deficits hardly surfaces in quantitative scales, although the NIH stroke scale scores neglect syndromes. The absence of a more detailed assessment of right hemisphere function obviates the appreciation of an extensive panoply of cognitive functions that may be improving or worsening during a particular neurological disease process. The common usage of "dominant" versus "non-dominant" hemisphere is completely biased towards language. The following illustrates how the right hemisphere is probably dominant for more (at least measurable by us thus far) items than the left hemisphere.

Simply stated, right hemisphere function may be crystallized into two principal properties, its dominance for both attention and emotion. Most of the disorders listed in the table below can be viewed as derivatives of these two major faculties. These syndromes may be considered as those that occur more commonly and those occurring less commonly, in addition to subtypes of the major conditions (Tables 9.1 and 9.2).

Evolutionary Insights into Right Hemisphere Disorders and Syndromes

Attention and emotional processing are the two principal systems that are right hemisphere dominant. Attention refers to the cerebral process that selects particular stimuli for further evaluation and not others, because of limited processing ability. It is a critical component of higher cortical functioning and prone to many alterations in health and disease. Aptly described by Koch, "Attention is evolution's answer to information overload. No brain can process all incoming information" [6]. The spectrum of the various levels of consciousness that may be considered is depicted in Fig. 9.1.

Clinical and animal studies point two interlinked attentional systems:

1. A general domain or domain-independent attentional control system that has a cortical component and a brainstem neurotransmitter modularity component
2. A modality-specific attentional system for objects, space, and faces for example

© Springer International Publishing Switzerland 2016
M. Hoffmann, *Cognitive, Conative and Behavioral Neurology*,
DOI 10.1007/978-3-319-33181-2_9

Table 9.1 Hemisphere dominance [1–5]

Right hemisphere dominance
• Attention
• Prosody
• Spatial construction
• Body image
• Melody
• Emotion
Left hemisphere dominance
• Language
• Praxis
• Analytical/mathematical
• Temporal sequencing of stimuli

Table 9.2 Right hemisphere syndromes

More common right hemisphere lesion disorders
Attentional disorders
Neglect syndromes
Anosognosias for hemiparesis
Visuospatial dysfunction
Aprosodias
Less common syndromes recognizable after right hemisphere lesions include:
Neglect
Motor neglect
Sensory neglect
Spatial neglect
Egocentric—neglect of their own space or personal space
Allocentric neglect, peripersonal, extrapersonal
Object-centered neglect
Environment-centered neglect
Representational neglect
Personal representational neglect
Anosognosias
Anosognosia for hemiparesis or hemiplegia
Anosodiaphoria
Somatoparaphrenia
Misoplegia
Asomatognosia
Autotopagnosia
Allesthesia
Allochiria
Synchiria
Allokinesia
Supernumerary phantom limb

(continued)

Table 9.2 (continued)

Xenomelia and apotemnophilia
Rubber hand illusion
Visuospatial function
Visuoperceptual
Visuomotor
Aprosodias
Motor
Sensory
Global
Transcortical subtypes
Other uncommon right hemisphere-related syndromes
Apraxia of eye opening
Response to next patient stimulation
Topographical disorientation (geographical disorientation) and planotopokinesia
Empathy alteration (loss of empathy or hyperempathy)
Gourmand syndrome
Hypergraphia, graphomimia, graphomania, and echographia
Nosagnosia overestimation
Delusional misidentification syndromes
Musical ability alterations (discussed under newly acquired cultural circuits chapter)

Together this expansive circuitry subserves the different attention-related processes that include:

- Overall arousal and responsiveness to the environment
- Selective attention whereby some stimuli are allocated priority over others
- Orientation to a stimulus or stimuli by the sensory organs (vision, tactile, auditory, olfactory)
- Sustained attention that provides ongoing vigilance
- Divided attention, the capability of attending to a number of different environmental events [7]

The evolutionary origins into these attentional control systems may be traced very far back, at least to the split prior to the invertebrates and vertebrates. The metazoans, inhabiting the Earth about 700 mya, would have benefited from a cognitive control system or process that developed hierarchies in action sequences during foraging.

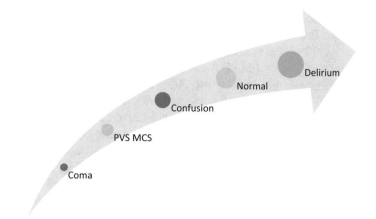

Fig. 9.1 Levels of consciousness may vary from coma to hyper-alertness. *PVS* persistent vegetative state, *MCS* minimally conscious state

The challenge posed by dynamic foraging in these early animals was likely the initiating force that required a sustained persistence of activities to achieve this. This overlaps considerably with the concept of working memory, which is sometimes also referred to as "working attention" by Baddeley [8]. Dopamine has been implicated in this mechanism described by Hills as the "persistence in action." This has been recognized in the vertebrate striatum where dopamine mediates the modulation of several intracellular G protein receptors, inducing both hyper- and hypoexcitability that transmits a downstream cascade triggering further neurotransmitters, most notably glutamate. The overall effect of dopamine, by these mechanisms, is understood to elevate the signal-to-noise ratio and result in the sustained persistence of a particular activity [9]. This may be the origin of the bottom-up (as opposed to top-down) attentional and arousal mechanisms in vertebrates.

The second component of the domain general component, the top-down influence of predominantly frontal cortex, occurred later in mammalian and subsequent anthropoid evolution. The development of the six-layer cortex and in anthropoids, the development of the granular cortical areas in the brain, and elaboration and extension of the cortical top-down component occurred.

The granular cortical development amongst anthropoids, specifically the medial prefrontal cortex, allowed the primate to choose amongst several stimuli and make a choice based on external signals from the sensory organs and cortices as well as from internal signals based on prior events and their outcome and motivations. The medial prefrontal cortex (MPFC) also allows choice when no external cue prime the organism at the time choice is required—hence sometimes the choice of action depends on the learnt outcome of an action from previous, often one single occurrence. Both direct and indirect action choice is therefore the premise of the MPFC. These latter contributing influences are due to the MPFC connections with the amygdala, hippocampus, and medial premotor areas [10]. Furthermore, connections to areas BA 9 and 10, the frontopolar cortex that is involved in autobiographical event recollection, and generate rules for subsequent actions, assigning outcomes to goals, may be represented. This is supported by both clinical lesion studies and direct cell recording in monkeys [11].

Being able to ignore many irrelevant stimuli is an important function of the anterior cingulated gyrus and is for example disrupted in psychiatric disease states such as schizophrenia. The anterior cingulate cortex (ACC) has extensive connections with the auditory association areas. It is notable that the auditory has the most connections with the prefrontal cortex (PFC) of any sensory modality and connects to every component of the PFC; the "auditory modality is spatially unconstrained and has a unique role in attention" [12]. The ACC also has particularly extensive connections to the amygdala (emotions) and hippocampus (memory) as well as the frontopolar cortex BA 10.

ACC hypofunction effects on the dorsolateral prefrontal cortex (DLPFC) are associated with working memory impairment; ACC hypofunction effects on BA10 are associated with disordered thought process, such as is evident with schizophrenia. Barbas et al. have suggested that "auditory signals are symbolic representations of organized thought." The extensive auditory connections of the ACC as well as the connections to the DLPFC (BA 9 and 46) and frontopolar cortex (FPC, BA 10) are mechanisms whereby the organization, coordination of thought, and the complexity of high-level cognitive processing are accomplished, including emotive and language information transmission. In addition the ACC is engaged when either physical or mental pain (depression) is present. A common clinical example is the inability to integrate emotional

memories with context in post-traumatic stress disorder (PTSD), and this may be due to the disturbance of the circuitry linking the ACC, medial temporal lobe, and amygdala [12]. The auditory modality has unrestricted spatial characteristics and unique connections to every PFC subcomponent, but particularly to the MPFC and FPC (BA10). The BA 10 connections are reciprocated to the association areas of the auditory cortex (Fig. 9.2). Interestingly the ACC (BA 32) deep cortical layers connecting to the upper auditory association cortex (AAC) and the FPC projections form the upper cortical layers impinged on mostly the middle cortical layers of the AAC considered to be associated with excitatory input and the terminations of the more superficial layers having a modularity role. The suppression of stimuli that may interfere with a task at hand is also important that leads to an increased error rate. The inhibitory control of the certain PFC pathways may have

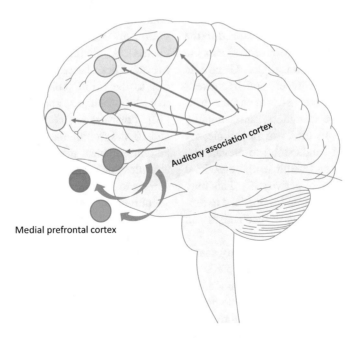

Auditory association cortex

Medial prefrontal cortex

Fig. 9.2 The primate auditory association cortex is unique in that it connects to all major prefrontal cortical Brodmann areas (BA): information based on research in rhesus monkeys. BA8—*pink*, BA9,46—*green*, BA12—*orange*, BA10—*yellow*, BA13—*purple*, *BA32—red*, BA25—*blue*. Figure based on information from Barbas H. Architecture and cortical connections of the prefrontal cortex in the rhesus monkey. Advances in Neurology 2000;84:87–110 and Barbas H, Bunce JG, Medalla M. Prefrontal pathways that control attention. In: Stuss DT, Knight RT. Principles of Frontal Lobe Function 2nd ed. Oxford University Press, Oxford 2013 and Barbas H, Zikopoulos B, Timbie C. Sensory pathways and emotional context for action in primate prefrontal cortex. Biol Psychiatry. 2011;69(12):1133–9

many different effects. For example, a decrease in DLPFC function and an increase in the AAC may lead to a noisy environment interfering with the ability to conduct a conversation for example [13].

Emotion

The other key RH function is emotion interfaces with the attentional circuitry. An event with particular emotional significance amplifies the salience of a stimulus that then leads to an attentional focus on that stimulus.

Similar to the posterior part of the orbitofrontal cortex, the ACC areas 32 and 25 (posteromedial PFC cortex) part of BA 24 have extensive amygdala connections. The ACC also has connections with the hypothalamus, basal forebrain, and brainstem components that are involved with autonomic control that allow the linking of emotionally significant events to deploy attentional resources [14]. Guidance by past experiences is enabled by the strong connections of the ACC with the hippocampal formation, particularly the CA1, subiculum areas, and entorhinal and parahippocal areas, which allow the ACC to play a key role in evaluating signals based on long-term episodic memory. In this way emotionally important stimuli are not only remembered, but initiate attention. In addition, pain captures attention and thalamic nuclei; in particular intralaminar and midline nuclei have particularly strong connections to the ACC and hence ACC activation. This activation occurs in response to not only pain itself but also viewing examples or images of situations associated with pain [15].

Classification of Right Hemisphere Functions and Related Lesion-Associated Syndromes

It may be readily appreciated that some syndromes notably increase in artistic talent and various forms of savant presentations are due to network disinhibitory effects. Similar to other lobes regions of the cortex, both topographical hypofunction and hyperfunction may be evident

as well as diaschisis-related syndromes or hodological hypofunction and hyperfunction.

Attention

There is right hemisphere dominance for both attention and arousal. The principal components of right hemisphere dysfunction can be related to disorders of attention. For example neglect is a disorder of spatial attention, which is also more profound after right versus left brain injury. The expansive cerebral network subserving spatial attention involves the posterior parietal cortex, the cingulate gyrus, frontal eye fields, subcortical components of thalamus, striatum, and superior colliculus and interacts with the ascending reticular activating system (Fig. 9.3). These may be depicted in a three-tier system with the frontal subcortical parietal cortical network referring to the top of the hierarchy, the mid-component being represented by the circuitry subserving various sensory modalities and leading to domain-specific or modality-specific deficits such as the neglect syndromes and the inferior component, the ascending neurotransmitter systems, providing modulation of various kinds but a key role being arousal (Fig. 9.4) [7].

The top-down component of the attentional circuitry may cause both domain-general and domain-specific attentional abnormalities. Frontal, posterior parietal, and medial temporal lesions can produce the common clinical syndromes such as the acute confusional state and various neuropsychiatric syndromes such as PTSD, schizophrenia, and autism. Other attentional disturbances include the various types of modality-specific neglect syndromes and many of the deficits of neuropsychiatric syndromes seen with PTSD, depression, and autism.

Impaired Prefrontal Attentional Control in Selected Neurological and Neuropsychiatric Conditions

The attentional circuitry of the DLPFC, ACC, and AAC is implicated in excitatory and inhibitory balance that is disrupted in schizophrenia for

Fig. 9.3 Attentional circuits: frontal subcortical, frontal eye field (FEF), posterior parietal (PP) region, ascending aminergic neurotransmitter activating system (ARAS). *A* amygdala, *C* cingulate, *OFC* orbitofrontal, *DLPR* dorsolateral prefrontal cortex, *DS* dorsal striatum, *VS* ventral striatum, *GP* globus pallidus, *T* thalamus

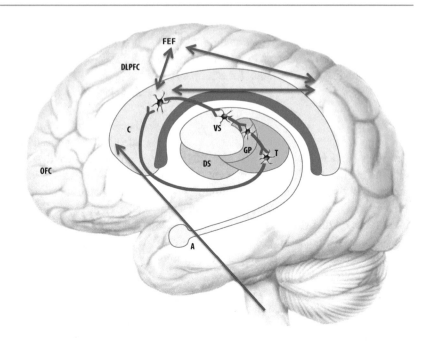

Fig. 9.4 Attentional circuits: top down, bottom up, middle modality specific

Clinical syndrome examples of lesions affecting these areas:

Akinetic mutism

Spatial neglect syndromes

Somnolence, coma

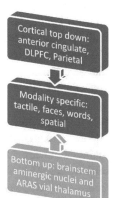

example. Here, ACC hypofunction with reduced ACC laminar-specific GABA neurons translates into excitatory glutamate transmission in ACC and DLPFC which explains the working memory impairment due to the DLPFC impairment, auditory hallucinations, and the dysmemory (episodic) due to hippocampal-to-ACC connectivity disruption. The ACC hypofunction and its connections to the FPC BA 10 that impair multitasking ability and the sequencing disability help explain the disordered thinking seen with schizophrenia [16].

Overall, the ACC plays a pivotal role in emotional communication, speech, language, and suppression of intrusive thoughts [17].

ACC-to-DLPFC and AAC connectivity is also abnormal in autism, probably genetic mutation based whereby inhibitory projections to the DLPFC mediate "noise reduction" which is amplified in those with autism and opposite to schizophrenia. In autism this translates into repetitive behavior due to perseveration within the attentional realm [18]. The ACC is also implicated in depression and PTSD. In the latter the increased arousal, dissociated behavior, and both hyperamnesias and amnesias are attributed to circuitry impairment linking the medial temporal region, amygdala, and ACC resulting in failure of appropriate integration of emotionally laden memories with the appropriate context [19].

The ACC impairment results in circuitry that may augment or decrease one or more inhibitory systems that impact the DLPFC. In schizophrenia, this is weakened and in autism strengthened. In depression there is a hyperactivity within the ACC adjacent to the rostral corpus callosum that may be causing increased internal focusing on negative thoughts and at the same time being unable to move attention to another stimulus or stimuli [20].

Altered Mental Status and Acute Confusional States

Also commonly referred to as delirium, ICU psychosis and clouding of consciousness of the cardinal feature is a disturbance of the attentional circuitry that is characterized by a fluctuating course. It is a state of altered arousal (hyperactive, hypoactive, or mixed), attentional, hallucinations, delusions, and disturbed circadian sleep–wake cycle. Epidemiological clinical studies reveal that delirium may be present in people presenting to emergency rooms in up to 10–16%, in hospitalized patients 14–56%, and in the ICU setting up to 82% [21]. It is a marker of underlying cognitive impairment as well as risk factor for subsequent mortality in the year post-delirium onset with people older than 65 having a 40% 12-month mortality rate with odds ratio of 1.6–2.0 versus controls. The main clinical identifiable causes include medication effects, infections both outside and within the neural axis, and vascular and metabolic derangements. Medications that have anticholinergic effects (opiates, benzodiazepines, antihistamines, beta blockers, dopamine agonists, cephalosporins, fluoroquinolones, antiseizure agents, and tricyclic antidepressant and many other neuropsychiatric drugs) are likely the most commonly implicated factor, with opiates and benzodiazepines the most common culprits. For example, an additional three medications during a hospitalization correlated with a three-fold delirium increase. The three most common are meds with anticholinergic properties, opiates, and benzodiazepines [22, 23].

Pathophysiology: Excess of Dopamine and Deficiency of Acetylcholine

Although first recognized as a marker of failing cognitive reserve mechanisms include anticholinergic activity, neuroinflammation, and glial activation with previous prior dementia; other brain pathologies and age-related cognitive impairment are the main predisposing factors for delirium that lead to network disruption, synaptic loss, and microglia activation. HLA-DR and CD68, both markers of microglial activation and pro-inflammatory IL-6 cytokines and GFAP (astrocyte activity marker), have been implicated pathophysiologically [24]. From a neuroimaging point of view leukoaraiosis and white matter integrity have been implicated and related to cognitive impairment in the year following the onset of delirium [25]. At a neurotransmitter level imbalances of the frontal cortical subcortical circuitry and their Ach, DA, and GABA are generally implicated. Although DA and Ach may be the final common pathophysiological mechanism NT interaction is complex and GABA-A receptor agonism for example inhibits dopamine release and with GABA agonist withdrawal a rebound hyperdopaminergic state may ensue with the dopamine elevation causing delirium which may be treated with antagonists that increase Ach and block dopamine [26].

Neglect Syndromes

Neglect syndromes may be defined as an altered or impoverished response when the left limbs or body are stimulated in the context of normal elementary neurological function (vision, hearing, tactile senses, motor system) and no significant aphasia.

Neglect syndromes represent a domain-specific attentional disorder and therefore are usually due to right hemisphere lesions, less often left hemisphere lesions. A clinical classification may be conceived as follows:

1. Motor neglect and related disorders
   Akinesia
   Motor extinction
   Allokinesia
   Impersistence
   Horizontal, vertical, angular line bisection impaired
2. Sensory neglect (inattention)
   Domains: visual, tactile, auditory neglect syndromes
   Spatial or personal
   Object centered
3. Asomatagnosia, anosognosia, and related disorders
4. Allesthesia
5. Allocentric (object)
6. Egocentric neglect (personal)
7. Neglect for items close up and those far away

Motor Neglect

With akinesia or hypokinesia, there is a failure of intent or intentional neglect that may pertain to the limb, eyes, or the head with relatively intact motor power or only mild weakness. When there is association with abulia or hypobulia forms and endogenous form and if it is in response to environmental stimuli it is termed exogenous akinesia or hypokinesia. The sudden call to motion or kinesis paradoxica occurring in Parkinson's in response to an alarming environmental stimulus or danger is an example of a sudden reversal of the endogenous form of akinesia [27]. Another subtype of akinesia termed motor extinction was described by Valenstein and Heilman whereby an instruction to move both arms for example at the same time, a contralesional akinesia, is noted, but not with each on its own [28]. The instruction to perform a par-

ticular act may be failed due to abandoning a sequence of writing, verbal narrative, and recitation of lists for example due to a motor impersistence. This may involve a limb posture or movement, the eyelids, the eyes, or a leg action. Allokinesia is akin to the sensory-based condition termed allesthesia, where the instruction to move one specified limb is not executed with, and instead the opposite limb is moved [29].

Testing includes observation of spontaneous movements of the person with respect to limb and axial musculature, verbal discourse, and scanning of the immediate environment for akinesia and abulia states. Motor neglect testing can be accomplished by line bisection tasks that should be about 10 cm long and in the horizontal, vertical, and angular (45°) planes. Cancellation tasks such as shapes, letters, or numbers can also be employed.

Sensory Neglect

The establishment of the degree of limb or hemisensory, elementary neurological deficit first needs to be ascertained and if profound hemianesthesia or hemianopia is present then sensory neglect testing is not possible. Thereafter bilateral simultaneous stimulation, with eyes closed, can be performed for tactile (touch both forearms at once) or gentle finger rubbing next to both ears can be performed. Hence one may have unimodal, bimodal, or trimodal hemineglect syndromes [30].

Spatial Neglect

Often observed during the instruction to draw items such as a flower, cube, house, or person, such patients may ignore parts of the picture or diagram that are in the contralateral space. During writing tasks a word, the page, or even the keyboard may be involved in the neglect syndrome. For example while writing, the left part of a word may be missing which is termed neglect paralexia; if only the right side of a page may be written up or only the right side of a keyboard may be used it is termed neglect paragraphia. These patients may fail to read the left part of a word, which is called neglect paralexia. They may write only on one side of the page or use only the right side of a typewriter that has been called neglect paragraphia [31].

Hemispatial Neglect May Pertain to Many Different Types and Things That Are Neglected (Fig. 9.5)

1. Egocentric—neglect of their own space or personal space
2. Allocentric
   • Peripersonal space neglect
   • Extrapersonal or reaching space neglect
3. Environmental neglect
4. Representational neglect—neglect for items recalled from memory

When testing for neglect using line bisection tasks, it may become apparent that some patients have not only a horizontal neglect or bias to the right but also an altitudinal or vertical neglect [32, 33]. Even a diagonal neglect or radial neglect syndromes have been described and the three in combination may also be possible [34]. Furthermore there may be a horizontal neglect for stimuli that are close to the body or only to those that are distant to the person. The converse may also be found where far-space neglect is present but not near-space neglect [35]. Halligan and Marshall described the neglect for near but

not far space in a man with right parietal stroke. Spatial maps for both peripersonal and extrapersonal space were first reported in monkeys. These may also be referred to as the grasping or reaching distance as opposed to the throwing or walking distance [36].

At least three different reference frames may be affected by neglect syndromes after right hemisphere lesions. These include viewer or retinotopic-centered, object-centered or allocentric, and environment-centered neglect.

Object-centered neglect refers to individual target-associated features whereby the left side of targets is relatively neglected. Rapcsak et al. tested patients with a page full of triangles, some of which had a gap in one of the lines on the left and others a gap on the right side of the triangle. People with object-centered neglect if asked to do a modified cancellation task of selecting triangles with gaps on one side select or cancel many more triangles with right-sided gaps as opposed to left-sided gaps [37, 38]. In addition the processing of the place-value information pertaining to multi-digit numbers may also be impaired in neglect patients [39].

Environment-centered neglect refers to the relationship between the environment and body and may be revealed by right parietal lesions. Measuring left-sided neglect in the upright position and while lying on the right side helps differentiate body-centered and environmental-centered neglect. This may also be tested by

Fig. 9.5 Neglect subtypes

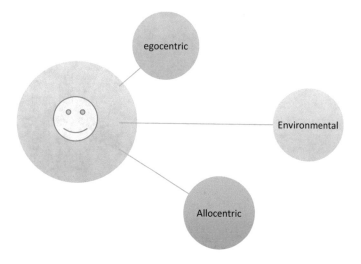

measuring visual reaction time in patients to right and left visual field stimulation and the tilting the head to the right or left through 90° to ascertain gravitational effects. The degree of attentional deficit was measured by the reaction times to the stimuli and attending to the left stimuli was relatively impaired to both retinal and gravitational reference frames [40]. In a particularly dangerous variety of spatial neglect syndrome, one of the author's patients who had a near-complete recovery after a large right hemisphere stroke secondary to right internal carotid dissection and subsequent occlusion left him with spatial neglect. He had been allowed to resume driving but his spouse became aware that when stopping at a 4-way stop, he would only look for cars approaching from the right-hand side, never the left side [41].

Neglect may occur for both the left side of person's view or the left side of the objects they are viewing. Viewer-centered neglect refers to the premise that an object may be appreciated both by the location and by its intrinsic spatial components. These object- or viewer-centered frames of reference may also be dissociated by right parietal lesions [42–44].

Representational Neglect

This type of neglect may be conveyed when recalling a scene from memory; right hemisphere lesion patients may omit details of the scene that were present on the left side of the image. This was reported by for example Bisiach and Luzzatti whose patients did not recall details of the left side of an imagined scene in the city of Milan, Italy, on emerging from the Milan cathedral that borders on the city square. Bisiach and Luzzatti questioned two people with neglect what they would see if standing on the steps of the Milan cathedral looking out onto Milan's very spacious city square, the Piazza del Duomo. Most of the features mentioned by both people were those on the right side from their point of reference with few items on the left side being noted (Fig. 9.6) [45–47].

Fig. 9.6 The Piazza del Duomo or Cathedral Square is the focal point of Milan and is dominated by the Milan Cathedral. Image with permission: Artist Catarina Belova/Shutterstock.com

Personal representational neglect refers to a relative impairment or inability to identify pictures of left-hand side compared to right-hand side [48, 49].

Bedside Testing May Be Performed by Line, Cancellation, and Object-Centered Cancellation Tasks

1. Line bisection tasks in the horizontal vertical and radial (45°) domains
2. Cancellation tasks including letters, numbers, shapes, object centered [50]
3. Drawing tasks; spontaneous, drawing a clock, drawing a person
4. Catherine Bergego scale—a semiquantitative behavioral assessment

Catherine Bergego scale is a ten-item behavioral scale, which may be used for the assessment of both anosognosia and neglect. Reliability and validity testing has been reported and confirmed by Azouvi et al. who found that it is more sensitive than pen and paper tests for neglect. The ten items are each scored from 0 to 3 (no neglect, or mild moderate, or severe neglect) yielding a total score out of 30 [51, 52].

Pathobiology of Neglect: Brain Lesions Associated with Neglect Syndromes

Mesencephalic reticular formation [53]
Thalamus [54].
Striatum—pulvinar [55].
Right inferior parietal lobe [56]
Right dorsolateral prefrontal cortex [57]
Cingulate gyrus [57]

Clinical Importance of Neglect

From clinical cases and case series some number of concepts about the neglect and the role of the right hemisphere have evolved:

Spatial attention is the fundamental concept underlying clinical neglect.

Although neglect may be present in association with a left hemisphere lesion, severity and duration tend to be less than with right hemisphere lesions.

There appears to be right hemisphere dominance for both attention and arousal.

The right hemisphere is responsible for attentional attributes to both hemispheres whereas the left hemisphere directs attention only contralaterally.

There are sensory, motor, and memory domains to attention resulting in spatial attentional or sensory neglect, spatial intention, or motor neglect and memory-related neglect syndromes (representational neglect) [44].

Right hemisphere lesions, most often middle cerebral artery territory stroke, are the most common causes of clinical neglect syndromes. They usually keep company with the other typical right hemisphere syndromes such as visuospatial dysfunction, anosognosia, and aprosodias (Fig. 9.7). Aprosodias and

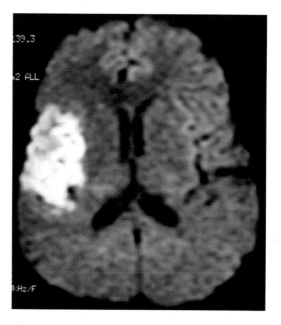

Fig. 9.7 Right middle cerebral artery branch infarction with trimodal hemineglect, motor neglect, visuospatial dysfunction, mild hemi-sensorimotor deficit, transient motor aprosodia, and anosognosia

anosognosias are often more transient conditions in cerebrovascular conditions whereas neglect may linger for much longer and has been correlated with significant factor in subsequent functional disability. Buxbaum et al. used a clinical measure, the Functional Independence Measure and Family Burden Questionnaire, to gain an appreciation of neglect subtypes in 166 right hemisphere stroke patients. Neglect was the most important predictor of adverse outcome and longer hospitalization. Overall neglect was present in almost half (48 %) and neglect subtypes included peripersonal neglect (27 %), perceptual neglect (21 %), motor neglect (17 %), and personal neglect (1 %) [49]. The prognosis differs according to these subtypes. In general, visual neglect recovers relatively rapidly (mean 8–9 weeks) in one study compared with the recovery of homonymous hemianopia and hemiparesis in one study [58].

Mechanisms Underlying Neglect and Related Disorders

Impaired Spatial Working Memory Hypothesis

This theory is supported by the finding of impairment in head and eye movements in the mapping of spatial locations and impaired gaze shifts that remap spatial locations already attended to (Fig. 9.8) [59].

Attentional Arousal Hypothesis

First proposed by Heilman and Valenstein it emphasizes the importance or emotional valence of incoming information which is triaged [57].

An Abnormal Exaggeration of the Troxler Effect?

The Troxler fading effect refers to neurons adapting to a stimulus and after sometime the stimulus is ignored, if deemed unimportant. Stimuli that

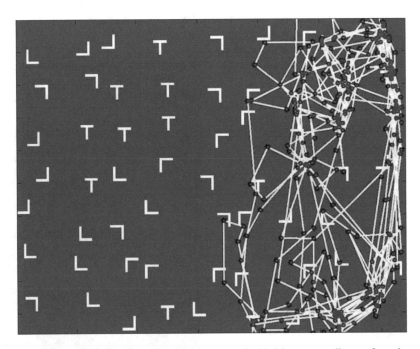

Fig. 9.8 Pathophysiological insights into neglect syndromes. Impaired working memory for identifying spatial locations. This figure indicates the scan tracks of a person with neglect after right parietal infarct. The instruction was to identify the "T"s amongst the "L's" which are the distractors. The diagram reveals that the person neglected the left-sided items as well as re-fixated on the "T's" on the right. Figure with permission: Husain M, Mannan S, Hodgson T et al. Impaired spatial working memory across saccades contributes to abnormal search in parietal neglect. Brain 2001;124:941–952

do not change or are unvarying cause them not to enter into awareness. It may be tested clinically by visual fixation on a point on a sheet of paper and about 5 cm away an "X" is marked. After about 20 s of fixating on the point the "X" is found to disappear. Troxler's fading may be viewed as a component of the sensory systems in general, and may be one of the mechanisms in explaining sensory neglect, perhaps by facilitation or inhibition of the neuronal circuitry after a lesion [60]. In clinical studies of patients with right parietal lesions accelerated Troxler fading can be documented. This is not seen in patients with frontal lesions, also part of the attentional circuitry [61].

Management and Treatment Options

A number of different treatment modalities have been reported and shown efficacy or improvement when studied, some in controlled trials, and others in case series or isolated case reports. Promising are the device-based therapies using transcranial magnetic stimulation and transcranial direct current stimulation, at this stage still experimental.

Behavioral therapy
Visual scanning training for reducing unilateral spatial neglect
Mirror therapy for unilateral spatial neglect
Restraint or forced use
Directing attention to contralesional space

Cross-sensory stimulation techniques
Caloric stimulation
Optokinetic stimulation
Neck vibration
Eye patching

Device-based therapies
Transcranial magnetic stimulation (TMS)
Transcranial direct current stimulation (t-DCS)

Neuropharmacological
Rotigotine
Dopamine agonists (other)

Amantadine
Apomorphine [62–70]

Personal Neglect Syndromes Asomatognosia, Anosognosia Variants, Subtypes, and Related Syndromes

Anosognosia

This refers to a specific right hemisphere-associated syndrome; the term is derived from the Greek, nosos, meaning disease, and gnosis, meaning knowledge, most often seen in the first few days after right hemisphere stroke, whereby there is a denial of illness or deficit such as hemiplegia. It may be partial denial admitting to some weakness, or gross underestimation of weakness or a frank denial of any weakness at all. For example in the acute stages or first after stroke, hemiplegic may claim to be able to walk without difficulty. It affects up to 58 % of patients with right hemisphere stroke [71] and others find much higher instances in up to 73 % of patients affected during the first week of stroke but and typically resolving relatively rapidly within days [72–74]. In other disease states, such as schizophrenia, it has been measured to affect 57–98 % of people with schizophrenia, at least by some estimates [75]. Other presentations include the disruptive patient. Yet others see the left limb, usually the arm, as foreign or representing a stranger lying in their bed and attempt to throw the limb out of bed sometimes causing them to fall out of bed as a consequence. It was first described by Oliver Sacks and described in detail together with other similar anecdotes in his book "The man who mistook his wife for a hat" [76]. The author has had experience of several patients with right hemisphere stroke, that were termed "disruptive" because and they had "fallen out of bed", sometimes more than once, due to a foreign, offensive limb lying in their bed. Some will recognize their limb but label it with a particular name, termed personification [77–81].

Clinical evaluation can be performed by simple, rapid bedside scales or semiquantitative measures depending on what specific deficit denial is

present. With anosognosia for hemiparesis for example the Bisiach scale may be employed:
Bisiach scale:

0 — Deficit spontaneously reported to general questioning
1 — Deficit reported only after specific question about limb strength
2 — Deficit acknowledged after examination demonstrated weakness
3 — Deficit denied even after demonstration of weakness or plegia [82]

Other quantitative metric measures that may be used for monitoring or for research include:

The Denial of Illness Scale [83]
Cutting's Anosognosia Questionnaire [71]
The Bimanual task of Cocchini [84]
The Catherine Bergego Scale [52]

Although anosognosia and hemineglect commonly appear together with right hemisphere lesions they remain independent syndromes. Some people with anosognosia due to stroke have been reported to be unable to appreciate similar deficits in others [85] that has been postulated to be due to a dual body map in the brain, our own, and one for others, in turn based on the mirror neuron system. Critical lesions within the right hemisphere, associated with anosognosia to date, have implicated the prefrontal, parietotemporal cortex, thalamus, and insula [86].

Other manifestations of the anosognosia family of disorders include what has been termed identification anosognosia. First described by Garcin et al., the patient with left hemiparesis, while travelling with his wife in the car, maintained that the arm on the steering wheel was his, and not that of his wife's, it being his right arm [87].

Rehabilitation and Management

In general, anosognosia, if it persists, is associated with a longer hospital stay and rehabilitative strategies also include vestibular caloric stimulation similar to those for hemineglect. One reported method included self-observation of motor behavior via video, with resolution reported in a single case study [88]. Others include various awareness interventions such as the comprehensive dynamic interactional model of awareness (CDIM) [89].

Other forms of personal neglect and related disorders and manifestations include the following:

Anosodiaphoria

An indifference or undue lack of concern for paretic arm, leg, or both also termed insouciance, from the French word "souciant" which means worrying and "in" referring to not. These patients have a distinct lack of insight into their deficit, that allows acknowledging that they have a problem but to markedly underestimate its severity and implications. This syndrome has been recorded to occur in up to 32 % of right hemisphere strokes, improving relatively rapidly and often due to insular lesions [90, 91].

Somatoparaphrenia

Aside from the denial of weakness in the context of acute left hemiplegia, some patients attribute the left limb or limbs to another person or animal that may be familiar to them. Both a denial of the limb ownership and attributing them to another are evident, sometimes termed the "positive variant" of anosognosia for hemiplegia [92].

Misoplegia

Although anosognosia and anosodiaphoria occur relatively commonly with acute-to-subacute right hemisphere lesions, less frequent manifestations of the personal neglect include misoplegia. To date only about half a dozen have been reported, in the English literature searches. First described by Critchley, it refers to a hatred, extreme dislike, and even acts of aggression towards the left limb or limbs [93, 94].

Asomatognosia

The denial of ownership of the left arm, less often the left leg after right hemisphere lesions, most often stroke. In a study by Feinberg et al., patients with asomatognosia had right temporoparietal and additionally medial frontal lesions appeared important in the development of the syndrome. This differed from the orbitofrontal lesions that were correlated with the development of somatoparaphrenia syndromes [95].

Autotopagnosia

This syndrome is associated with left, not right, parietal lesions and refers to an inability to localize stimuli on a limb or different parts of the body with somatotopagnosia suggested as a more appropriate term. This occurs in the context of preserved motor, proprioceptive, and semantic ability [96].

Allesthesia

A response whereby there is a mislocation of sensory stimuli within the same limb when stimulated.

Allochiria

There is a mislocation of the sensory stimulus to the opposite arm or leg. This may be in the visual, auditory, and tactile domains. Lesions have been shown to be in the right temporal, parietal, and mesial frontal regions [97–100].

Synchiria

Represents one form of dyschiria, the generic term. For example synchiria is diagnosed when a stimulus to one side of the body is perceived on both sides of the body [101].

Allokinesia

A request to move a limb is followed by moving the opposite limb to the one requested. It represents the motor counterpart in a sense of the sensory syndrome of allochiria [102]. Described to date only in monkeys, no clinical reports have been located in the English medical literature; however clinical examples have on rare occasion been observed [Hoffmann M, Unpublished material].

Supernumerary Phantom Limb

Also referred to as phantom third hand this condition is an illusory sensation of the hand or arm, rarely the leg in a position that is different to that of the actual paretic or paralyzed limb. In a related illusion, Ehrenwald reported a case of a patient describing multiple phantom feelings that was described as a "nest of hands" [103].

Xenomelia

This refers to the expressed desire by a patient with a right superior parietal lesion for amputation of healthy left arm or leg. The right superior parietal lobe RSPL integrates multimodal sensory inputs to create body image. When the RSPL is lesioned, this is thought to lead to an abnormal response whereby the tactile function of the arm for example remains but is not incorporated into the cohesive body image—hence the request for amputation [104]. Other terms or similar syndromes include apotemnophilia and body integration disorder.

Rubber Hand Illusion

The rubber hand illusion refers to the phenomenon that when a person watches the stroking of an adjacent rubber hand at the same time the person's own (hidden) hand is stroked which causes the relocation of the stroking sensation to the rubber hand. This has been interpreted as motor intention overriding the visual and sensory signaling of the movement. The tactile sensation is transferred to the "alien" hand and supports the hypothesis that there is a three-component interaction of the sensations of tactility, proprioception, and vision. Vision may have dominance over the other sensations for both consciousness and the body representation [105–107].

Visuospatial Function

Although a wide variety of lesion sites in the brain may be associated with visuospatial dysfunction, the right posterior brain region is particularly important in this function. This has both sensory and motor properties and may be divided into:

1. Visuoperceptual—test by judgment of line orientation
2. Visuomotor or visuoconstructual ability

A variety of diagrams may be copied or instructed to draw, typically tested copying two-dimensional images of examiner's flower, intersecting pentagons or clock drawing for easy and moderate difficulty tasks. A more difficult task and perhaps a surrogate for three-dimensional image representations can be done by copying the examiner's cube or drawing a person.

Metric testing can be performed using the following:

Rey-Osterrieth Complex Figure Test which is scored according to 18 items each, scored 0, 1, or 2 for a maximum of 36 both for copying and after 15- or 30-min delay from memory.
Taylor Complex Figure, 18 items each, scored 0, 1, or 2 for a maximum of 36
RBANS figure, scored for 20 items [108]

Prosody and Aprosodias

In Darwin's view the first protolanguage was thought to be musical in nature, driven by sexual selection, akin to current-day birdsong.

Motherese, paretense, and infant-directed speech is often used as an example of the hyperprosodic manner of speaking that mothers use for communication with infants. Pet-directed speech is similar and demonstrates the communication value of prosody in everyday speech. Emotional communication dominance by the right hemisphere was first enunciated by Hughlings Jackson in 1915 [109]. About 60 years later Heilman et al. reported the association of emotional inflection and intonations of linguistically neutral sentences or phrases and alteration in response to production, repetition, and comprehension abnormalities in relation to right hemisphere lesions as opposed to left hemisphere or normal controls [110]. Mesulam cited a particularly poignant example of Chinese Mandarin and the word "Ma" that may signify vastly different meanings depending on the prosodic expression, specifically intonation. Mandarin, Thai, and Chinese are tonal languages whereby segments of speech tones are altered for a particular meaning. Taiwanese people with right hemisphere lesions had differed in their motor aprosodia compared to English people with similar aprosodias [111, 112] (Table 9.3).

With right hemisphere dominance for emotional processing, in the language context this comprises of prosody, the melodic intonation associated with speaking, and is regarded as a key paralinguistic component of language. There are instances when prosody, together with non-verbal communication through facial expressions, body language, and mannerisms, may override the literal messages of verbal discourse. Ekman's research has proposed that humans are capable of making over 10,000 differing facial expressions, including what has been termed "micro-expressions" that have been used as possible indicators of lying as an example. Pathological states, often due to stroke, involve

Table 9.3 Mandarin tonal language example [7, 170]

Mandarin	English meaning
Ma	Mother
Ma	Numbness
Ma	Horse
Ma	Curse

the right middle cerebral artery associated with an aprosodic state, usually transient and in association with other common right hemisphere syndromes such as neglect and anosognosia [113].

Language communication may be understood in terms of five principal components and is a function of both hemispheres. More specifically, the role of each hemisphere is understood to include the following:

Right Hemisphere

Prosody or the emotional valence attached to words, grammar, and sentences achieved varying loudness, cadence, accents, timbre, melody, and tone during speaking. Prosody may also be understood in terms of emotional, intellectual, intrinsic, affective, and inarticulate components [114]. Emotional prosody incorporates the valencies of fear, anger, sadness, or happiness; intellectual prosody accents particular words that may then indicate the opposite to the literal meaning. Affective prosody allows the accentuation of a word that may similarly convert it to the opposite literal meaning and inarticulate prosody includes paralinguistic speech elements such as grunts, sighs, or whistling sounds that amplify or downgrade a particular emotion. The right hemisphere is also dominant for facial expressions and body language that accompany spoken language [115].

Left Hemisphere

Control phonemes, the smallest unit of a word, lexicon (words) sequencing, and syntax (grammar) that determines word order are strung together into phrases and sentences.

Clinical Disorders of Prosody

Hyperprosodia
There may be an overlap with normality and brain lesions, most often of the right hemisphere, which may cause more overt abnormalities. For example hyperprosodia can be appreciated amongst people that are particularly expressive during speech and is most obvious in the context of motherese, infant-, or pet-directed speech. Hyperprosodia can also be a component of a manic presentation.

Aprosodia
May be appreciated after right hemisphere stroke, left hemisphere Broca's dysphasia, and Parkinsonian syndromes. Further clinical subtyping may be done according to the aprosodia battery and classification system proposed by Ross et al. that closely resembles that of the Boston Geschwind Aphasia classification [116].

Dysprosody
Certain changes in prosody particularly of intonation, accents, sentence structure, and enunciation and intonation may result in changes in a persons voice so that they may sound foreign—hence the syndrome of foreign accent syndrome [117].

Clinical Evaluation of Aprosodia and Dysprosodia

The motor or expressive aspects of affective prosody can be assessed during the patient interview, by noting whether the person's voice is noted (intonation present or flat monotone) particularly when engaging in emotionally relevant topics or questions.

Repetition of affective prosody is evaluated by asking the person to repeat a sentence in a happy, sad, or angry voice.

The comprehensive component of prosody can be evaluated by asking the person to evaluate the examiner's sentence said in a happy, sad, or angry voice. Testing should be done so that the patient cannot see facial expressions or gestures for example by standing behind the person. Hence the following syndromes can be deduced:

Sensory aprosodia—good affective prosody during speech and gesturing but auditory appreciation of affective prosody is impaired and emotional gesturing appreciated visually is impaired.

Table 9.4 Clinical aprosodia subtype classification after Ross et al. [171]

Affective prosody			Gestures		Aprosodia
Spontaneous	Repetition	Comprehension	Spontaneous	Comprehension	Aprosodia type
Poor	Poor	Good	Poor	Good	Motor
Poor	Good	Good	Poor	Good	Transcortical motor
Poor	Good	Poor	Poor	Poor	Transcortical mixed
Poor	Poor	Poor	Poor	Poor	Global
Good	Poor	Poor	Good	Poor	Sensory
Good	Good	Poor	Good	Poor	Transcortical sensory
Good	Poor	Good	Good	Good	Conduction
Good	Good	Good	Good	Poor	Agesic

Expressive or motor aprosodia is characterized by monotone speech with relative paucity of spontaneous gesturing. In addition affective prosody repetition is impaired but affective prosody (auditory) comprehension as well as emotional gesturing by visual comprehension remain intact.

With global aprosodia there is an overall lack of affect through prosody and gesturing, a flattened affect, as well as impairment of comprehension and repetition of affective prosody and visual comprehension of emotional gesturing [118, 119] (Table 9.4).

Prosody: A Brief Bedside Test

Spontaneous Prosody

Does person impart affect conversation through prosody (1) and gestures (1)?

Scoring	
Present	2
Absent	0
Total	/2

Repetition of Affective Prosody

A sentence ("today is a sunny and windy day") devoid of emotional words is said in a

Happy
Angry
Sad

tone of voice by the examiner which the person repeats in the same affective intonation

Scoring	
Present	1
Absent	0
Total	/3

Comprehension of Affective Prosody

The examiner says the sentence ("today is a sunny and windy day"), free of emotional words, in a:

Happy
Angry
Sad

tone of voice, whereafter the person needs to identify the emotion.

Scoring	
Present	1
Absent	0
Total	/3
Grand total	/8

Other Uncommon Right Hemisphere-Related Syndromes

Apraxia of Eye Opening

Eye closure with inability to open to command but may open spontaneously, seen rarely after right hemisphere lesions. Although sometimes due to a different process associated with Parkinson' disease, progressive supranuclear palsy, cortico-basal degeneration, and Shy-Drager disease, right hemisphere lesions including subcortical stroke or hemorrhage may cause a transient involuntary eye closure [120–122].

Response to Next Patient Stimulation

Described in 11 of 134 consecutive patients with relatively large right hemisphere strokes by the authors. Patients responded to questions themselves when these were directed at patients near them or adjacent to them and postulated to be form of perseveration [123].

Topographical Disorientation (Geographical Disorientation) and Planotopokinesia

Although this syndrome may occur with lesions in either hemisphere, particularly the medial hippocampal region, persistent retrograde (as opposed to anterograde TD) has been correlated with right medial occipitotemporal impairment [124]. TD may be the predominant or presenting symptom of some disease processes such as limbic encephalitis. Hirayama et al. reported the case of a woman who had a relatively sudden onset of TD for familiar places with magnetic resonance imaging revealing abnormalities in both anteromedial temporal lobes, the right posterior parahippocampal region, and right retrosplenial and right inferior precuneus areas with anti-Hu antibody detected [125]. In the author's stroke experience, a TD was the heralding event or symptom of herpes simplex encephalitis with the only other abnormality clinical of a mild pyrexia (personal database, unpublished), and it refers to a loss of orientation in places or space. The term planotopokinesia refers to a similar entity, with the composite word derived from the Greek word planos or wandering, topos or place, and kinesis referring to motion.

Empathy Alteration (Loss of Empathy or Hyperempathy)

Relative loss of empathy is one of the most common components of frontotemporal lobe degeneration and dementia of the behavioral variety, seen with right frontotemporal lobe hypometabolism and atrophy. The opposite syndrome, one of hyperempathy, has recently been reported for the first time in a patient requiring right amygdalo-hippocampectomy for treatment-resistant epilepsy [126].

Gourmand Syndrome

Described in patients with right anterior brain lesions that refers to a particular preoccupation with food and dining. The disorder has a correlation with impulse control disorders and the serotonergic influence of the right hemisphere, which likely differs to that of the left hemisphere serotonergic influence, according to the authors [127].

Hypergraphia, Graphomimia, Graphomania, and Echographia

Although more commonly presenting in association with the Geschwind-Gastaut temporal lobe syndrome or epileptic induced it may also be a manifestation of complex partial seizures [128, 129]. A related or variant condition may be that of compulsive painting [130]. Yamadori et al. described a variant syndrome in a person with a large right hemisphere infarct with "inappropriate and permanent writing behavior" but with abnormal spatial construction of the writing itself. This case was interpreted as a disinhibition syndrome secondary to release of left hemisphere writing activity. Hence this form of hypergraphia was referred to as graphomimia [131]. Graphomania, different to the above syndromes with …, was reported in a frontal and callosum glioma with other accompanying symptoms and signs being attentional impairment and expressive dysphasia.

"Spontaneous and induced writing was abundant and incoercible." The description of graphomania was suggested to differentiate the syndrome from other similar behaviors of excessive writing such as hypergraphism and echographia, the latter being a type of field-dependent behavior [132].

Nosagnosia Overestimation

Exaggeration of the strength of unaffected limb may be present in association with right hemisphere syndromes, anosognosia and hemineglect. The lack of insight of being able to adequately perform bilateral limb tasks may outlast the left-sided neglect and paresis.

Nimmo-Smith et al., on the basis of their study, recommended questioning in regard to both a bimanual and bipedal task ability as a bedside assessment of this condition [133].

Buccal Hemineglect

This may constitute an underestimated syndrome, perhaps frequently undiagnosed or misdiagnosed. In the study by Andre et al., the vast majority occurred in relation to right hemisphere lesions associated with other forms of neglect. The presentations included dysphagia, retention of food in the left buccal space, salivation from the left side of the mouth, choking, as well as dysgeusia for some of the primary tastants such as salt, sweet, and sour. Experimentally, in monkeys, area 6 may be a hub area for this function as well as connection to the thalamus and parietal cortex [134].

Delusional Misidentification Syndromes or Content-Specific Delusion Syndromes

These take the form whereby a person incorrectly identifies or duplicates persons, places, objects, or even events which may be learned by self-report or from family members or friends. Many different DMIS have been reported and the specific delusion is then matched according to one's known to assist in the diagnosis. These include the following:

Capgras Syndrome

The belief by the person that a familiar individual or even the person themselves had been replaced by an imposter (hypoidentification).

Fregoli's Syndrome

An individual, familiar to the person, is actually impersonating and is presenting themselves as a stranger (hyperidentification).

Intermetamorphosis

Two people, both familiar to the person, have interchanged identities with one another.

Reduplicative Paramnesias

(a) Place reduplication
    The person is of the belief that there are two identical places sharing the same name but located in geographically distinct areas.
(b) Chimeric
    The person claims to be in their house when in fact they are lying in hospital, which may be considered to have been transformed into their own home.
(c) Extravagant spatial localization
    The person claims to be in a place different to the one at the time of interview, usually a place familiar to the person.

De Clerambault's Syndrome

The belief that one is being loved by an individual belonging to higher socioeconomic status than self.

Doppelgänger Syndrome

The person has a duplicate or twin.

Othello Syndrome

The belief that the spouse is being unfaithful.

Ekbom Syndrome (Parasitosis)

There is a belief of being infested or infected with insects.

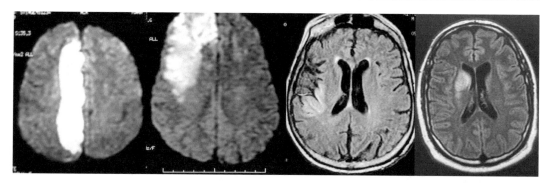

Fig. 9.9 Right hemisphere stroke lesions associated with content-specific delusions. From *left* to *right*: Right anterior cerebral territory, right middle and anterior cerebral territory, right middle cerebral branch occlusion, and right caudate nucleus infarction

Cotard's Syndrome
The belief that one is actually demised.

Dorian Gray Syndrome
The belief that one is immune to aging and is not actually aging.

Lycanthropy
The belief of one that there is a periodic transformation into an animal, most commonly a wolf.

Incubus Syndrome
The person is visited by phantom lover.

Picture Syndrome
Persons in the news such as in newspapers and on television are present in the home of the individual.

Phantom Boarder
There is a guest in the person's home who is not welcomed there by the individual [135–147].

In a series of 260 misidentification syndromes, most were diagnosed with Capgras syndrome ($n=174$) misidentifying other persons, Fregoli's syndrome ($n=18$), reduplicative paramnesia ($n=17$), and intermetamorphosis ($n=11$). Although almost half (127) were attributed to schizophrenia stroke, traumatic brain injury and dementia were other common clinical diagnoses [148].

The lesion in those presenting acutely is usually associated with a right frontal lesion but several reports of subcortical lesions in particular the right caudate nucleus have been implicated (Fig. 9.9). The incidence of CSD in a psychiatric population was described to be 1–5 % and in a case series of acute and subacute stroke population ($n=1476$) there were 31 instances of CSD (one patient may have had more than one CSD) with an overall incidence of 2 %. CSD occurred in approximately one in nine patients after right hemisphere; particularly right frontal stroke and CSD were often accompanied by other right hemisphere cognitive disorders. Reduplicative paramnesia was the most frequently recorded entity followed by Capgras and Fregoli's syndrome [149, 150]. Although usually a transient disturbance, some success has been recorded with mirtazapine use [151].

Emotion Disorders

Although the brain's emotional circuitry is extensive and widespread involving many cortical and subcortical regions, the temporal lobe serves as an important hub. Our understanding of the much more expansive neural circuitry has prompted a reappraisal of the critical, necessary, and other involved brain regions in emotional assessment, regulation, and impairments secondary to disease processes (Fig. 9.10).

Evolutionary Insights

The brain has a two-tier system to help cope with survival. Primates and humans inherited vision as the predominant sense which is closely linked to

Medial view   **Lateral view**

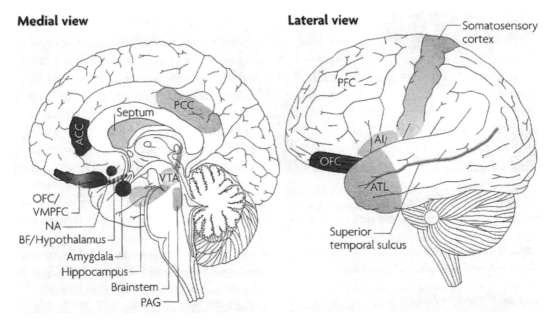

Fig. 9.10 Emotional brain: Core (*red colored*) and extended brain (*orange colored*) regions. Core emotional brain: *OFC* orbitofrontal cortex, *VMPFC* ventromedial prefrontal cortex, *ACC* anterior cingulate cortex, *BF* basal forebrain, *NA* nucleus accumbens. Extended emotional brain: *PAG* periaqueductal gray matter, *ATL* anterior temporal lobe, *AI* anterior insula, *PCC* posterior cingulate cortex, *VTA* ventral tegmental area. Figure with permission: Pessoa L. On the relationship between emotion and cognition. Nature Reviews Neuroscience 2008;9:148–158

the emotional brain centers. These enable a rapid and non-conscious reaction to environmental stimuli or predators that can be a life-saving, being a more ancient and "unconscious," system which developed earlier in evolution concerned with more elementary functions and survival, not the least of which is dealing with the vast sensory information that requires processing. This is a standard infrastructure amongst vertebrates in general. The conscious component became a later elaboration of the primate brain (Fig. 9.11) [152]. The occipitotemporal fasciculus is a particularly large fiber tract in humans, part of the inferior longitudinal fasciculus that allows rapid transfer of visual information to the anterior temporal lobe and amygdaloid complex for environmental threat or predator evaluation and human social and emotional salience processing (Fig. 9.12) [153].

As humans we are particularly cooperative as a species and our coordinated behavior is in marked contrast to all other animals [154]. Metacognition, a function of the frontopolar

cortex (BA 1), represents the apex of higher cortical function ability. This region may be divided into medial lateral and anterior regions within BA 10, mediating self-reflection, introspection, monitoring and processing of internal states with the externally acquired information, episodic memory subfunctions such as source memory and prospective memory, and contextual and retrieval verification (Fig. 9.13) [155, 156].

Shea et al. have proposed a two-level metacognitive control system, a system 1 and system 2. System 1 (also called cognitively lean system) of metacognition that operates non-consciously or implicitly and is hypothesized to be involved in processing information from the senses for intrapersonal cognition relevant to many animals (Fig. 9.14). System 2 (also termed cognitively rich system) is thought to be unique to humans and is concerned with processing high-level information amongst several conspecifics or people, also termed suprapersonal control (Fig. 9.15). Shea et al. conjecture that the suprapersonal coordination metacognitive system antedated

a Visual pathways

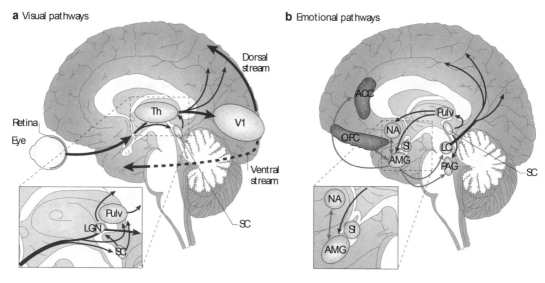

b Emotional pathways

Fig. 9.11 Linking vision (**a**) and emotion (**b**): Cortical and subcortical pathways. *Th* thalamus, *V1* primary visual area, *SC* superior colliculus, *LGN* lateral geniculate nucleus, *Pulv* pulvinar, *OFC* orbitofrontal cortex, *ACC* anterior cingulate cortex, *NA* nucleus accumbens, *AMG* amygdala, *SI* substantia innominata, *PAG* periaqueductal gray matter, *LO* locus coeruleus. Figure with permission: Tamietto M, de Gelder B. Neural bases of nonconscious perception of emotional signals. Nature Neuroscience Reviews 2010;11:697–709

Fig. 9.12 Occipito-temporal pathway. Figure with permission: Catani M. Jones DK, Donato R, ffytche DH. Occipito-temporal connections in the human brain. Brain 2003;126:2093–2107

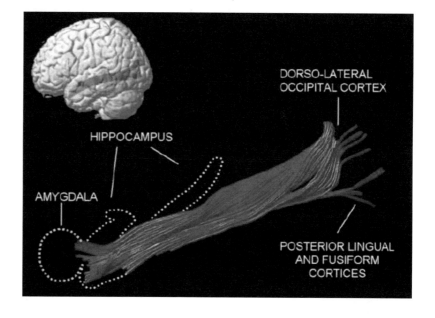

system 1 or that which controlled the intrapersonal cognitive processes that underlie emotional intelligence and inhibitory control [157]. In line with these hypotheses is that one of the core frontal functions of disinhibition was a critical development in human evolution.

Clinical

Emotional intelligence (EI) defined differently according to the various clinical brain specialties such as psychology, neurology, and psychiatry. EI has been shown to be important for personal suc-

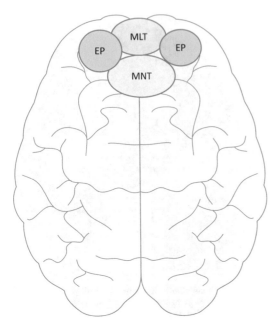

Fig. 9.13 Frontopolar cortex subregion activation patterns; regional cerebral blood flow and blood oxygen level dependent. *MLT* multitasking, *MNT* mentalizing, *EP* episodic memory. Figure adapted from Gilbert SJ, Spengler S, Simons JS et al. Functional Specialization within Rostral Prefrontal Cortex (Area 10): A Meta-analysis. Journal of Cognitive Neuroscience 2006;18:6: 932–948

cess and career success and in navigating the social complexities in day-to-day living [158]. EI impairment may be apparent in both apparently healthy people and after brain illnesses such as stroke, frontotemporal lobe dementia, Alzheimer's disease, and multiple sclerosis [159, 160]. Baron for example has proposed that EI comprises of the group of abilities that allow one to

1. Understand one's own emotions and be able to express feelings
2. Understand how others that one interacts with feel and how one relates to them
3. Manage and be in control of one's own emotions
4. Use one's emotions in adapting to one's environment
5. Generate positive emotions and use them for self-motivation in facing challenges

The neural circuitry for EI involves orbitofrontal, anterior cingulate, insula, and amygdaloid complex regions [161, 162]. Processes that injure these areas in particular include traumatic brain injury, stroke, multiple sclerosis, and frontotemporal lobe disorders. A holistic clinical brain injury assessment should

Fig. 9.14 System one (lean) meta-cognitive system that allows improved oversight of sensory signals and the optimum output in terms of effector organs. Contention scheduling refers to the solving of appropriate output after receiving competing signals. Shea N, Boldt A, Bang D, Yeung N, Heyes C, Frith CD. Suprapersonal cognitive control and metacognition. Trends in Cognitive Sciences 2014;18:186–193

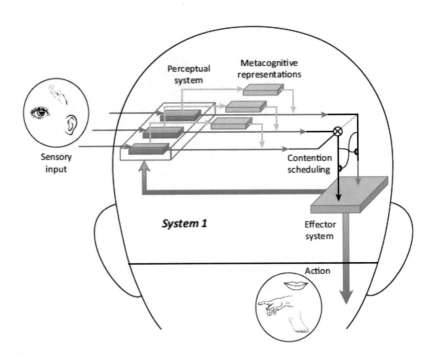

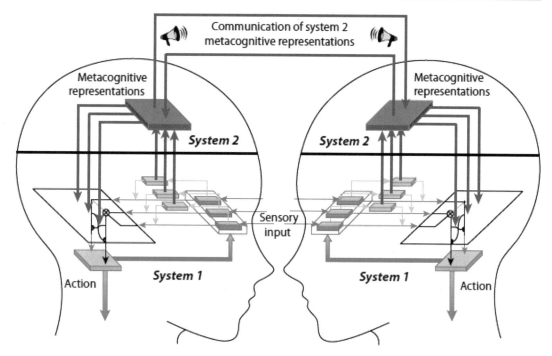

Fig. 9.15 System 2 (rich) meta-cognitive multi-agent cognitive control system. Cooperation between two people. Information from system 1 is available verbal broadcasting translating into increased reliability of the sensory signals. Figure with permission: Shea N, Boldt A, Bang D, Yeung N, Heyes C, Frith CD. Suprapersonal cognitive control and metacognition. Trends in Cognitive Sciences 2014;18:186–193

therefore involve all the clinical brain specialties, neurological, neuropsychiatric, and psychological, and speech and language. EI assessment has already been extensively embraced in the corporate arena in view of its relationship with not only career success but also productivity and institutional health [163] and EI is amenable for behavioral intervention programs [164, 165]. In a study involving stroke patients, many different brain lesions affected EI scores, including with frontal, temporal, subcortical, and subtentorial stroke lesions. The most significant abnormalities were however associated with the frontal and temporal region lesions, supporting Pessoa's extended emotional brain circuitry [166, 167].

Involuntary Emotional Expression Disorder

Involuntary emotional expression disorder (IEED) refers to clinically recognized syndromes that are characterized by intermittent inappropriate laughing and/or crying as well as intermittent emotional lability. Lesion studies have implicated the frontal primary motor (BA 4), premotor (BA 6), supplementary motor (BA 8), dorsal anterior cingulate gyrus, posterior insular, and parietal regions and corticopontine projections to the amygdala, hypothalamus, and periaqueductal gray matter. In addition seizure activity of these circuits may result in laughing (gelastic) or crying episodes (dyscrastic) [168]. Dextromethorphan is a sigma-1 receptor agonist and noncompetitive N-methyl-d-aspartate (NMDA) receptor antagonist. Quinidine is a competitive inhibitor of CYP2D6 which increases plasma levels of dextromethorphan. In placebo, double blinded studies using validated IEED scales, the Center for Neurologic Study-Lability Scale in multiple sclerosis and amyotrophic lateral sclerosis patients, the combination drug was superior to placebo [169].

References

1. Corballis PM, Funnell MG, Gazzaniga MS. Hemispheric asymmetries for simple visual judgements in the split brain. Neuropsychologia. 2002;40:401–10.
2. Levy-Agresti J, Sperry RW. Differential perceptual capacities in major and minor hemispheries. Proc Natl Acad Sci U S A. 1968;61:1151.
3. Bear DM. Hemispheric specialization and the neurology of emotion. Arch Neurol. 1983;40:195–202.
4. Ross ED, Homan RW, Buck R. Differential hemispheric specialization of primary and social emotions. Neuropsychiatr Neurosurg Behav Neurol. 1994;7:1–19.
5. Joseph R. The right cerebral hemisphere: emotion, music, visuospatial skills, body image, dreams and awareness. J Clin Psychol. 1988;44:630–73.
6. Koch C. Consciousness. Cambridge, MA: MIT Press; 2012.
7. Mesulam M-M. Principles of behavioral and cognitive neurology. 2nd ed. Oxford: Oxford University Press; 2000.
8. Baddeley A. Working memory: theories, models, and controversies. Annu Rev Psychol. 2012;63:1–29.
9. Hills TT. The evolutionary origins of cognitive control. Top Cogn Sci. 2011;3:231–7.
10. Passingham RE, Wise SP. The neurobiology of the prefrontal cortex. Anatomy, evolution and the origin of insight. Oxford: Oxford University Press; 2012.
11. Tsujimoto S, Genovesio A, Wise SP. Frontal pole cortex: encoding ends at the end the endbrain. Trend Cogn Sci. 2011;15:169–76.
12. Barbas H, Bunce JG, Medalla M. Prefrontal pathways that control attention. In: Stuss DT, Knight RT, editors. Principles of frontal lobe function. 2nd ed. Oxford: Oxford University Press; 2013.
13. Knight RT, Staines WR, Swick D, Chao LL. Prefrontal cortex regulates inhibition and excitation in distributed neural networks. Acta Psychol (Amst). 1999;101:159–78.
14. Ghashghaei HT, Barbas H. Pathways for emotions: Interactions of prefrontal and anterior temporal pathways in the amygdala of the rhesus monkey. Neuroscience. 2002;115:1261–79.
15. Ghashghaei HT, Barbas H. Neural interactions between the basal forebrain and functionally distinct prefrontal cortices in the rhesus monkey. Neuroscience. 2001;103:593–614.
16. Kerns JG, Cohen JD, MacDonald 3rd AW, Johnson MK, Stenger VA, Aizenstein H, et al. Decreased conflict and error related activity in the anterior cingulate cortex in subjects with schizophrenia. Am J Psychiatry. 2005;162:1833–9.
17. Honey GD, Fletcher PC. Investigating principles of human brain function underlying working memory: what insights from schizophrenia? Neuroscience. 2006;139:59–71.
18. Medalla M, Barbas H. Anterior cingulated synapses in prefrontal areas 10 and 46 suggest differential influence in cognitive control. J Neurosci. 2010;30:16068–81.
19. Medalla M, Barbas H. Synapses with inhibitory neurons differentiate anterior cingulate from dorsolateral prefrontal pathways associated with cognitive control. Neuron. 2009;61:609–20.
20. Mayberg HS. Limbic-cortical dysregulation: a proposed model of depression. J Neuropsychiatry Clin Neurosci. 1997;9(3):471–81.
21. Ely EW, Shintani A, Truman B, Speroff T, Gordon SM, Harrell FE, et al. Delirium as a predictor of mortality in mechanically ventilated patients in the intensive care unit. JAMA. 2004;291:1753–62.
22. Alagiakrishnan K, WIens CA. An approach to drug induced delirium in the elderly. Postgrad Med J. 2004;80:383–93.
23. Inouye SK, Charpentier PA. Precipitating factors for delirium in hospitalized elderly persons: predictive model and interrelationship with baseline vulnerability. JAMA. 1996;275:852–7.
24. Munster BC, Aronica E, Zwinderman AH, Eikelenboom P, Cunningham C, Rooij SE. Neuroinflammation in delirium: a postmortem case-control study. Rejuvenation Res. 2011;14:615–22.
25. Morandi A, Rogers BP, Gunther ML, Merkle K, Pandharipande P, Girard TD, et al. The relationship between delirium duration, white matter integrity, and cognitive impairment in intensive care unit survivors as determined by diffusion tensor imaging: the VISIONS prospective cohort magnetic resonance imaging study. Crit Care Med. 2012;40:2182–9.
26. Maldonado JR. Neuropathogenesis of delirium: review of current etiologic theories and common pathways. Am J Geriatr Psychiatry. 2013;21(12):1190–222.
27. Oguro H, Ward R, Bracewel M, Hindle J, Rafal R. Automatic activation of motor programs by object affordances in patients with Parkinson's disease. Neurosci Lett. 2009;463(1):35–6.
28. Valenstein E, Heilman KM. Unilateral hypokinesia and motor extinction. Neurology. 1981;31(4):445–8.
29. Burcham KJ, Corwin JV, Stoll ML, Reep RL. Disconnection of medial agranular and posterior parietal cortex produces multimodal neglect in rats. Behav Brain Res. 1997;86(1):41–7.
30. Binder J, Marshall R, Lazar R, Benjamin J, Mohr JP. Distinct syndromes of hemineglect. Arch Neurol. 1992;49(11):1187–94.
31. Valenstein E, Heilman KM. Apraxic agraphia with neglect-induced paragraphia. Arch Neurol. 1979;36(8):506–8.
32. Shelton PA, Bowers D, Heilman KM. Peripersonal and vertical neglect. Brain. 1990;113:191–205.
33. Rapcsak SZ, Cimino CR, Heilman KM. Altitudinal neglect. Neurology. 1988;38(2):277–81.

34. Mark VW, Heilman KM. Diagonal spatial neglect. J Neurol Neurosurg Psychiatry. 1998;65(3):348–52.

35. Vuilleumier P, Valenza N, Mayer E, Reverdin A, Landis T. Near and far visual space in unilateral neglect. Ann Neurol. 1998;43(3):406–10.

36. Halligan PW, Marshall JC. Left neglect for near but not far space in man. Nature. 1991;350(6318): 498–500.

37. Rapcsak SZ, Verfaellie M, Fleet WS, Heilman KM. Selective attention in hemispatial neglect. Arch Neurol. 1989;46(2):178–82.

38. Behrmann M, Moscovitch M. Object-centered neglect in patients with unilateral neglect: effects of left-right coordinates of objects. J Cogn Neurosci. 1994;6(1):1–16.

39. Klein E, Moeller K, Zinsberger D, Zauner H, Wood G, Willmes K, et al. Object-based neglect in number processing. Behav Brain Funct. 2013;9:5. doi:10.1186/1744-9081-9-5.

40. Ladavas E. Is the hemispatial deficit produced by right parietal damage associated with retinal or gravitational coordinates. Brain. 1987;110:167–80.

41. Calvanio R, Petrone PN, Levine DN. Left visual spatial neglect is both environment-centered and body-centered. Neurology. 1987;37(7):1179–83.

42. Chatterjee A. Picturing unilateral spatial neglect: viewer versus object centred reference frames. J Neurol Neurosurg Psychiatry. 1994;57:1236–40.

43. Farah MJ, Brun JL, Wong AB, Wallace MA, Carpenter PA. Frames of reference for allocating attention to space: evidence from the neglect syndrome. Neuropsychologia. 1990;28:335–47.

44. Hillis AE, Rapp B, Benzing L, Caramazza A. Dissociable coordinate frames of unilateral spatial neglect: "viewer-centered" neglect. Brain Cogn. 1998;37(3):491–526.

45. Bisiach E, Luzzatti C. Unilateral neglect for representational space. Cortex. 1978;14:129–33.

46. Denis M, Beschin N, Logie RH, Della SS. Visual perception and verbal descriptions as sources for generating mental representations: evidence from representational neglect. Cogn Neuropsychol. 2002;19:97–112.

47. Guariglia C, Palermo L, Piccardi L, Iaria G, Incoccia C. Neglecting the left side of a city square but not the left side of its clock: prevalence and characteristics of representational neglect. PLoS One. 2013;8(7), e67390. doi:10.1371/journal.pone.0067390.

48. Coslett HB. Evidence for a disturbance of the body schema in neglect. Brain Cogn. 1998;37(3):527–44.

49. Buxbaum LJ, Ferraro MK, Veramonti T, Farne A, Whyte J, Ladavas E, et al. Hemispatial neglect: subtypes, neuroanatomy, and disability. Neurology. 2004;62(5):749–56.

50. Albert ML. A simple test of visual neglect. Neurology. 1973;23(6):658–64.

51. Azouvi P, Olivier S, de Montety G, Samuel C, Louis-Dreyfus A, Tesio L. Behavioral assessment of unilateral neglect: study of the psychometric properties of the Catherine Bergego Scale. Arch Phys Med Rehabil. 2003;84:51–7.

52. Bergego C, Azouvi P, Samuel C. Validation d'une e´chelle d'e´valuation fonctionnelle de l'he´mine´gligence dans la vie quotidienne: l'e´chelle CB. Ann Readapt Med Phys. 1980;38:183–9.

53. Watson RT, Heilman M. Thalamic neglect. Neurology. 1979;29:690–4.

54. Vallar G. Extrapersonal visual unilateral spatial neglect and its neuroanatomy. Neuroimage. 2001;14:S52–8.

55. Wilke M, Turchi J, Smith K, Mishkin M, Leopold DA. Pulvinar inactivation disrupts selection of movement plans. J Neurosci. 2010;30(25):8650–9.

56. Critchley M. The parietal lobes. New York, NY: Hafner; 1966.

57. Heilman KM, Valenstein E. Frontal lobe neglect in man. Neurology. 1972;22:660–4.

58. Jehkonen M, Ahonen JP, Dastidar P, Koivisto AM, Laippala P, Vilkki J, et al. Visual neglect as a predictor of functional outcome one year after stroke. Acta Neurol Scand. 2000;101(3):195–201.

59. Husain M, Mannan S, Hodgson T, Wojciulik E, Driver J, Kennard C. Impaired spatial working memory across saccades contributes to abnormal search in parietal neglect. Brain. 2001;124:941–52.

60. Troxler D, Himly K, Schmidt JA. Über das Verschwinden gegebener Gegenstände innerhalb unseres Gesichtskreises [On the disappearance of given objects from our visual field]. Ophthalmol Bibl. 1804;2(2):1–53.

61. Mennemeier MS, Chatterjee A, Watson RT, Wertman E, Carter LP, Heilman KM. Contributions of the parietal and frontal lobes to sustained attention and habituation. Neuropsychologia. 1994;32(6): 703–16.

62. Corbetta M. Hemispatial neglect: clinic, pathogenesis, and treatment. Semin Neurol. 2014;34(5):514–23.

63. Matano A, Iosa M, Guariglia C, Pizzamiglio L, Paolucci S. Does outcome of neuropsychological treatment in patients with unilateral spatial neglect after stroke affect functional outcome? Eur J Phys Rehabil Med. 2015.

64. Ng MJ, Singh P, Pandian JD, Arora R, Kaur P, et al. Mirror therapy in unilateral neglect after stroke (MUST trial): a randomized controlled trial. Neurology. 2014;83(11):1012–7.

65. Arai T, Ohi H, Sasak H, Nobuto H, Tanaka K. Heimspatial sunglasses: effect on unilateral spatial neglect. Arch Phsy Med Rehabil. 1997;78:230–2.

66. Gorgoraptis N, Mah YH, Machner B, Singh-Curry V, Malhotra P, Hadji-Michael M, et al. The effects of the dopamine agonist rotigotine on hemispatial neglect following stroke. Brain. 2012;135(Pt 8):2478–91.

67. Buxbaum LJ, Ferraro M, Whyte J, Gershkoff A. Coslett HB Amantadine treatment of hemispatial neglect: a double-blind, placebo-controlled study. Am J Phys Med Rehabil. 2007;86(7):527–37.

68. Jacquin-Courtois S. Hemi-spatial neglect rehabilitation using non-invasive brain stimulation: or how to modulate the disconnection syndrome? Ann Phys Rehabil Med. 2015;pii:S1877-0657(15)00471-6. doi:10.1016/j.rehab.2015.07.388.

69. Goedert KM, Zhang JY, Barrett AM. Prism adaptation and spatial neglect: the need for dose-finding studies. Front Hum Neurosci. 2015;9:243. doi:10.3389/fnhum.2015.00243. eCollection 2015.

70. Guilbert A, Clément S, Moroni C. Hearing and music in unilateral spatial neglect neurorehabilitation. Front Psychol. 2014;5:1503. doi:10.3389/fpsyg.2014.01503.

71. Cutting J. Study of anosognosia. J Neurol Neurosurg Psychiatry. 1978;41:548–55.

72. Pedersen PM, Henrik MA, Jorgensen HS, Nakayama H, Raaschou HO, Olsen TS. Frequency, determinants and consequences of anosognosia in acute stroke. J Neuro Rehabil. 1996;10:243–50.

73. Jehkonen M, Ahonen JP, Dastidar P, Koivisto AM, Laippala P, Vilkk J, et al. Predictors of discharge to home during the first year after right hemisphere stroke. Acta Neurol Scand. 2001;104:134–6.

74. Maeshima S, Dohi N, Funahashi K, Mnakai K, Itakura T, Komai N. Rehabilitation of patients with anosognosia for hemiplegia due to intracerebral haemorrhage. Brain Inj. 1997;11:691–7.

75. Lehrer DS, Lorenz J. Anosognosia in schizophrenia: hidden in plain sight. Innov Clin Neurosci. 2014; 11(5-6):10–7.

76. Sacks O. The man who mistook his wife for a hat. Br J Psychiatry. 1995;166(1):130–1.

77. Vuilleumier P. Anosognosia: the neurology of beliefs and uncertainties. Cortex. 2004;40:9–17.

78. Heilman KM, Barrett AM, Adair JC. Possible mechanisms of anosognosia: a defect in self-awareness. Philos Trans R Soc B Biol Sci. 1998;353(1377): 1903–9.

79. Ramachandran VS, Blakeslee S. Phantoms in the brain: probing the mysteries of the human mind. New York, NY: Quill; 1999. p. 113–57.

80. Baier B, Karnath HO. Incidence and diagnosis of anosognosia for hemiparesis revisited. J Neurol Neurosurg Psychiatry. 2005;76:358–61.

81. Moro V, Pernigo S, Zapparoli P, Cordioli Z, Aglioti SM. Phenomenology and neural correlates of implicit and emergent motor awareness in patients with anosognosia for hemiplegia. Behav Brain Res. 2011;225:259–69.

82. Bisiach E, Vallar G, Perani D, Papagno C, Berti A. Unawareness of disease following lesions of the right hemisphere: anosognosia for hemiplegia and anosgnosia for hemianopia. Neuropsychologica. 1986;24:471–82.

83. Starkstein Starkstein SE, Fedoroff JP, Price TR, Leiguarda R, Robinson RG. Anosognosia in patients with cerebrovascular lesions. A study of causative factors. Stroke. 1992;23:1446–53.

84. Cocchini G, Beschin N, Fotopoulou A, Della Sala S. Explicit and implicit anosognosia or upper limb motor impairment. Neuropsychologia. 2010;48(5): 1489–94.

85. Ramachandran VS, Rogers-Ramachandran D. Denial of disabilities in anosognosia. Nature. 1996;382:501.

86. Orfei MD, Robinson RG, Prigatano GP, Starkstein S, Rüsch N, Bria P, et al. Anosognosia for hemiplegia after stroke is a multifaceted phenomenon: a systematic review of the literature. Brain. 2007;130:3075–90.

87. Garcin R, Varay A, Dimo H. Documentporu servier a l'etude des troubles du schema corporea. Rev Neurol. 1938;69:498–510.

88. Fotopoulou A, Rudd A, Holmes P, Kopelman M. Self observation reinstates motor awareness in anosognosia after hemiplegia. Neuropsychologia. 2009;47:1256–60.

89. Kortte KB, Hillis AE. Recent trends in rehabilitation interventions for visual neglect and anosognosia for hemiplegia following right hemisphere stroke. Future Neurol. 2011;6:33–43.

90. Appelros P, Karlsson GM, Seiger A, Nydevik I. Neglect and anosognosia after first-ever stroke: incidence and relationship to disability. J Rehabil Med. 2002;34:215–22.

91. Critchley M. The parietal lobes. London: Hafner Press; 1953.

92. Paulig M, Weber M, Garbelotto S. Somatoparaphrenia. A positive variant of anosognosia for hemiplegia. Nervenarzt. 2000;71(2):123–9.

93. Loetscher T, Regard M, Brugger P. Misoplegia: a review of the literature and a case without hemiplegia. J Neurol Neurosurg Psychiatry. 2006;77:1099–100.

94. Critchley M. Personification of paralysed limbs in hemiplegics. BMJ. 1955;2(4934):284–6. and Critchley M. Misoplegia, or hatred of hemiplegia. Mt Sinai J Med. 1974;41:82–7.

95. Feinberg T, Venneri A, Simone AM, Fan Y, Northoff G. The neuroanatomy of asomatognosia and somatoparaphrenia. J Neurol Neurosurg Psychiatry. 2010;81:276–81.

96. Buxbaum LJ, Coslett HB. Specialised structural descriptions for human body parts: evidence from autotopagnosia. Cogn Neuropsychol. 2001;18: 289–306.

97. Lepore M, Conson M, Grossi D, Trojano L. On the different mechanisms of spatial transpositions: a case of representational allochiria in clock drawing. Neuropsychologia. 2003;4:1290–5.

98. Venneri A, Pentore R, Cobelli M, Nichelli P, Shanks MF. Translocation of the embodied self without visuospatial neglect. Neuropsychologia. 2012;50(5):973–8.

99. Halligan PW, Marshall J, Wade D. Left on the right: Allochiria in a case of left visuo-spatial neglect. J Neurol Neurosurg Psychiatry. 1992;55:717–9.

100. Young RR, Benson DF. Where is the lesion in allochiria. Arch Neurol. 1992;49:348–9.

101. Meador KJ, Allen ME, Adams RJ, Loring DW. Allochiria vs allesthesia. Is there a misperception? Arch Neurol. 1991;48(5):546–9.

102. Heilman KM, Valenstein E, Day A, Watson R. Frontal lobe neglect in monkeys. Neurology. 1995;45(6):1205–10.

103. Ehrenwald H. Anosgnosie und depersonalisation. Der Nervenartzt. 1930;4:681–8.

104. McGeoch PD, Brang D, Song T, Lee RR, Huang M, Ramachandran VS. Xenomelia: a new right parietal lobe syndrome. J Neurol Neurosurg Psychiatry. 2011;82:1314–9.

105. Pasqualotto A, Proulx MJ. Two-dimensional rubber-hand illusion: the Dorian gray hand illusion. Multisens Res. 2015;28(1-2):101–10.

106. Ehrsson HH, Spence C, Passingham RE. That's my hand! Activity in premotor cortex reflects feeling of ownership of a limb. Science. 2004;305(5685):875–7.

107. Botvinick M, Cohen J. Rubber hands 'feel' touch that eyes see. Nature. 1998;391:756.

108. Swindell CS, Holland AL, Fromm D, Greenhouse JB. Characteristics of recovery of drawing ability in left and right brain damaged patients. Brain Cogn. 1988;7:16–30.

109. Hughlings Jackson J. On affections of speech from diseases of the brain. Brain. 1915;38:106–74.

110. Heilman KM, Scholes R, Watson RT. Auditory affective agnosia: disturbed comprehension of affective speech. J Neurol Neurosurg Psychiatry. 1975;38:69–72.

111. Edmondson JA, Ross ED, Chan JL, Seibert GB. The effect of right brain damage on acoustical measures of affective prosody in Taiwanese patients. J Phon. 1987;15:219–33.

112. Ross ED. Hemispheric specializations for emotions, affective aspects of language and communication and the cognitive control of display behaviors in humans. Prog Brain Res. 1996;107:583–94.

113. Ekman P. Facial expression and emotion. Am Psychol. 1993;48(4):384–92 [Review] and [Ekman P. Emotions revealed. Recognizing faces and feelings to improve communication and emotional life. St Martin's Griffin, New York. 2003.].

114. Bowers D, Bauer RM, Heilman KM. the nonverbal affect lexicon: theoretical perspectives from neuropsychological studies of affect perception. Neuropsychology. 1993;7:433–44.

115. Blonder LX, Bowers D, Heilman KM. The role of the right hemisphere in emtinoal communciation. Brain. 1991;114:1115–27.

116. Ross ED, Orbelo DM, Cartwright J, Hansel S, Burgard M, Testa JA, et al. Affective-prosodic deficits in schizophrenia: profiles of patients with brain damage and comparison to schizophrenic symptoms. J Neurol Neurosurg Psychiatry. 2001;70:597–604.

117. Monrad-Krohn GH. Dysprosody or altered 'melody of language'. Brain. 1948;70:405–15.

118. Darby DG. Sensory aprosodia: a clinical clue to lesions of the inferior division of the right middle cerebral artery? Neurology. 1993;43:567–72.

119. Gorelick PB, Ross ED. The aprosodias: further functional-anatomic evidence for the organization of affective language in the right hemisphere. J Neurol Neurosurg Psychiatry. 1987;50:553–60.

120. Johnston JC, Rosenbaum DM, Picone CM, Grotta JC. Apraxia of eyelid opening secondary to right hemisphere infarction. Ann Neurol. 1989;25(6):622–4.

121. Lin Y-H, Liou L-M, Lai C-L, Chang Y-P. Right putamen hemorrhage manifesting as apraxia of eyelid opening. Neuropsychiatr Dis Treat. 2013;9:1495–7.

122. De Renzi E, Gentilini M, Bazolli C. Eyelid movement disorders and motor impersistence in acute hemisphere disease. Neurology. 1986;36(3):414–8.

123. Bogousslavsky J, Regli F. Response-to-next-patient-stimulation: a right hemisphere syndrome. Neurology. 1988;38(8):1225–7.

124. Barrash J. A historical review of topographical disorientation and its neuroanatomical correlates. J Clin Exp Neuropsychol. 1998;20(6):807–27.

125. Hirayama K, Taguchi Y, Sato M, Tsukamoto T. Limbic encephalitis presenting with topographical disorientation and amnesia. J Neurol Neurosurg Psychiatry. 2003;74(1):110–2.

126. Richard-Mornas A, Mazzietti A, Koenig O, Borg C, Convers P, Thomas-Antérion C. Emergence of hyper empathy after right amygdalohippocampectomy. Neurocase. 2013;20:666–70. doi:10.1080/13554794.2013.826695.

127. Regard M, Landis T. "Gourmand syndrome": eating passion associated with right anterior lesions. Neurology. 1998;50(3):831.

128. Yamadori A, Mori E, Tabuchi M, Kudo Y, Mitani Y. Hypergraphia: a right hemisphere syndrome. J Neurol Neurosurg Psychiatry. 1986;49(10):1160–4.

129. Carota A, Annoni JM, Combremont P, Clarke S, Bogousslavsky J. Hypergraphia, verbal aspontaneity and post-stroke depression secondary to right cingulate and corpus callosum infarction. J Neurol. 2003;250(4):508–10.

130. Panico A, Parmegriani A, Trimble MR. Compulsive painting: a variant of hypergraphia? Neurology. 1996;9(3):177–80.

131. Gil R, Neau JP, Aubert I, Fabre C, Agbo C, Tantot AM. Anosognosic graphomimia: an uncommon variety of hypergraphia in right sylvian infarction. Rev Neurol (Paris). 1995;151(3):198–201.

132. Cambier J, Masson C, Benammou S, Graphomania RB. Compulsive graphic activity as a manifestation of fronto-callosal glioma]. Rev Neurol (Paris). 1988;144(3):158–64.

133. Nimmo-Smith I, Marcel AJ, Tegnér R. A diagnostic test of unawareness of bilateral motor task abilities in anosognosia for hemiplegia. J Neurol Neurosurg Psychiatry. 2005;76(8):1167–9.

134. Andre JM, Beis JM, Morin N, Paysant J. Buccal hemineglect. Arch Neurol. 2000;57(12):1734–41.

135. Mallory PF, Richardson ED. Frontal lobes and content specific delusions. J Neuropsychiatry. 1994;6:455–66.

136. Feinberg TE. Delusional misidentification. Psychiatr Clin N Am. 2005;28:665–6683.

137. Ramachandran VS. Consciousness and body image: lessons from phantom limbs, Capgras syndrome and

pain asymbolia. Philos Trans R Soc Lond B. 1998;353:1851–9.

138. Sinkman A. The syndrome of capgras. Psychiatry. 2008;71(4):371–7.

139. Collins MN. Capgras' syndrome with organic disorders. Postgrad Med J. 1990;66:1064–7.

140. Hakim H. Pathogenesis of reduplicative paramnesia. J Neurol Neurosurg Psychiatry. 1988;51:839–41.

141. Hirstein W. Capgras syndrome: a novel probe for understanding the neural representation of the identity and familiarity of persons. Proc R Soc Lond B. 1997;264:437–44.

142. Moriyama Y. Fregoli syndrome accompanied with Prosopagnosia in a women with a 40-year history of schizophrenia. Keio J Med. 2007;56(4): 130–4.

143. Kapur N, Turner A, King C. Reduplicative paramnesia: possible anatomical and neuropsychological mechanisms. J Neurol Neurosurg Psychiatry. 1988;51:579–81.

144. Feinberg TE, Roane DM. Misidentification syndromes. In: Feinberg TE, Farah MJH, editors. Behavioral neurology and neuropsychology. New York, NY: McGraw Hill; 1997.

145. Aziz MA, Razik GN, Donn JE. Dangerousness and management of delusional misidentification syndrome. Psychopathology. 2005;38(2):97–102.

146. Devinsky O. Delusional misidentification s and duplications: right brain lesions, left brain delusions. Neurology. 2009;72:80–7.

147. Larner AJ. Delusion of pregnancy in frontotemporal lobar degeneration with motor neurone diseases. Behav Neurol. 2008;19:199–200.

148. Forstl H, Almeida OP, Owen AM, Burns A, Howard R. Psychiatric, neurological and medical aspects of misidentification syndromes: a review of 260 cases. Psychol Med. 1991;21(4):905–10.

149. Hoffmann M, Taylor. Content specific delusions after stroke. Fourth international vascular dementia conference, Porto, Portugal, 28–31 Oct, 2005.

150. McMurtray AM, Sultzer DL, Monserratt L, Yeo T, Mendez MF. Content-specific delusions from right caudate lacunar stroke: association with prefrontal hypometabolism. J Neuropsychiatry Clin Neurosci. 2008;20(1):62–7.

151. Khouzam HR. Capgras syndrome responding to the antidepressant mirtazapine. Compr Ther. 2002;28(3): 238–40.

152. Kaas J. Convergences in the modular and areal organization of the forebrain of mammals: implications for the reconstruction of forebrain evolution. Brain Behav Evol. 2002;59(5-6):262–72.

153. Catani M, Jones DK, Donato R, Ffytche DH. Occipito-temporal connections in the human brain. Brain. 2003;126:2093–107.

154. Tomasello M. Why we cooperate. Cambridge, MA: MIT Press; 2009.

155. Gilbert SJ, Spengler S, Simons JS, Steele JD, Lawrie SM, Frith CD, et al. Functional specialization within the rostral prefrontal cortex (area 10): a meta-analysis. J Cogn Neurosci. 2006;18:932–48.

156. Christoff K, Gabrieli JDE. The frontopolar cortex and human cognition: evidence for a rostrocaudal hierarchical organization within the human prefrontal cortex. Psychobiology. 2000;28:168–86.

157. Shea N, Boldt A, Bang D, Yeung N, Heyes C, Frith CD. Suprapersonal cognitive control and metacognition. Trends Cogn Sci. 2014;18:186–93.

158. Goleman DP. Emotional intelligence: why it can matter more than IQ for character, health and lifelong achievement. New York, NY: Bantam Books; 1995.

159. Mendez MF, Lauterbach EC, Sampson SM. An evidence-based review of the psychopathology of frontotemporal dementia: a report of the ANPA committee 16 on research. J Neuropsychiatry Clin Neurosci. 2008;20(2):130–49.

160. Van der Zee J, Sleegers K, Van Broeckhoven C. The Alzheimer disease frontotemporal lobar degeneration spectrum. Neurology. 2008;71:1191–7.

161. Bar-On R, Tranel D, Denburg NL, Bechara A. Exploring the neurological substrate of emotional and social intelligence. Brain. 2003;126: 1790–800.

162. Shamay-Tsoory SG, Tomer R, Goldsher D, Berger BD, Aharon-Peretz J. Impairment in cognitive and affective empathy in patients with brain lesions: anatomical and cognitive correlates. J Clin Exp Neuropsychol. 2004;26(8):1113–27.

163. Loehr J, Schwartz T. The making of a corporate athlete. Harv Bus Rev. 2001;79:120–8. and Gilkey R, Kilts C. Cognitive fitness. Harv Bus Rev. 2007;85:53–54.

164. Cherniss C, Adler M. Promoting emotional intelligence in organizations. Alexandria, VA: American Society for Training and Development; 2000.

165. Goleman D. Working with emotional intelligence. London: Bloomsbury; 1999.

166. Hoffmann M, Benes-Cases L, Hoffmann B, Chen R. Emotional intelligence impairment is common after stroke and associated with diverse lesions. BMC Neurol. 2010;10:103. doi:10.1186/1471-2377-10-103.

167. Pessoa L. On the relationship between emotion and cognition. Nat Rev Neurosci. 2008;9:148–58.

168. Lauterbach EC, Cummings JL, Kuppuswamy PS. Toward a more precise, clinically-informed pathophysiology of pathological laughing and crying. Neurosci Biobehav Rev. 2013;37(8):1893–916.

169. Pioro EP. Review of dextromethorphan 20 mg/quinidine 10 mg (NUEDEXTA® for pseudobulbar affect. Neurol Ther. 2014;3(1):15–28.

170. Hughes CP, Chan JL, Su MS. Aprosodia in Chinese patients with right cerebral hemisphere lesions. Arch Neurol. 1983;40:732–6.

171. Ross ED. Modulation of affect and nonverbal communication by the right hemisphere. In: Mesulam MM, editor. Principles of behavioural neurology. Philadelphia, PA: FA Davis and Company; 1985.

Language, Aphasias, and Related Disorders

10

Language consists of ideas that we convert into linguistic construction through an output that may be vocal or graphic. Ideas can also be transmitted by other means such as music, images, and visual art. An example of the latter was the imagistic transmission of concepts represented by the Franco Cantabrian cave art, dated to 40,000 years ago, for example. Archeological evidence is that visual art forms long preceded language. The earliest evidence of visual art we have dates to the Blombos cave findings estimated at 100,000 years ago [1].

Evolutionary Overview of Language and Speech Origins in Humans

There is no direct fossilized or other archeological finds related to human speech and language. Archeological fossil evidence of bipedalism however provides some insights. Bipedalism, walking, and running evolved about three million years before language.

This was associated with the development of a larger brain and increased complexity with respect of sensorimotor control. Rhythm is essential for walking, running, and, in general, complex coordination. Cortical subcortical circuits are involved in motor learning and in the regulation of motor control of speech, syntax, and cognition. In this respect walking was arguably the initiating event in the evolution of human speech and language. The adaptations for both walking and endurance running are dependent on the subcortical basal nuclei sequencing circuitry. The subcortical structures, basal ganglia, and the cerebellum expression are regulated by FOXP2 "language gene" [2]. Similarly, effective bipedalism requires rhythm or a linking between the auditory and motor system and also the possible foundations of musicality. Unsurprisingly this may be the basis of music therapy in motor and speech disorders in a variety of neurological diseases today [3].

Comparative primate studies suggest that this evolved from vocalizations of our ancestral primates, a gestural type of communication or through musical protolanguage. It has been estimated that only 14 distinct calls exist amongst monkeys and baboons and 50,000 words among humans. Vervet monkeys are known to have different vocal signals for alerting the troop to a leopard (loud barking), a snake (chutter), and an eagle (double-syllable cough) [4]. These have been interpreted by Mithen to be holistic as they conform to a message and manipulative in that they recommend fleeing for example. Gibbons, also referred to as the singing apes, are known singing together with a female song component that may last between 6 and 80 min, a male reply phrase (coda), and this is thought to facilitate their pair bonding and their monogamous relationship.

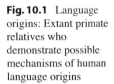

Fig. 10.1 Language origins: Extant primate relatives who demonstrate possible mechanisms of human language origins

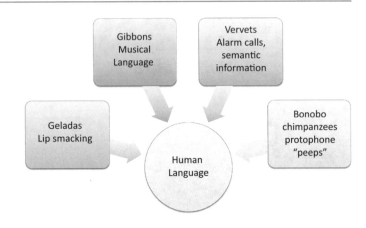

Gibbons are considered by Mithen to have holistic, musical, manipulative, and multimodal communication that may represent part of the Hmmmm theory of language evolution [5]. The gelada baboon lip smacking behavior as well as a musical communication assisted in their social interaction. They have been observed to synchronize their own vocalizations with other members in their group using various melodies and rhythms. Bergman's observations of Geladas revealed that a rapidly opening and closing of the mouth resulting in gelada "wobbles" or rapid variations in rhythm, pitch, and volume was akin to some aspects of human speech [6] (Fig. 10.1).

Bonobos chimpanzees produce a type of call known as a "peep," a high-pitched short-duration call produced with a closed mouth that is made during a range of their natural activities. Termed a protophone, these may represent a transition from the fixed vocal calls of animals to the much more flexible vocalizations of humans. Considered to reflect a dimensional aspect of the emotional both positive and negative (happy and sad) components of language, these are features of humans, infants, as well as bonobos. This stage would have occurred about 6–10 million years ago, in a shared common ancestor of great apes and humans [7]. The social theories of language evolutions are supported by the long-term observations of baboon populations by Cheney and Seyfarth for example, who suggest that baboons appear to have a basic theory of mind capability and able to have many social interactive concepts but no words for such [8].

The Development of Cortical Neural Circuitry

Hints at the complexity of neural mechanisms and reorganization can be appreciated from comparative extant primate between humans, chimpanzees, and macaques with investigations of fiber tractography. The marked expansion in the fiber connectivity that occurred in the sensorimotor connections in particular the arcuate fasciculus and its components bear testimony to this [9, 10] (Fig. 10.2).

The arcuate fasciculus (AF) circuitry is considerably more complex than was first appreciated. Aboitiz and Frey et al. in particular have contributed to the details of the arcuate fasciculus tract complex [10, 11]. The AF runs form the posterior superior temporal gyrus to both Brodmann areas (BA) 44 and 45 as well as the dorsolateral prefrontal cortical areas of BA 6 and 8. In addition the supramarginal gyrus component of the inferior parietal lobe projects to BA 44 and the intraparietal sulcus areas (ventral posterior component) receives input from the superior temporal lobe gyrus through the middle longitudinal fasciculus and the inferior longitudinal fasciculus, and finally a ventrally situated tract from the anterior temporal cortex to the prefrontal cortical BA 45 and 47 traverses via the extreme capsule as well as the uncinate fasciculus which has bilateral representation [12] (Fig. 10.3).

The current theories of language evolutionary attempt to explain the connections of the acoustic,

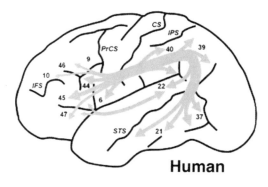

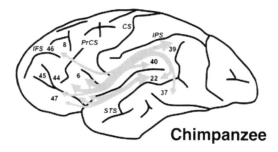

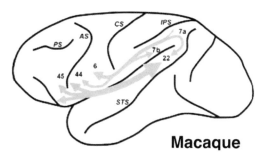

Fig. 10.2 Human language evolution: White matter tract expansion. A comparative analysis of the arcuate fasciculus in macaques, chimpanzees, and humans. Figure with permission: Rilling JK, Glasser MF, Preuss TM, et al. Evolution of the arcuate fasciculus revealed with comparative DTI. Nature Neuroscience 2008;11:426

symbolic concepts and motor programs with subcortical basal ganglia, premotor, supplementary, and motor areas (Table 10.1). Perhaps all are relevant and explain one or more of the complex steps or processes in language evolution. Cortical-subcortical circuits are integral to motor control, cognition, and syntax. The discovery of the FOX P2 gene (so-called language gene) was shown to be key to regulation of and embryogenesis of the basal ganglia, deletions of which cause speech impairment [2, 11].

Gestural protolanguage may be regarded as an intermediate stage in language evolution and in Arbib's model of the transition of gesture to speech, one of the driving forces may have been that of increased social complexity (kin selection theory). This may have been represented by the phase of *Homo ergaster* 1.8 million years ago and the one million years of relative stasis before another rapid brain expansion phase approximately 600 kya. During this period the so-called mimetic culture existed whereby mimesis (imitation) and a nonlinguistic type of communication likely existed and was seen as a bridge between the ape and modern human type of communication [21]. This was a holistic type of communication and the first stage of what Mithen later termed the Hmmmm communication [15].

Changes brought about by climate and subsequent foraging and sociality amongst early hominids expanded upon the existent vocalizations and gestures in place which were holistic and manipulative. Perhaps gestures and musical like vocalizations increased.

For the hominid communication became more integrated for a more complex communication type to become a Hmmmm (holistic, mimetic, manipulative, multimodal).

Gestural protolanguage: Because humans move the hands and arms, and show facial expression during conversation, it has been postulated that a visual manual communication system preceded language. Gesture types are many and include pantomime (objects/actions acted out without speech), co-speech gestures (occur during speech), deictic gestures (pointing to an object during speech), iconic gestures (spatial representation of an object during speech), emblematic gestures (cultural associated meanings of finger-associated gestures), and metaphorical/lexical (co-speech utterance) [22].

Presumably, gestures were initially substitutes for spoken words and later cross-modal transference between auditory, tactile, and visual domains occurred by virtue of the mirror neuron system. The cross-modal integration of these neural mechanisms forms the basis of the Arbi and Rizzolatti model of language evolution. Arbib suggested that gesture speech transitioned to a

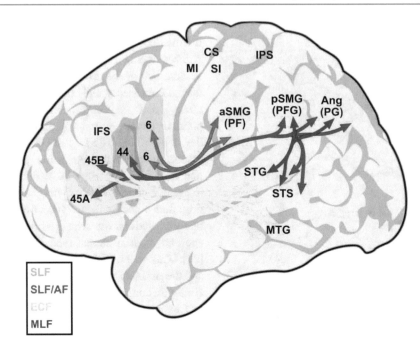

Fig. 10.3 Language evolution: Tripartite input into Broca's region: The arcuate fasciculus, superior longitudinal, middle longitudinal fasciculus, and extreme capsule fasciculus. Superior longitudinal fasciculus, arcuate fasciculus (*red* and *green*), middle longitudinal fasciculus (*blue*), extreme capsule fasciculus (*yellow*). Figure with permission: Aboitiz F. Gestures, vocalizations and memory in language origins. Frontiers in Evolutionary Neuroscience 2012, February doi:10.3389/fnevo.2012. 00002

Table 10.1 Language evolution theories [13–20]

Building on earlier language theories of Wallace (divine attribute), Chomsky (the whole is greater than the sum of its parts), Gould (language exaption theory), and Pinker (language instinct), the following may be considered current contemporary language theories:
1. Gestural protolanguage [Fitch]
2. Extended mirror system hypothesis [Rizolatti, Arbib]
3. Musical protolanguage (Hmmmm) [Mithen, Blacking, Wray, Darwin]
4. Lexical protolanguage [Fitch]
5. Synesthetic bootstrapping [Ramachandran]
6. Vocal grooming hypothesis [Dunbar]

system that was augmented by some vocalization, one that was predominantly a vocal system. In brief the extended mirror neuron system hypothesis involved several stages:

Stages 1–3 occurred within primates leading to the last common ancestor with present-day chimpanzees

S1—grasping
S2—mirror system for grasping

S3—simple imitation (shared with chimpanzees, not macaques)

Stages 4–7 hypothetically from the last common ancestor to modern humans

S4—complex imitation (after chimpanzees)
S5—protosign (open repertoire)
S6—protospeech (cortical vocal control via collateralization)
S7—modern language

Arbib suggested that the mechanism of the step from gesture to speech was due to collateralization, from the motor limb to the neighboring human oromotor vocal control areas, a cultural development occurring in the last 100 ky (Fig. 10.4).

The vocal grooming hypothesis of Dunbar and Aiello was based on the gossip concept and size of the group. Grooming over 50 individuals was facilitated by language rather than tactile grooming—hence "vocal grooming." Gelada monkeys

Fig. 10.4 Gestural control circuits "collateralized" speech control area. Control of contiguous body regions mapping on the motor cortical areas and gestural control circuits "collateralizing" speech control areas. Figure with permission: Fitch WT. The Evolution of Language. Cambridge University Press, Cambridge, 2010

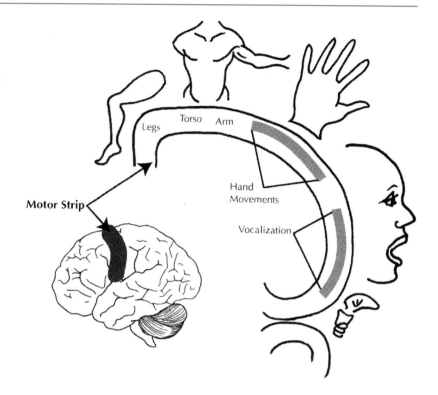

are thought to have this form of vocal grooming by using rhythm and melody during grooming. Vocal grooming may also be related to the origin of music more so than language. Singing promotes bonding and feelings of contentment and pleasure akin to the tactile grooming effects, both oxytocin mediated [23, 24].

Musical Protolanguage (Prosodic Protolanguage)

Music can be both vocal (speech, song) and gestural (dance, sign language) and can express emotion and arouse emotion, in different and perhaps more effective ways than language. The hyperprosody and effects on emotional states of infant direct speech give us some insights into aspects of language evolution from a human developmental point of view. Human infants have a particular sensitivity to melody, tempo, and rhythms of speech that is evident long before language abilities. This may be regarded as a type of mental circuitry that underlies the prosody of our

language and the subsequent development of the neural circuitry of language may be integral to this [Fitch]. Music is a powerful means of social bonding, and facilitates cooperative behavior with one of the neurobiological mechanisms likely being oxytocin release in the basal forebrain [25]. Neanderthals, more so than early modern humans, may have used musical ability for communication that is the basis of Mithen's hmmmmm model (holistic, musical, mimetic, manipulative, multimodality). In his conceptualization these components existed in earlier primates but were combined in modern humans. Primate gestures and vocalizations were already manipulative and holistic whereas later early hominid communications systems were elaborated to include mimetic, musical components that were conceived as Mithen's hmmmm communication system. In the Hmmmm model, each component already evident within the primate line ultimately became integrated and working together within the early hominins [15].

Both music and language are pervasive in humans today and share many similarities as they

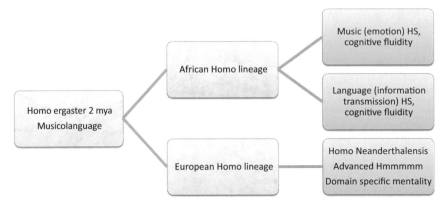

Fig. 10.5 Evolution of music and language. Adapted from Mithen S. The Singing Neanderthals. Weidenfeld and Nicolson London 2005 and Blacking J. How Musical is Man? Seattle University of Washington Press 1973

can be transmitted vocally, in physical form (gesture body movement), and in writing. In addition they have constructs that are combinatorial and hierarchical and allow recursion. This would argue for a common precursor earlier musicolanguage that existed prior to *Homo ergaster* about 2 mya which subsequently differentiated into analytical language and musical abilities [16, 26, 27] (Fig. 10.5).

Lexical

The lexical protolanguage theory is where an utterance of a single word or even several words is combined resulting in an extensive lexicon of meaningful words but lacking syntax. Onomatapoeia may represent a bridging or a protolanguage from earlier nonlinguistic hominins to modern language. Most people are intuitively capable of sound synesthesia whereby they can recognize the link between the sound of a word and some of the physical characteristics of an animal for example. In extant traditional societies whose lives are intimately linked with nature and animals, onomatopoeia is common for names used for animals for example. The Peruvian rainforest tribe approximately a third of names for 206 different birds have an onomatopoeic source. This is explainable by sound synesthesia or a mapping of one sense onto another sense such as the size of aan animal and the associated sound. For example i' sound is associated with small

whereas 'u-a-o' is associated bigger objects explained by the sound made by the tongue and lips relating to the size of the object and a larger oral cavity [27].

Bickerton proposed a word-based protolanguage drawing on experience from pidgin languages in which there is a lexicon but little grammar [28]. Jackendorff posited a stage where function words and relational words allowed combination of phrases to be into more complex structures [29]. Modern syntax is viewed as the final act or achievement in language evolution. Humans, no matter what culture, possess the ability to arrange the language signals into more complex, yet meaningful, syntactically constructed signals or language. Rather than being biologically based this is most likely a result of cultural concepts termed "grammaticalization" in which verbs and nouns are metamorphosed into prepositions and other function words [13, 30].

Syntax and the use of grammar are regarded as a function of more complex cognition. In particular thinking or inner speech has been envisaged as a type of motor activity and a function of the frontal cortex, in particular Broca's area [31, 32].

Synesthetic Boot Strapping [18]

The non-arbitrary, cross activation of brain circuits for visual information and auditory information and linking to the motor programs within Broca's regions is explainable by the mirror neuron system. A linking between manual gesture, lip, and tongue movement was first reported by Darwin, explainable by the adjacent cortical areas of mouth and hand. Similar to the sensory-based synesthesia, there is also co-activation of motor circuitry termed synkinesis, which likely played a role in the transformation of gestural protolanguage into modern language. In their view concepts may represent in brain circuits similar to how percepts are and cross activation of these brain circuits is postulated to be the basis for metaphorical thinking and related to the higher synesthesia incidence in famous scientists and artists both visual and literary. The more extensive "cross wiring" gives them increased opportunity for metaphorical thinking. This mechanism is also implicated in the synesthesia origin of language. A number of cross-modal computations may be envisaged in early hominins:

1. Visual-auditory mapping: This is demonstrated by the bouba kiki effect which refers to the non-arbitrary mapping between the visual shape aspect of a particular object and associated speech sounds, first described by Köhler [33].
2. Audio-visual mapping to motor vocalization mapping within Broca's area.
3. Broca's area to motor manual gestural areas.

The inferior parietal lobule, a hub of the mirror neuron circuit, would allow visuo-auditory mapping and the inferior frontal lobe, a hub for the motor mirror neurons for tongue and lip movements, allowing these abstractions to occur. In the final phase there is an envisaged "autocatalytic bootstrapping culminating in the emergence of a vocal protolanguage" [18, 19].

Descent of the Larynx, Shorter Trachea, and Pharynx Development

Comparative studies of the red deer and giraffes provided insights. Red deer lower their larynxes to achieve lower resonance calls with the presumed effect of size exaggeration However others con-

tend that a more important reason to lower the larynx is to shorten the trachea with consequent reduced need for lung compression for phonation pressure. For example giraffes don't vocalize. Laryngeal lowering has the effect of increasing vocal tract length, reducing the frequencies of resonances, and causing an animal to sound bigger. This was the postulate suggested by Fitch in his size exaggeration hypothesis [34]. Humans as opposed to the red deer for example have a permanent lowering of the larynx that requires raising of the larynx during swallowing and increased risk of choking, currently ranked number four of accidental deaths in the USA [35, 36]. The development of a pharynx allowed improved vowel sound formation.

A number of other significant exaptions of the neuroanatomy and neural circuitry had to evolve for the development of modern language. A more mechanical alteration of the pharyngeal and laryngeal anatomical and neuromuscular apparatus was associated with a descent of the tongue within the pharynx, with the tongue position occupying an approximately 1:1 pharyngeal to oral components of the supralaryngeal aspect of the vocal tract. This allows the production of "quantal sound units" (Fig. 10.6).

Breathing Control

A third major development was the more precise control of breathing. Human syllable production or the units of sounds are about tenfold faster in comparison to nonhuman primates, an important component of modern language and for transferring information between conspecifics [37]. In particular the intercostal muscle, innervated by thoracic nerves, allows the production of extended phrases during the expiration of quiet breathing. In addition the fine-tuning and control of subglottal pressure allow intonation and pitch control. Without such control, only limited intonation and short exclamations would have been possible. Enlarged vertebral canals suggest improved breathing control. Other factors are basocranial flexion suggesting descent of the larynx, higher pharynx, hyoid bone, and hypoglos-

Fig. 10.6 Descent of the larynx: The size exaggeration hypothesis. Figure with permission: Fitch WT. "Vocal tract length and formant frequency dispersion correlate with bodysize in rhesus macaques," Journal of the Acoustical Society of America 1997;102(2):1213–1222

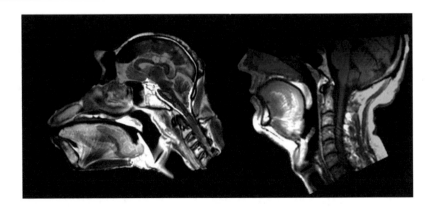

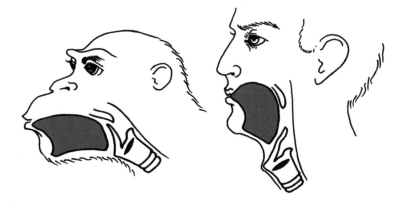

sal canal enlargement for improved tongue motor control. Cross-sectional thoracic canal area in mm² measured at T10 in *Australopithecus africanus* was 96.6, *Homo ergaster* 123.4, and *Homo sapiens* 223.4. On the basis of these thoracic vertebral canal measurements it has been postulated that in the fossil hominid ancestor (KNM-WT 15,000), *Homo ergaster* species was not able to achieve this degree of control. Estimates of the evolution of this precision control of respirations are during the period 1.6–0.1 mya [38].

It Is Difficult to Separate Thought and Language

We may think in words silently, named inner speech (endophasia), or speak or write it, external speech (exophasia). Thoughts and words were aptly described as the "mental commerce" of the mind by Gardiner [39]. Similarly Brain referred to language and "words as the guardians of our

thoughts." An abstract thought can be retained in mind by words, or a mathematical symbol or symbols that represent it [40]. The key differentiating features of human language in comparison to great apes are the vast repertoire of symbolic words, the complex grammar, and recursion. The latter refers to the ability to embed relational concepts within a phrase or sentence ("John thought that Mary thought that he knew the facts already"). The use of hand actions and gestures is closely related to speaking. However, written language (reading, writing) is a more recent evolutionary development and is acquired through formal tuition and imitation.

How Are Speech and Language Different?

Language impairment is due to a problem of left brain lesions, rarely the right hemisphere, and speech is usually an extracerebral problem concerning the articulatory structures such as the cranial musculature or those of phonation (vocal cords).

Neurobiology, Definitions of Language Components, and Related Activities

1. Thoughts—mental activity that may take place both in a linguistic and nonlinguistic form.
2. Endophasia—inner or internal language.
3. Exophasia—spoken or written language.
4. Ideas—the encoding of thoughts into sounds and sometimes utterances. These ideas pertain to events, objects, actions, and employing verbal symbols for transmission to another person.
5. Language—the communicable part of a person's knowledge. A signaling system used by people to communicate with one another.
6. Speech—oral communication by coordinated muscular activity and its neural control.
7. Grammar—the rules applying to the vocal sounds are combined to form words (lexicon), phrases, and sentences. Grammar may be divided into the following:
   Phonology (phonemes or sounds)
   Morphology (morphemes or words
   Grapheme—the smallest meaningful unit in writing
   Semantics—the symbolic meaning associated with a word or phrase
   Syntax (sentence construction)

The concept of a lemma refers to a psycholinguistic conceptualization of a concept with a specific meaning but no specific sound associated with it. With the utterance of a word, the suggestion is that we are turning thoughts into sounds, termed lexicalization [41]. The further elaboration of transmission of thoughts entails melodic intonation of speech, or prosody and pragmatics, which refers to the intricate arrangement of speech symbols such as words and prepositions [42].

Functional magnetic resonance imaging (fMRI, PET, magnetoencephalography) of language shows the participatory circuits and the

both hemispheres are active during language (left more than the right) as well as subcortical components (basal nuclei, thalamus, and the anterior and inferior temporal lobe (Fig. 10.7). Although much of our understanding comes from lesion models, mostly due to stroke, the newer imaging techniques such as MR- and CT-based perfusion scanning have shown that perfusion is a much better predictor of the aphasic syndrome or deficit compared to tissue impairment, usually DWI [43] (Fig. 10.8).

Language Dominance and Left Handedness

About 10 % of population is left handed and has been associated with different cerebral language organization to right-handed people. Left-handedness is associated with left hemisphere dominance in approximately 70 % and right brain dominance of about 30 % for language. Furthermore, left-handedness is associated with a greater likelihood of developing aphasia with stroke suggesting increased bilateral language representation. Recovery of aphasia is usually better in left-handed people. Insightful observations by Ribot and Pitres of aphasia in polyglots can be summarized by the following:

Ribot's law: better performance in the native language than newly acquired language
Pitres law: the language used more frequently prior to the stroke, recovers better

For example a case study by Obler and Albert, a Russian immigrant with global aphasia for the more recently acquired Hebrew language, had only mild anomia for his childhood-acquired Russian language [44]. Studies of aphasia in people that learnt two languages at the same time appear to affect both languages similarly— termed compound bilinguals. Polyglots likely involve their right hemisphere more when learning languages acquired later in life, which may account for the discrepancy [45].

Crossed aphasia refers to the rare instance whereby right-handed people have aphasia after

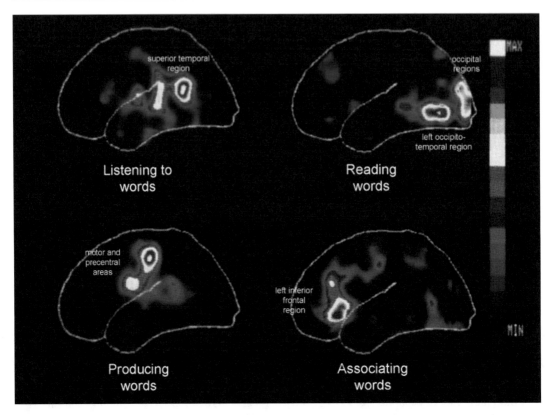

Fig. 10.7 Functional imaging of language networks. Viewing words correlate with predominantly occipital activation, listening with temporal lobe activation, speaking with Broca's area, and adjacent frontal cortical area activation and word generation with dorsolateral prefrontal cortex activation. Figure with permission: Scientific American Library. Posner MI, Raichle ME. Images of Mind. Scientific American Library, 1994, New York

damage to the right hemisphere as opposed to the left hemisphere which occurs in about 99 % of the time [46].

Functional imaging has revealed a widespread activation of not only perisylvian language cortex of Broca's and Wernicke's areas but also anterior temporal, inferior temporal, and occipital cortex. In summary these include the following:

Auditory: Heschl's gyri bilaterally, superior temporal lobe gyrus
Visual: left inferior occipital lobe gyrus
Phonologic networks: left superior temporal lobe gyrus, supramarginal gyrus
Semantic evaluation: left middle temporal lobe gyrus, angular gyrus
Short-term memory (auditory verbal): left supramarginal and superior temporal gyrus
Word retrieval: left frontal opercular, middle and inferior temporal gyrus, angular gyrus

Phonologic output: left superior temporal lobe gyrus and supramarginal gyrus
Speech production: Broca's area, putamen bilaterally, and motor cranial nerve nuclei
Written output: superior parietal region, left angular gyrus, middle frontal (Exner's area)
Sentence assembly: left frontal operculum
Syntax assembly: left dorsolateral prefrontal cortex L DLPFC
Conceptual and semantic components are integrated by the anterior temporal lobes bilaterally [47]

Clinical Aphasias

The most common clinical aphasia classification is the Boston Classification of Geschwind, Benson, and Alexander of eight different aphasic syndromes (Fig. 10.9) [48]. These classic aphasia

Fig. 10.8 Diffusion-weighted imaging and perfusion imaging mismatch in acute stroke expressive aphasia (*left images*) and global aphasia (*right images*)

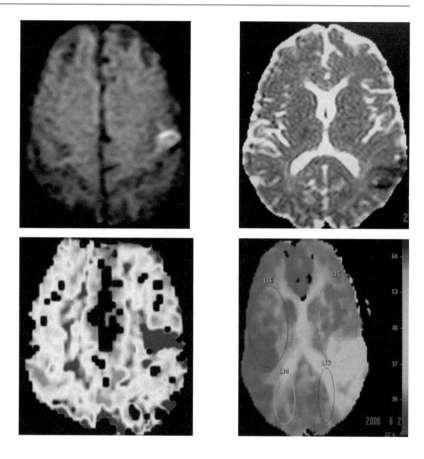

Fig. 10.9 Clinical aphasia subtypes

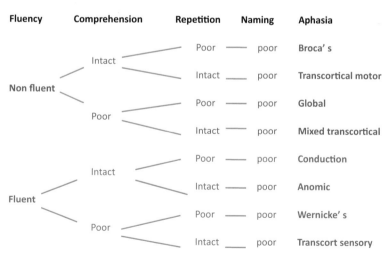

syndromes are vascular syndromes that follow a particular artery occlusion, for example Broca's and superior division of MCA and Wernicke's due to inferior division of the MCA, underscoring why the latter is so rare. PWI better accounts for the patients' clinical deficits as compared to DWI. This is because the former and not the latter is able to reveal dysfunctional tissue due to hypoperfusion.

Neuroimaging of vascular aphasic syndromes therefore is best by PWI if possible. Functional imaging techniques such as PET, fMRI, and magnetoencephalography reveal that both hemispheres are activated in language tasks albeit the left more than the right. There is also activation seen in areas outside the generally accepted language circuitry such as the inferior temporal, and anterior temporal lobe, basal nuclei, and thalamus. Aphasia can also be analyzed by at least two different methods. The neurobiological process attempts to decipher specific cognitive mechanisms that underlie language processed that include semantics, modality independent lexical access, orthographic and phonological representations, and articulatory programming. The other method is by analyzing clinical aphasia syndromes after lesion studies mostly due to stroke [49].

Lesion studies in neurology gave us the initial brain localization associated with major aphasic syndromes pioneered by Lichtheim, Broca, Wernicke, Geschwind, and others. However modern cerebral connectomics and functional MRI studies point to much more extensive brain circuitry involved in language comprehension and output than a perhaps more restricted concept of speech areas such as motor posterior inferior frontal and receptive language, temporal lobe regions [50–53]. For example, a direct cortical surface recording study in neurosurgical patients revealed that Broca's area is silent during speech production whereas the motor cortex is active. The interpretation of this new data is that Broca's area should be conceived as a coordinating center (hub) of the transfer of information from large-scale cerebral (cortical and subcortical) networks that are required for spoken language. Broca's area is therefore not the site of articulation but a hub for forwarding transmission of neural information to motor areas [54]. Lieberman on the other hand rejects the traditional Broca-Wernicke brain-language model as incorrect emphasizing the critical importance of the frontal subcortical circuitry in speech and language processing. Furthermore he maintained that aphasia does not occur unless there is subcortical involvement

(nor cortical such as Broca's area), requiring involvement of the basal ganglia and their connecting fiber tracts [55, 56]. Nevertheless a wealth of over a hundred years of clinical lesion reports have allowed us to identify eight or more principal language disorders that occur commonly especially after cerebral infarction, but also due to other lesions such as tumors, intracerebral hemorrhage, trauma, and ictal aphasia for example.

The major aphasic syndromes are conveniently presented and discussed under the following subheadings:

1. Core linguistic features
2. Associated cognitive neurological signs
3. Associated elementary neurological signs
4. Neuroradiology and typical lesion location
5. Prognosis

Clinical Aphasia Subtype Classification

Perisylvian aphasias
  Broca's
  Wernicke's
  Conduction
  Extra perisylvian
  Transcortical motor
  Transcortical sensory
  Transcortical mixed
  Very large and very small lesions causing aphasia
  Global aphasia
  Anomic aphasia
  Subcortical aphasia
  Thalamic
  White matter
  Internal capsule
  Caudate
  Other aphasias and related syndromes
  Mutism
  Aphemia
  Foreign accent syndrome
  Forced hyperphasia
  Dynamic aphasia of Luria

Pure word deafness

Gerstmann's syndrome

Angular gyrus syndrome

Allied speech disorders

Dysarthria

Aphonia

Disorders of written language and newer cultural abilities (separate chapter)

Reading impairment

Alexia with agraphia

Central alexia

Semantic paralexia

Alexia without agraphia

Agraphias

Acalculias

The Clinical Aphasia Syndromes May Be Conveniently Parcellated into Four Principal Functions Yielding a Clinical Classification System of Eight Different Types

These include fluency, comprehension, and repetition. All have naming deficits, so that the mildest form will present only with dysnomia. However this is an overly simplistic approach as people with Broca's aphasia have components of comprehension deficits and those with Wernicke's aphasia may have components of expressive dysphasia. Conceptualization of the aphasic syndromes may also be assisted by considering them as follows:

1. Global aphasia: Broca's + Wernicke's
2. Transcortical motor aphasia (TCM): Broca's aphasia but with preserved repetition
3. Transcortical sensory aphasia (TCS): Wernicke's aphasia but with preserved repetition
4. Transcortical mixed aphasia (TMA): Global aphasia with preserved repetition or TCM + TCS
5. Conduction aphasia: Opposite to transcortical aphasia with impaired and spared repetition, respectively

The Three Perisylvian Aphasias Are Relatively Common and Often Due to Cerebral Infarcts of Embolic Etiology

Broca's dysphasia (expressive aphasia)

Core linguistic features

Sparse output (10–12 words per minute)

Impaired repetition

Loss of melody (dysprosodic)

Short phrase length

Poor syntactic output, incorrect sentence formulation using words and phrases

Decrease in the number of prepositions

Decrease in grammatical modifiers

Verb use is usually more impaired than nouns

Associated cognitive neurological signs

Writing impairment

Reading impaired more than auditory comprehension. This is also referred to as a third alexia.

Apraxia of the normal arm and leg (left) is common in Broca's aphasia with left ideomotor apraxia on imitation that occurs in approximately 2/3 of patients.

Ideomotor apraxia of the right arm may be detected if no overriding weakness is present.

Facial apraxia is common.

Associated elementary neurological signs

Right hemiparesis, usually in Broadbent' distribution or right arm weaker than right leg

Neuroradiology and typical lesion location

Lesion studies—Left inferior frontal gyrus lesion, most often infarction, involving the pars triangularis and pars opercularis or Brodmann's area 44 and 45

Prognosis

May be a Baby Broca's (lesion restricted to cortical BA 44 and 45) or Big Broca's (larger lesion of frontoparietal operculum) with former recovering completely after days to weeks and the latter persisting

Usually evolve to less severe syndrome and to dysnomic aphasia [57, 58]

Wernicke's Aphasia

Core linguistic features

Predominant finding is impaired comprehension and repetition.

Fluency is retained and may be increased with logorrhea, usually nonsensical.

Increased speech output may be in the order of 200 words or more per minute.

Speech output is interspersed with frequent paraphasias.

The paraphasias may take the form of semantic, literal, phonemic, or neologistic.

Semantic paraphasias—a real word that has semantic relationship to the target word (boat and ship).

Phonemic or literal paraphasias are defined as the substitution of a word that preserves half the syllables or components of the word intended to be spoken. This may also apply to non-words.

Neologistic paraphasia—less than 50 % of phonemes in common with the target, usually with different initial phoneme.

Jargon aphasia is meaningless and not comprehensible but seems to have meaning to the person speaking it.

Associated cognitive neurological signs

There is an apparent unawareness of the language deficit and may even include euphoric disposition.

This is in marked contradistinction to Broca's aphasia where depression and emotional lability may be associated.

Written and spoken language impairments are not similarly affected.

Patients usually have a more profound word deafness than word blindness, although the reverse may also occur.

One aspect of auditory comprehension, the correct performance of axial movements such as eyes and trunk elicited by commands such as close your eyes, stand to attention, and turn around, may be spared.

Associated elementary neurological deficits

There is a notable lack of long tract signs with usually no accompanying hemiparesis or hemihypoesthesia.

The only elementary neurological sign is a frequent right upper quadrantanopia and minor and transient faciobrachial sensory loss.

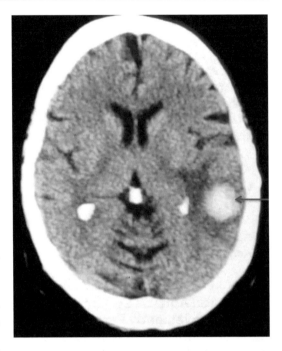

Fig. 10.10 Wernicke's aphasia due to a left temporoparietal intracerebral hemorrhage (*arrow*)

Neuroradiology and typical lesion location

Lesion studies—left superior posterio-temporal gyrus (BA 22)

May also involve the left supramarginal gyrus (BA 39) and angular gyrus (BA 40) (Fig. 10.10)

Prognosis

May evolve through conduction aphasia or anomic aphasia [59, 60]

Conduction Aphasia

Core Linguistic Features

Repetition is markedly difficult. However although unable to repeat a word the person is able to accurately describe the word. Bedside testing progresses from single-word repetition to phrases and sentences as a rough approximation of mild, moderate, and severe categories.

Fluency is retained but with paraphasias with mostly literal, some semantic, and occasional neologisms in a typical patient.

Comprehension normal

Writing is impaired but less so than spoken language.

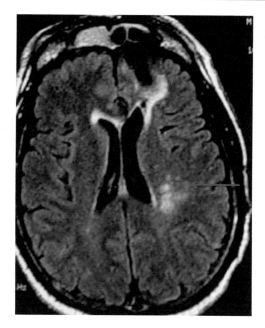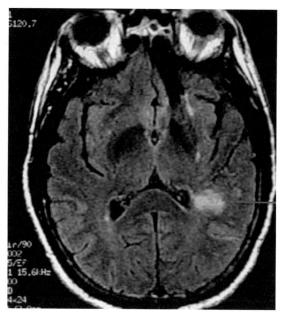

Fig. 10.11 Conduction aphasia: A typical infarct location of subjacent white matter tracts to the planum temporale (*red arrows*)

Associated Cognitive Neurological Signs

Occasional ideomotor and buccolingual facial apraxia

Associated Elementary Neurological Signs

Occasional right hemisensory impairment or heminanopia, hemiparesis is unusual [61]

Neuroradiology and Typical Lesion Location

Left supramarginal gyrus (BA 40), arcuate fasciculus, white matter connections to insula, and primary auditory cortex. The arcuate fasciculus connects Wernicke's and the frontal cortex, specifically the premotor and motor areas, and not Wernicke's and Broca's areas as previously held [62] (Fig. 10.11). In a case of left middle cerebral artery infarction of the parietal lobe, with sparing of both Wernicke's and Broca's areas, a DTI evaluation of the arcuate fasciculus showed a much smaller tract on the left compared to the spared right hemisphere. This was also evaluated by the anisotropy index of 0.43 ± 0.10 of the left side compared to 0.46 ± 0.10 of the right uninvolved fiber tract [63] (Fig. 10.12). Functional activation studies show that the area critically associated with clinical conduction aphasia is part of the superior temporal lobe, the planum temporale area, also a hub for phonological working memory [64].

Prognosis

Case report studies have reported excellent recovery from conduction aphasia in the weeks to months after stroke, for example, even with persistent abnormality of the arcuate fasciculus by functional MRI with diffusion tensor imaging [65].

Global Aphasia

This may be viewed as a combination of both receptive and expressive aphasias and due to relatively large left hemisphere lesions often due to large middle cerebral artery infarctions.

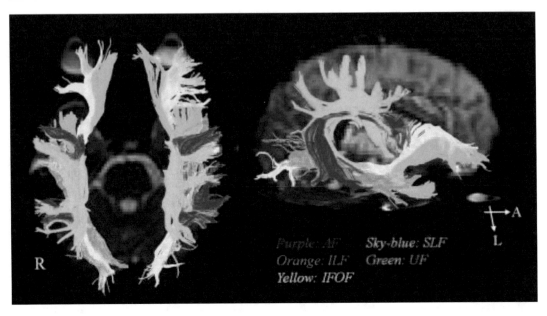

Fig. 10.12 Aphasias and arcuate fasciculus imaging by DTI. Figure with permission: Jang SH. Diffusion Tensor Imaging Studies on Arcuate Fasciculus in Stroke Patients: A Review. Frontiers in Human Neuroscience 2013;7:749. doi: 10.3389/fnhum.2013.00749

Core Linguistic Features

Marked impairment in verbal output
    Comprehension impaired
    Repetition impaired
    Anomia
    Agraphia
    Alexia

Associated Cognitive Neurological Signs

Often beginning with initial muteness and evolving to some degree of phonation, frequently with ability to produce simple words, even occasionally short phrases.

Comprehension is generally better than verbal output and these patients often show the ability to interpret gestures and other non-language communication such as intonations and facial or body part communication. Associated neuropsychiatric syndromes such as agitation and depression occur relatively frequently. Interestingly sparing of function such as musical ability may be evident [66].

Associated Elementary Neurological Signs

Right hemiparesis
    Right hemihypesthesia
    Right homonymous heminanopia
    At times large left hemisphere stroke, even with right-handedness, may be remarkably devoid of language deficits or recovery very rapidly, within days.

Neuroradiology

Most often due to cerebral infarction or intracerebral hemorrhage with extensive left middle cerebral artery territory infarct due to ICA or MCA stem occlusion.

Prognosis

Evolution is most frequently into Broca's aphasia or anomic aphasia, less often Wernicke's aphasia [67, 68].

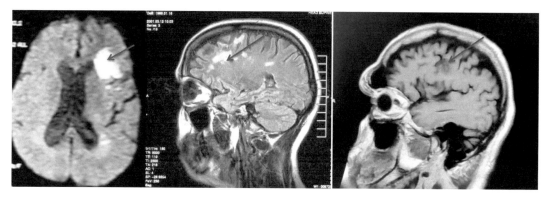

Fig. 10.13 Transcortical motor aphasia lesions depicting the suprasylvian aphasia centers (*red arrows*)

The Transcortical Aphasic Syndromes Are Usually Due to Watershed-Type Lesions and Are Due to Lesions Located More Peripherally Than the Perisylvian Language Syndromes

The overall recovery of transcortical aphasias has a relatively good prognosis and in some case series has an excellent recovery trajectory to normal language in the weeks to months after stroke. The very nature of these aphasias may suggest the stroke mechanism and prompt initiating investigations to detect large vessel cervicocephalic or intracranial stenosis implied by this diagnosis [69].

Transcortical Motor Aphasia: Motor Aphasia with Relatively Preserved Repetition

Core linguistic features
  Non-fluent aphasia
  Preserved repetition
  Preserved comprehension
  May have echolalia
  Associated cognitive neurological signs
  Abulia
  Frontal lobe release signs
  Associated elementary neurological findings
Atypical right hemiparesis with right leg weaker than arm. This is the reverse of Broadbent's hemiparesis distribution typical of middle cerebral artery infarctions.
  Neuroradiology and typical lesion location

Left ACA territory lesion, anterior to Broca's area
  Deep subcortical frontal lesions
  Medial frontal lesions of supplementary motor area (Fig. 10.13)
  Prognosis
  Generally good to excellent and improves to anomic aphasia and often normality

Transcortical Sensory Aphasia

This aphasia is also associated with a relatively good prognosis and recovery that was tracked by one of the largest contemporary aphasia registries using the Western Aphasia Battery for diagnosis and monitoring. The WAB allows both aphasia quotient evaluation and aphasia subtype characterization into eight different aphasias including transcortical subtypes [70]. It is postulated to be caused by disruption of the left hemisphere phonological processing areas (posterior superior temporal lobe) and the lexical to semantic evaluation circuit in the inferior parietal lobe [71].
  Core linguistic features
  Fluent paraphasic speech
  Semantic paraphasias
  Impaired comprehension (auditory and written)
  Preserved repetition
  It may be summarized as Wernicke's aphasia syndrome but with preserved repetition
  Associated cognitive neurological signs

Angular gyrus syndrome

Gerstmann's syndrome

Echolalia

Associated elementary neurological signs

Right visual field defect (quadrantanopia)

Neuroradiology and typical lesion location

White matter of posterior periventricular region that connects to the inferior lateral occipital temporal cortex [72]

Inferior temporal gyrus (hemorrhage more likely than bland infarct)

Thalamic lesions

Prognosis

Generally good to excellent and improves to anomic aphasia and often normality

Transcortical Mixed Aphasia Is Also Called Isolation of the Speech Area

These syndromes are typically related to more extensive watershed brain lesions affecting the left hemisphere with sparing of the perisylvian language areas and circuitry. The perisylvian areas have effectively been "isolated" from the left hemisphere (Figs. 10.14 and 10.15). The person has impaired language function in all domains including dysfluent speech, comprehension impairment, and dysnomia but is able to repeat relatively well. In a sense this can be considered the opposite to conduction aphasia where the person has relatively preserved language ability but with a remarkable inability to repeat.

Core linguistic features

Mute or nearly so

Cannot comprehend

Repetition retained

Reading and writing impaired

May be summarized as global aphasia with preserved repetition

Associated cognitive signs

Echolalia

Associated elementary neurological signs

Abulia, paratonia, other frontal lobe signs

Right hemiparesis, sensory loss

Right homonymous hemianopia or quadrantanopia

Neuroradiology and typical lesion locations

Frontal and temporal lesions, multiple lesions, or large border-zone infarction

Anterior thalamic lesions

Prognosis

May evolve into transcortical motor aphasia or anomia [73–75]

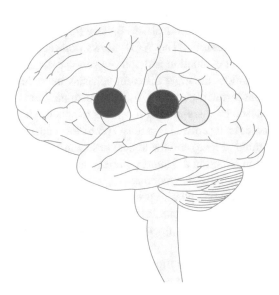

Fig. 10.14 Broca's (*blue*), conduction (*red*), and Wernicke's (*yellow*) aphasia approximate lesion sites

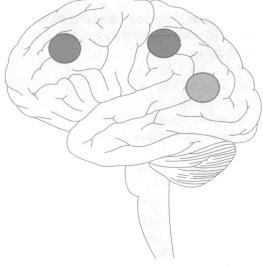

Fig. 10.15 Transcortical motor (*pink*), transcortical sensory (*orange*), and transcortical mixed (*green*) aphasia approximate lesion sites

Other More Restricted Language Syndromes, Overlapping and Related Syndrome

1. Anomias
2. Subcortical aphasias
3. Aphemia
4. Pure word deafness and auditory agnosia
5. Gerstmann's syndrome
6. Angular gyrus syndrome
7. Mutism
8. Dysarthria
9. Dysphonia
10. Ictal aphasia
11. Foreign accent syndrome
12. Acquired stuttering
13. Echolalia and palilalia
14. Dynamic aphasia
15. Festination

In addition to the above entities, the syndromes of alexia, agraphias, and acalculias can be considered in this category but will be discussed separately in the chapter of newly acquired cultural circuits and their disorders.

Anomias

Definition

These are relatively mild aphasic syndromes that have good comprehension, fluency, and repetition but impaired confrontation naming and word generation tests (categories of animals or words). Anomias represent the most common language deficit in aphasic syndromes and occur in all aphasic syndromes to a greater or lesser degree. Nouns, verbs, and adjectives are all affected but nouns are most susceptible and circumlocution is common as a method of countering the deficit. Anomia occurs in a wide variety of neurological illnesses such as dementia, delirium, raised intracranial pressure, subarachnoid hemorrhage, encephalitis, toxic metabolic encephalopathy, and stroke but also overlaps with normality. When there is some comprehension deficit, but good repetition, transcortical sensory aphasia and

anomic aphasia may be regarded as existing on a continuum with the differentiating feature restricted to the degree of comprehension that exists. In general, all aphasia syndromes recover to become an anomic aphasia of varying degree. It is very sensitive to the overall neurological status but is not specific. Subtypes of anomic aphasia include the following:

Word Production Anomia

Inability to produce the correct name but retention of knowledge of the correct one. Therefore there is retained ability to recognize the correct word and there is a distinct benefit from cueing such as priming the patient with the first phoneme. It may accompany all anterior and posterior aphasic syndromes as well as diffuse neurological illness, medication effects, and metabolic abnormality

Word Selection Anomia

With this category of anomia there is an inability in selecting the correct word despite the prompting afforded by cueing. There is retention of knowledge of how to use the particular object for example as well as retention of the ability to single out the object when given the correct name to the patient. The patient is unable to name the object no matter what sensory modality, whether visual, auditory, and tactile, is used for the presentation.

Semantic Anomia

With this dysnomia subtype, the patient does not respond to cueing and in addition is not able to single out the object when given the correct name. This may be regarded therefore as a combination of both word production and word selection anomia, or a two-way deficit.

In summary, when one cannot name but is enabled with cue it is word production anomia. When one cannot name but still can select correct object from a group, it is word selection anomia. When one can neither name even if aided by cueing or selecting from a group even when given the right name it is semantic anomia. The symbolism of the word and the object has been severed [76, 77].

Disconnection Anomias

A fourth category of anomic aphasias are referred to as the group of disconnection anomia comprising of three subtypes: modality-specific, category-specific, and callosal forms. The lesion location is in the posterior white matter.

Modality-Specific Anomia

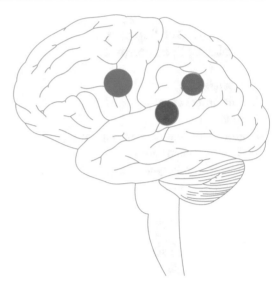

Fig. 10.16 Word production (*blue*), word selection (*red*), and semantic anomia (*purple*) lesions

The responsible lesion interrupts the networks between the language and one of the specific sensory areas of the cortex. Anomia is impaired for the specific sensory modality that may pertain to colors, objects, faces, or letters. Hence the diagnoses of visual anomia may be a color anomia or object anomia for example. These are similar to object or color agnosia. With color agnosia the person cannot recognize but can name and in color anomia the person can recognize but cannot name. Typical lesion location is in left posterior white matter.

Category-Specific Anomia

An inability in the naming certain body parts has been described, for example, as in the selective anomia for facial expressions [78, 79]

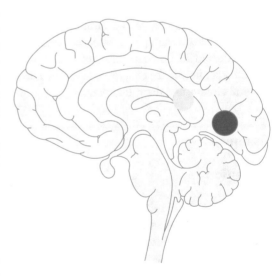

Fig. 10.17 Disconnection anomia lesions. Callosal anomia (*yellow*) and category and modality specific (*blue*)

Callosal Anomia

With the intact right hemisphere sensory cortex, objects can be adequately evaluated by size, shape, texture, and weight. However the transmission to the left hemisphere is impaired by a callosal lesion which renders the left hemisphere unable to name the object [80–82] (Figs. 10.16 and 10.17).

Aphemia (Pure Word Mutism, Apraxia of Speech)

Core linguistic features

Acquired inability to speak, mutism in acute stage then non-fluent speech

Regarded as a restricted form of Broca's aphasia

Preserved comprehension and written language is normal

Evaluation of written sentences indicates they are well constructed

Elementary neurological findings

Right facial weakness

May have transient right hemiparesis

Neuroradiology

A number of regions have been implicated

A restricted lesion within Broca's area, either small cortical lesion or subcortical lesion

Left frontal opercular region

Premotor area

Left medial frontal cortex and cingulated gyrus

Supplementary motor cortex

Prognosis

May evolve into a transcortical motor aphasia

Sometimes "foreign accent syndrome"

Anomia then normality [83–86] (Fig. 10.18)

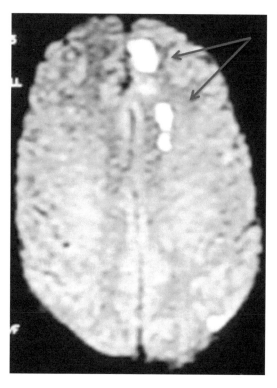

Fig. 10.18 Aphemia infarct locations

Pure Word Deafness and Auditory Agnosia

Pure word deafness may be viewed as a restricted form of Wernicke's aphasia and Wernicke's aphasia may improve and evolve into a PWD and subsequently a dysnomia.

Core linguistic components

Spontaneous speech, reading, and writing normal

Cannot comprehend verbal speech

Pure tones such as telephone ringing, door bell ring, dog barking, and environmental sounds remain intact

Repetition is impaired

Naming may vary from normal to mild dysnomia

Nonverbal communication remains intact

Associated cognitive neurological signs

Minor aphasic deficits may include paraphasias

Associated elementary neurological signs

Usually no long tract signs are associated

Neuroradiology and typical lesion location

Bilateral temporal lobe lesions are necessary for this syndrome

Pure word deafness and auditory agnosia may be viewed as a disconnection of Wernicke's area (BA 22), which is not affected, from both auditory cortices (Heschl's gyri)

If there is destruction of both Heschl's gyri, cortical deafness results with inability to hear words as well as environmental sounds

Unilateral lesions have been reported with disconnection due to subcortical lesions that involve the dominant temporal lobe involving ipsilateral auditory radiations and contralateral auditory radiation via the corpus callosum

Prognosis

Improves to a dysnomia [87–89]

Examples of two different, albeit rare, mechanisms that led to a presentation of pure word deafness. One was due to bilateral idiopathic thrombotic thrombocytopenic purpura, with bilateral temporal infarcts, and another due to bilateral cerebral metastases from lung adenocarcinoma (Fig. 10.19).

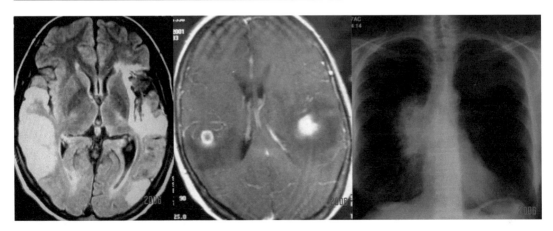

Fig. 10.19 Pure word deafness lesion location examples: Bilateral temopral lobe infarctions secondary to idiopathic, thrombotic, thrombocytopenic purpura (*left image*), and bilateral temporal metastatic deposits (*center image*) secondary to lung adenocarcinoma (*right image*)

Auditory agnosia refers to an inability to identify environmental sounds that may be due to unilateral (right posterior temporal lobe) or bilateral superior temporal lobe lesions. It may be independent of pure word deafness or both may be present. Normal ear function and connecting tracts have to be ascertained first [90].

Subcortical Aphasia

A common, relatively mild, mixed aphasia that generally has a good prognosis and has a correlation with small vessel cerebrovascular disease mechanisms. In one recent study it accounted for 12 % of aphasic disorders in one of the largest case series reported [91].

Core Linguistic Features

Acute mutism that evolves into hypophonia and
  subsequent further improvement
Dysarthria, slow, impaired articulation
Paraphasias noted by the observations that repetition tends to abolish them
Comprehension, naming, and reading generally
  mildly impaired

Neuroradiology and Lesion Locations

Left-sided infarcts or hemorrhages in the caudate, putamen, thalamus, supplementary motor area, and anterior cerebral artery territory (Fig. 10.20). White matter language tract infarction may also result in subcortical aphasia particularly the insular white matter and the superior temporal cortex to inferior frontal area connections. An additional postulated mechanism is a diaschisis deactivation that is transient, usually over days to weeks.

Prognosis

Generally good, improvement to normality is usual [92–95].

Other Language-Related Syndromes of the Left Hemisphere

1. Gerstmann's syndrome
2. Angular gyrus syndrome
3. Mutism
4. Dysarthria
5. Dysphonia
6. Ictal aphasia

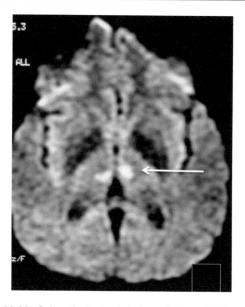 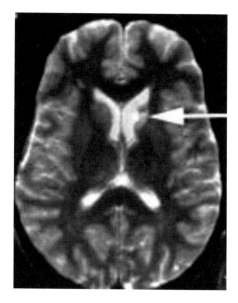

Fig. 10.20 Subcortical aphasia lesions, thalamus (*right image*, *arrow*), and caudate nucleus (*left image*, *arrow*)

7. Foreign accent syndrome
8. Acquired stuttering
9. Echolalia and palilalia
10. Frontal dynamic aphasia of Luria
11. Progressive aphasias
12. Cortical sensory loss

Gerstmann's Syndrome

The tetrad of signs first described by Josef Gerstman in 1930, comprises of:

Finger agnosia
Dysgraphia
Dyscalculia
Right left disorientation

The combination of these superficially disparate deficits may be explained by the important role of fingers in counting and praxis. The parietal lobes are at the crossroads of the sensory modalities and enable cross-modal associations of tactile, visual, and acoustic perceptions. In addition the intraparietal sulcus is the hub for numerosity. Lesion studies have consistently shown the correlation of these deficits with left inferior parietal lobe, recently bolstered by the transcranial magnetic stimulation virtual lesion of this area and associated with the components of Gerstmann's syndrome. Relatively discrete lesions of this region are uncommon but a case of cerebral amyloid angiopathy-related hemorrhage is presented in Fig. 10.21 [96, 97].

The Angular Gyrus Syndrome

Is seen with a more extensive lesion than with Gerstmann's syndrome and has additional deficits, namely:

Finger agnosia
Dysgraphia
Dyscalculia
Right left disorientation
Dysnomia
Alexia

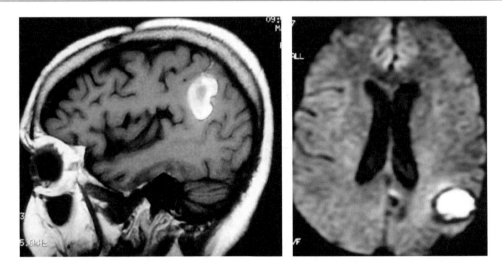

Fig. 10.21 A person with a relatively "pure" Gerstmann's syndrome due to small intracerebral hemorrhage likely amyloid angiopathy related

Commonly Confused with Dysphasia Are Mutism, Dysarthria, and Dysphonia

Mutism is a common syndrome in acute neurological illness and often improves rapidly over days to weeks depending on the underlying process. There are many different causes including:

Mesencephalic lesions
Frontal lesions
Left thalamic disturbance
Supplementary motor lesions
Acute damage of the posterior inferior frontal region (Broca's area)
Severe laryngitis
Cranial nerve damage

Dysarthria

A condition whereby the speech is altered that causes problematic communication but is not due to the cortical language networks and hence writing is unaffected. It may be due to subcortical or basal ganglia or cerebellar abnormalities. Subtypes include:

Paretic (stroke myasthenia, cranial nerve dysfunction)
Ataxic (cerebellar lesions)
Extrapyramidal (Parkinson's, Huntington's, choreiform disorders)
Dystonic
Hyperkinetic
Hypokinetic
Spastic

Dysphonia

Impaired communication due to phonation difficulty but with preservation of written language, fluency, comprehension, and repetition. The verbalizations and sounds that emanate may be difficult to understand and misunderstood for a dysphasia. Different forms may be recognized depending on the causative neurological or laryngeal process.

Aphonia and spastic dysphonias may occur. The latter refers to a focal dystonia that involves the laryngeal muscles. The response to botulinum toxin injection into the thyroarytenoid muscle is significant and alleviates symptoms in approximately in large case series [98].

Stuttering, Palilalia, and Echolalia

Although stuttering is usually congenital, it may be acquired secondary to neurological disease affecting Broca's region. Involuntary repetition of syllables, words, and sometimes phrases is referred to as palilalia and echolalia, respectively, and is encountered with frontal lobe lesions, schizophrenia, Tourette's syndrome, Parkinson's, and seizures [99, 100].

Ictal Aphasia Syndromes

Transient interruptions of language or speech may be the sole epileptic manifestation, sometimes the only sign of nonconvulsive status epilepsy. The syndromes include the Landau Kleffner syndrome, SMART syndrome (stroke-like migraine attacks after radiation therapy), and more rarely lesions of the temporal lobes [101, 102].

Foreign Accent Syndrome

A relatively rare speech disorder that manifests with a disordered accent, associated with predominantly motor aphasias due to a number of different pathologies but most commonly due to stroke but neurodegenerative diseases and multiple sclerosis related, has also been described. It may be important to recognize as the person may have fully recovered from a previous dysphasic syndrome. The most commonly reported accents have included English to French, American to Irish, as well as Brazilian to North American and Japanese to Korean. It has been ascribed to impairment of the motor speech planning circuitry that involves the cerebellum, premotor cortex, and motor cranial nerves. These in turn result in an alteration of intonation, rhythm, and stress of articulation of vowels, consonants, and syllables [103–107].

Frontal Dynamic Aphasia of Luria

Described by Luria [108], this syndrome is noted by a marked reduction in spontaneous speech but with preserved comprehension, naming, and repetition. Echoing and perseveration may be associated. Two forms may be distinguished, with left inferior frontal lesions and bilateral frontal or frontal subcortical lesions. The postulated neurobiology is an impairment at the level of conceptual preparation that takes place at the prelinguistic level [109].

Progressive Aphasias

These present with progressive language and speech impairment and may be associated with frontotemporal lobe degeneration or Alzheimer's disease. The three principal subtypes include the following:

Nonfluent, agrammatic primary progressive aphasia is associated with tau pathology
Semantic variant progressive aphasia correlates with TDP43 pathology
Logopenic variant progressive aphasia correlates with AD pathology [110–112]

Trajectories of Recovery in Aphasia Syndromes. Some Patterns of Reported Recovery

Recovery is generally more favorable in left versus right handers
Usually excellent recovery often occurs with dysnomias, and transcortical syndromes
Poorer recovery is seen with global, severe Broca's and Wernicke's subtypes
Global aphasia usually recovers to a Broca's subtype
Wernicke's may recover into a transcortical sensory, pure word deafness, or dysnomia

Conduction aphasia generally recovers to an anomic aphasia

Anomic aphasia and transcortical motor aphasia represent a continuum, with the main differentiating feature being the degree of comprehension

Testing for Aphasia

Bedside Testing

The principal elements of language evaluation include the assessment of spontaneous speech for fluency (nonfluent aphasia is less than 50 words per minute and fluent aphasia is usually 100–200 words per minute), comprehension for spoken language, repetition, naming, reading, and writing. This can easily be accomplished at the bedside or in the clinic with a semiquantitative test battery as follows:

Speech and Language Assessment

Scoring: 1 for each error, 0 is normal. Any point lost in any of the eight items constitutes an abnormality for that category.

Language	
1. Naming: Name three objects (pen, watch, ID card) and name three colors	/6
2. Fluency: Grade as fluent (0), nonfluent (1), mute (2)	/2
3. Comprehension: Close your eyes, squeeze my hand. Score 1 for each failure	/2
4. Repetition: "Today is a sunny and windy day." Nil (2), partial (1), all (0)	/2
5. Write a sentence. What is your job? Contain subject, verb, and make sense	/3
6. Reading: "Close your eyes." No words read (2), partial (1), or all words (0)	/2
Motor speech	
7. Dysarthria; slurring of words in interview. Nil (0), mild (1), marked slurring (2)	/2
8. Hypophonia; normal (0), voice softer than normal (1), barely audible (2)	/2
Total score	/21

Adapted from reference [91]

Metric Testing

Boston Diagnostic Aphasia Examination (Version 3)

Several brief, abbreviated, and more comprehensive language assessment tools have been developed, each with a different focus. Boston Diagnostic Aphasia Examination (BADE version 3) for example focuses on the components of language more from a neurobiological perspective rather than a lesion mechanism. It is a comprehensive and time-demanding test that may take 3–4 h and consisting of 34 subtests but has a short form option that can be completed in approximately 1 h. An aphasia severity score is provided as well as the number of subtest scores to help monitor and guide therapies [113].

The Western Aphasia Battery (WAB-Revised)

Now in its revised version it has a different focus providing a clinical aphasia classification through metric assessments. These include the expressive, receptive, global transcortical (motor sensory, mixed), and anomic aphasia subtypes. In addition both an aphasia quotient (100 is the maximum score) and a cortical quotient may be calculated to monitor progress or deterioration over time and assist in the rehabilitative process [114].

Boston Naming Test (BNT Version 2)

This language screening test consists of line drawings that are graded into 20 easy, 20 moderately difficult, and 20 more difficult items. It has two versions, a 60-time and a short 15-item one. Both come with a cueing option with four semantic cueing items on the back of each line drawing. Population, education, and age-specific norms are available and typical normal adult score for the 60-item test is 52+/3 [109].

Two other relatively frequent tests used in aphasia research and clinical reports include the Aachen Aphasia Battery and the Porch Index of Communicative Ability (PICA) [115, 116].

Treatment

Accurate the determination of aphasia subtype, facilitates monitoring of the language impairment allowing tracking of changes and whether deterioration or improvement is occurring. This will allow treatment interventions to be appropriately evaluated. A number of different treatment modalities and approaches have been considered. These range from standard aphasia therapies to pharmacological, constraint-related, and more recently device-related therapies such as TMS and t-DCS [117]. Although vascular aphasic syndromes have dominated our understanding of aphasia, another approach based on specific language tasks such as lexical semantics, articulatory planning, phonologic and orthographic representation, and modality-independent lexical access has been delineated by Hillis et al. [47]. This may have more relevance in treatment strategies for nonvascular syndromes such as the progressive aphasia syndromes such as FTD, TBI, and other pathologies such as brain tumor and multiple sclerosis related [118]. Aphasia therapies may be grouped into the following entities:

1. Intensive speech-language aphasia therapy
2. Pharmacological (Nootropics, Dopamine, Methylphenidate, Donepezil)
3. Melodic intonation therapy
4. Gestural therapy
5. Constraint-induced therapy for aphasia
6. Transcranial magnetic stimulation therapy
7. Transcranial direct current stimulation
8. Wernicke's aphasia treatment proposals
9. Right hemisphere engagement

Intensive Speech-Language Therapy

Speech-language therapy has been the cornerstone of aphasia treatment and more recent studies have shown that intensive aphasia therapy may show significant improvements if applied. Aphasia therapy has shown to be of benefit if intense sessions are given, such as 2-h sessions occurring 4× per week [119]. A recent meta-analysis concluded that treated versus untreated patients in the early post-stroke period rendered an effect size of 1.83 [120]. In a study of German-speaking patients with chronic aphasia duration of up to 12 months, intensive language therapy of 9 h per week for 6–8 weeks was associated with significant psychometric improvement by the Aachen Aphasia test scores in about 2/3 of patients [121, 122].

Pharmacotherapy

A number of case reports, case series, and controlled trials have repeatedly shown benefit from stimulatory-type therapy with a variety of different agents. Although not all studies have been positive, the aphasia populations studied and the controls and outcome measures have been very heterogenous [123, 124].

Nootropics such as piracetam, evaluated by placebo-controlled trials, has been shown to provide effective recovery from aphasia when treatment ensues in the early stroke period but not during the chronic aphasia period. Stimulating drugs that act on the tri-monoamine systems such as the noradrenaline, dopamine, serotonin, and acetylcholine ascending neurotransmitter systems have shown general, but with variable efficacy. A number of studies have used bromocriptine in both acute and chronic aphasias with efficacy demonstrated in case series, open-label, as well as placebo-controlled trials. It appears most effective in nonfluent aphasias. Other reports with positive results have used acetylcholinesterase inhibitors such as Donepezil as well as Dexamfetamine and Memantine [125].

Melodic Intonation Therapy

Based on the observations that people with profound motor aphasia may sing remarkably well, this therapy uses a melodic intonation of words or phrases or a employing a left limb rhythmicity to facilitate fluency. These have been shown to engage right frontotemporal (superior temporal, inferior frontal, premotor, and sensorimotor cortical areas) circuitry, one of the two recognized recovery mechanisms in aphasia, the other being left hemisphere perilesional cortical involvement or recruitment. First developed in 1973 by Albert, many studies have since demonstrated its efficacy in nonfluent stroke patients. Intensive MIT therapy in chronic stroke patients confers language and speech improvements assessed by a variety of methods including the Boston Diagnostic Aphasia Examination as well as demonstrating measurable homotopic white matter fiber tract plasticity [126–129].

Device-Based Therapies: Transcranial Magnetic Stimulation (TMS) and Direct Current Transcranial Stimulation (d-TCS)

Research from vascular aphasias and f-MRI evaluation indicates that cerebral language recovery and reorganization occur in at least two stages. A reduction of left hemisphere activation of left hemisphere language-associated networks initially (acute phase) is followed by homologous right hemisphere activation in weeks to months after the insult [30]. Both stimulation and activity by TMS and d-TCS modalities have already been successful using these pathophysiological insights. One of the mechanisms of MIT has been postulated to be decreasing the hyperactive right homologous Broca's area that in turn may be inhibiting the lesioned left Broca's area. TMS

and t-DCS are able to stimulate or inhibit brain areas depending on the frequency and placement of the stimulation. These noninvasive devices are able to stimulate activity in the lesioned cortical area or in dysfunctional circuitry as well as being able to suppress overactive areas that may impede functional improvement.

After a stroke for example relatively excitable cortex can be found in the unaffected hemisphere, depicted in red and hypoactive cortex within the affected hemisphere depicted in blue. Presumed mechanisms include inhibition across the corpus callosum (yellow arrow) from the intact to the lesioned hemisphere. Stimulating the hypoactive region by high-frequency rTMS as well as anodal tDCS correlates with improvement of motor function. Also, inhibiting the hyperactive unaffected hemisphere with low-frequency rTMS as well as cathodal tDCS facilitates motor recovery and function (Fig. 10.22). Controlled trials have shown improvement in word finding retrieval, as well as reaction in naming tests [130–135].

Gestural Therapy

This therapy is based on the evolutionary origins whereby gestural control circuits "collateralized" speech control areas, due to their close contiguous association in the motor cortex Gestural con-

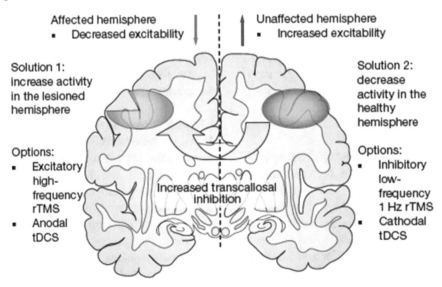

Fig. 10.22 Device-based therapies: tDCS and TMS. Figure with permission: Fregni F, Pascual-Leone A. Technology Insight: noninvasive brain stimulation in neurology: perspectives on the therapeutic potential of rTMS and tDCS. Nature Clinical Practice Neurology 2007;3:383–393

trol circuits "collateralized" speech control area. Control of contiguous body regions mapping on the motor cortical areas and gestural control circuits "collateralizing" speech control areas.

The most commonly encountered dysphasic syndromes include dysnomias and word retrieval impairment. The literature on the treatment of word retrieval has included a number of positive studies with respect to the aphasia population. In general, people with aphasia, if exposed to an enriched environment which may include semantic phonologic activities, errorless naming, orthographic cues, or gestural facilitation, improve word retrieval. In a comparison study gestural therapy (GES) and errorless naming therapy (ENT) were both effective in both ENT and GES led to improvements in naming, including both verbal production and word retrieval [136–138].

Mirror Neuron System Role in Rehabilitation

Similar activity of the frontoparietal temporal mirror neuron system (MNS) circuitry is found due to observation of a task that is similar to the one used for a particular action. This has particular applicability in stroke and severe motor impairment as the circuitry can be activated merely by observation. The various MNS-based rehabilitation techniques include imitation, motor imagery, and action observation and has been shown to improve motor aphasia in addition to primarily motor limb weakness [139, 140].

Situational Therapy for Wernicke's Aphasia

Wernicke's aphasia is particularly challenging in rehabilitation and no specific therapy exists. Clinical observations however suggest that people with Wernicke's aphasia can communicate quite well by facial expressions, gestures, and actions. Altschuler et al. proposed that non-language communication facilitation should be engaged in their rehabilitation. A proposal is simulated or situational therapy by engaging them in real-life situations such as grocery shopping, and catching a train or bus. No clinical trial however has yet been undertaken [141].

Constraint-Induced Therapy for Aphasia

Constraint-induced therapy for aphasia (CIMT) is analogous to the constraint therapy for motor disability after stroke. The therapy involves forcing use of affected limbs, and at the same time ceasing what is termed compensatory strategies. Presumably use-dependent reorganization occurs at the cortical level in response to this therapy. In chronic aphasia patients this is achieved by countering compensatory strategies that they may use such as pointing, gesturing, and writing [142, 143].

Right Hemisphere Engagement

The benefit of singing with respect to aphasia may be due to aspects of rhythm and lyric. In a comparison study of 15 patients with chronic motor aphasia, standard speech therapy, singing therapy, or rhythmic therapy were evaluated. All three groups showed improvement but in different domains. The rhythmic and singing therapy groups improved with respect to formulaic (standard pattern) phrases more than the singing group, but the latter improved more in the no-formulaic speech arena. These results support the use of both standard therapy that is thought to engage the perilesional cortical areas and formulaic phrase training (aside from singing therapy), which likely engages the right hemisphere circuitry [144].

References

1. Henshilwood C, d'Errico F, van Nierkerk KL, Coquinot Y, Jacobs Z, Lauritzen SE, et al. A 100000 year old ochre processing workshop at the Blombos Cave, South Africa. Science. 2011;334:219–21.
2. Lieberman P. The evolution of human speech. Anthropology. 2007;48:39–66.
3. Thaut MH, McIntosh KW, McIntosh GC, Hoernberg V. Auditory rhythmicity enhances movement and speech motor control in patients with Parkinson's disease. Funct Neurol. 2001;16:163–7.
4. Seyfarth RM, Cheney DL, Marler P. Monkey responses to three different alarm calls: evidence of predator classification and semantic communication. Science. 1980;210(4471):801–3.
5. Mithen S. The prehistory of the mind: the cognitive origins of art, religion and science. London: Thames and Hudson; 1996.

6. Bergman TJ. Speech like vocalized lip smacking in geladas. Curr Biol. 2013;23:R268–9.

7. Clay Z, Archold J, Zuberbuhler K. Functional flexibility in wild bonobo vocal behavior. Peer J. 2015; 3:e1124. doi:10.7717/peerj.1124.

8. Cheney DL, Seyfarth RM. Baboon metaphysics. The evolution of the social mind. Chicago, IL: The University of Chicago Press; 2008.

9. Rilling JK, Glasser MF, Preuss TM, Ma X, Zhao T, Hu X, et al. Evolution of the arcuate fasciculus revealed with comparative DTI. Nat Neurosci. 2008;11:426–8.

10. Frey S, Campbell JS, Pike GB, Petrides M. Dissociating the human language pathways with high angular resolution diffusion fiber tractography. J Neurosci. 2008;28:11435–44.

11. Aboitiz F. Gestures, vocalizations and memory in language origins. Front Evol Neurosci. 2012;4:2. doi:10.3389/fnevo.2012.00002.

12. Hickok G, Poeppel D. The corticoal organization of speech processing. Nat Rev Neurosci. 2007;8: 393–402.

13. Fitch WT. The evolution of language. Cambridge: Cambridge University Press; 2010.

14. Rizzolatti G, Arbib MA. Language within our grasp. Trends Neurosci. 1998;21(5):188–94.

15. Mithen S. The singing Neanderthals. London: Weidenfeld and Nicolson; 2005.

16. Wray A. Protolanguage as a holistic system for social interaction. Lang Commun. 1998;18:47067.

17. Blacking J. How musical is man? Seattle, WA: Seattle University of Washington Press; 1973.

18. Ramachandran VS. The tell tale brain. New York, NY: WW Norton and Company; 2011.

19. Darwin C. The descent of man, and selection in relation to sex. London: John Murray; 1871.

20. Dunbar R. Psychology. Evolution of the social brain. Science. 2003;302(5648):1160–1.

21. Donald M. Origins of the modern mind. Cambridge, MA: Harvard University Press; 1991.

22. Fitch WT. Evolutionary developmental biology and human language evolution: constraints on adaptation. Evol Biol. 2012;39(4):613–37.

23. Dunbar R. The social role of touch in humans and primates: behavioural function and neurobiological mechanisms. Neurosci Biobehav Rev. 2010;34(2):260–8.

24. Aiello LC. Terrestiality bipedalism and the origin of language. Proc Br Acad. 1996;88:269–89.

25. Freeman W. A neurobiological role for music in social bonding. In: Wallin NL, Merker B, Brown S, editors. The origins of music. Cambridge, MA: MIT Press; 2000. p. 411–24.

26. Rosseau JJ. Essay on the origin of languages (1781). Chicago, IL: University of Chicago Press; 1966.

27. Jespersen O. Language: its nature, development and origin. New York, NY: WW Norton & Co; 1922.

28. Bickerton D. Adam's tongue. New York, NY: Hill and Wang; 2009.

29. Jackendoff R. Possible stages in the evolution of the language capacity. Trends Cogn Sci. 1999;3(7): 272–9.

30. Lieven E, Behrens H, Speares J, Tomasello M. Early syntactic creativity: a usage-based approach. J Child Lang. 2003;30(2):333–70.

31. Lieberman P. On the nature and evolution of the neural bases of human language. Am J Phys Anthropol. 2002;45:36–62.

32. Ardila A. A proposed neurological interpretation of language evolution. Behav Neurol. 2015;2015:872487. doi:10.1155/2015/872487.

33. Köhler W. Gestalt psychology. New York, NY: Liveright; 1929.

34. Fitch WT. Vocal tract length and formant frequency dispersion correlate with body size in rhesus macaques. J Acoust Soc Am. 1997;102(2):1213–22.

35. Owren MJ, Seyfarth RM, Cheney DL. The acoustic features of vowel-like grunt calls in chacma baboons (Papio cyncephalus ursinus): implications for production processes and functions. J Acoust Soc Am. 1997;101:2951–63.

36. Lieberman DE. The evolution of the human head. Cambridge, MA: Harvard University Press; 2011.

37. Lieberman P, Laitman JT, Reidenberg JS, Gannon PJ. The anatomy, physiology, acoustics and perception of speech—essential elements in analysis of the evolution of human speech. J Hum Evol. 1992;23:447–67.

38. MacLarnon AM, Hewitt GP. The evolution of human speech: the role of enhanced breathing control. Am J Phys Anthropol. 1999;109:341–63.

39. Gardiner AH. The theory of speech and language. Westport, CT: Greenwood Press; 1979.

40. Brain R. Speech disorders. Aphasia, apraxia and agnosia. Oxford: Butterworth-Heinemann; 1967.

41. Caramazza A. How many levels of processing are there in lexical access? Cogn Neuropsychol. 1997; 14:177–208.

42. Rousseaux M, Daveluy W, Kozlowski O. Communication in conversation with stroke patients. J Neurol. 2010;257:1099–107.

43. Hillis AE, Barker PB, Beauchamp NJ, Gordon B, Wityk RJ. MR perfusion imaging reveals regions of hypoperfusion associated with aphasia and neglect. Neurology. 2000;55(6):782–8.

44. Obler LK, Albert ML. Influence of aging on recovery from aphasia in polyglots. Brain Lang. 1977;4(3):460–3.

45. Wuillemin D, Richardson B, Lynch J. Right hemisphere involvement in processing later learned languages in multilinguals. Brain Lang. 1994; 46:620–36.

46. Mariën P, Paghera B, De Deyn PP, Vignolo LA. Adult crossed aphasia in dextrals revisited. Review. Cortex. 2004;40(1):41–74.

47. Hillis AE. Aphasia. Progress in the last quarter of a century. Neurology. 2007;69:200–13.

48. Albert ML, Helm-Estabrooks N. Diagnosis and treatment of aphasia. Part I. JAMA. 1988; 259(7):1043–7.

49. Alexander MP, Hillis AE. Aphasia. Handb Clin Neurol. 2008;88:287–309.

50. Lichtheim L. On aphasia. Brain. 1885;7:433–84.

51. Broca P. Perte de la parole, ramollissement chronique et destruction partielle du lobe anterieur gauche du cerveau. Bull Soc Anthropol Paris. 1861;2:235–8.

52. Wernicke C. Der aphasische symptomenkomplex. Breslau: Cohn & Wigert; 1874.

53. Geschwind N. Disconnexion syndromes in animals and man. Brain. 1965;88:237–94.

54. Flinker A, Korzeniewska A, Shestyuk AY, et al. Redefining the role of Broca's area in speech. Proc Natl Acad Sci U S A. 2015;112:2871–5.

55. Naeser MA. Arch Neurol 1982;39:2 and [Lieberman P. Synapses, language and being human. Science 2013;342:944–45]

56. Lieberman P. The evolution of human speech. Its anatomical and neural basis. Curr Anthropol. 2007;48:39–66.

57. Mohr JP, Pessin MS, Finkelstein S, et al. Broca's aphasia: pathologic and clinical features. Neurology. 1978;28:311–24.

58. Sirigu A, Cohen L, Zalla T. Distinct frontal regions for processing syntax and story grammar. Cortex. 1998;34:771–8.

59. Lazar RM, Marshall RM, Prell GD, Pile-Spellman J. The experience of Wernicke's aphasia. Neurology. 2000;55:1222–4.

60. Godefroy O, Dubois C, Debachy B, Leclerc M, Kreisler A. Vascular aphasia: main characteristics of patients hospitalized in acute stroke units. Stroke. 2002;33:702–5.

61. Anderson JM, Gilmore R, Roper S, Crosson B, Bauer RM, Nadeau S, et al. Conduction aphasia and the arcuate fasciculus: are examination of Wernicke Geschwind model. Brain Lang. 1999;70:1–12.

62. Ardila A. A review of conduction aphasia. Curr Neurol Neurosci Rep. 2010;10(6):499–503.

63. Yamada K, Nagakane Y, Mizuno T, Hosomi A, Nakagawa M, Nishimura T. MR tractography depicting damage to the arcuate fasciculus in a patient with conduction aphasia. Neurology. 2007;68:789.

64. Buchsbaum BR, Baldo J, Okada K, Berman KF, Dronkers N, D'Esposito M, et al. Conduction aphasia, sensory-motor integration, and phonological short-term memory – an aggregate analysis of lesion and fMRI data. Brain Lang. 2011;119(3):119–28.

65. Kwon HG, Jang SH. Excellent recovery of aphasia in a patient with complete injury of the arcuate fasciculus in the dominant hemisphere. NeuroRehabilitation. 2011;29(4):401–4.

66. Basso A, Capitani E. Spared musical abilities in a conductor with global aphasia and ideomotor apraxia. J Neurol Neurosurg Psychiatry. 1985;48(5):407–12.

67. Mazzocchi F, Vignolo LA. Localization of lesions in aphasia: clinical CT scan correlates in stroke patients. Cortex. 1979;15:627–53.

68. Kertesz A, Harlock W, Coates R. Computed tomographic localization, lesion size and prognosis in aphasia and non verbal impairment. Brain Lang. 1979;8:34–50.

69. Flamand-Roze C, Cauquil-Michon C, Roze E, Souillard-Scemama R, Maintigneux L, Ducreux D,

70. et al. Aphasia in border-zone infarcts has a specific initial pattern and good long-term prognosis. Eur J Neurol. 2011;18(12):1397–401.

70. Pedersen PM, Vinter K, Olsen TS. Aphasia after stroke: type, severity and prognosis. The Copenhagen aphasia study. Cerebrovasc Dis. 2004;17(1):35–43.

71. Boatman D, Gordon B, Hart J, Selnes O, Miglioretti D, Lenz F. Transcortical sensory aphasia: revisited and revised. Brain. 2000;123:1634–42.

72. Alexander MP, Hiltbrunner B, Fischer RS. Distributed anatomy of transcortical sensory aphasia. Arch Neurol. 1989;46(8):885–92.

73. Heilman KM, Tucker DM, Valenstein E. A case of mixed transcortical aphasia with intact naming. Brain. 1976;99(3):415–26.

74. Geschwind N, Quadfasel FA, Segarra JM. Isolation of speech area. Neuropsychologia. 1968;6:327–40.

75. Bogousslavsky J, Regli F, Assal G. Isolation of speech area from focal brain ischemia. Stroke. 1985;16:441–3.

76. Geschwind N. The varieties of naming errors. Cortex. 1967;3:97–112.

77. Goodglass H, Wingfield A. Word finding deficits in aphasia: brain-behavior relations and clinical symptomatology. In: Goodglass H, Wingfield A, editors. Anomia: neuroanatomical and cognitive correlates. San Diego, CA: Academic; 1997.

78. Rapsack SZ, Conner JF, Rubens AB. Anomia for facial expressions: neuropsychological mechanisms and anatomical correlates. Brain Lang. 1993;45:233–52.

79. Geschwind N, Fusillo M. Color naming defects in association with alexia. Arch Neurol. 1966;15:137–46.

80. Benson DF, Geschwind N. Aphasia and related disorders. In: Mesulam MM, editor. Principles of behavioral neurology. Philadelphia, FA: Davis Company; 1985. p. 193–238.

81. Tranel D, Damasio H, Damasio HR. On the neurology of naming. In: Goodglass H, Wingfield A, editors. Anomia: neuroanatomical and cognitive correlates. San Diego, CA: Academic; 1997.

82. Howard D, Patterson K, Franklin S, Orchard-Lisle V, Moton J. Treatment of word retrieval deficits in aphasia. A comparison of two therapy methods. Brain. 1985;108:817–29.

83. Graff-Radford J, Jones DT, Strand EA, et al. The neuroanatomy of pure apraxia of speech in stroke. Brain Lang. 2014;129:43–6.

84. Dronkers NF. A new brain region for coordinating speech articulation. Nature. 1996;384:159–61.

85. Hillis AE, Work M, Barker PB, Jacobs MA, Breese EL, Mauer K. Re-examining the brain regions crucial for orchestrating speech articulation. Brain. 2004;127:1479–87.

86. Rheims S, Nighoghossian N, Hermier M. Aphemia related to a premotor cortex infarction. Eur Neurol. 2006;55:225–6.

87. Slevc LR, Martin RC, Hamilton AC, Joanisse MF. Speech perception, rapid temporal processing, and

the left hemisphere: a case study of unilateral pure word deafness. Neuropsychologia. 2011;49(2):216–30.

88. Tanaka Y, Yamadori A, Mori E. Pure word deafness following bilateral lesions. A psychophysical analysis. Brain. 1987;110:381–403.

89. Buchman AS, Garron DC, Trost-Cardamone JE, Wichter MD, Schwartz M. Word deafness: one hundred years later. J Neurol Neurosurg Psychiatry. 1986;49(5):489–99.

90. Tranel D, Damasio A. The agnosias and apraxias in neurology. In: Bradley WG, editor. Clinical practice in neurology. 2nd ed. Stoneham MA: Butterworth; 1996.

91. Hoffmann M, Chen R. The spectrum of aphasia subtypes and etiology in subacute stroke. J Stroke Cerebrovasc Dis. 2013;22(8):1385–92.

92. Mega MS, Alexander MP. Subcortical aphasia: the core profile of capsulostrial infarction. Neurology. 1994;44:1824–9.

93. Mohr JP, Watters WC, Duncan GW. Thalamic hemorrhage and aphasia. Brain Lang. 1975;2:3–17.

94. Tomić G, Stojanović M, Pavlović A. Speech and language disorders secondary to diffuse subcortical vascular lesions: neurolinguistic and acoustic analysis. A case report. J Neurol Sci. 2009;283(1-2):163–9.

95. Bonilha L, Fridriksson J. Subcortical damage and white matter disconnection associated with nonfluent speech. Brain. 2009;132, e108. doi:10.1093/brain/awn200.

96. Gerstman J. Zur symptomatologie der herderkrankungen in der ubergansregion der unterren parietal und mittleren okzipitalhirnwindung. Dtsch Z Nervenheilk. 1930;116:46–9.

97. Rusconi E, Pinel P, Dehaene S, Kleinschmidt A. The enigma of Gerstmann's syndrome revisited: a telling tale of the vicissitudes of neuropsychology. Brain. 2010;133:320–32.

98. Blitzer A, Brin MF, Stewart CF. Botulinum toxin management of spasmodic dysphonia (laryngeal dystonia): a 12-year experience in more than 900 patients. Laryngoscope. 2015;125(8):1751–7.

99. Critchley M. On Palilalia. J Neurol Psychopathol. 1927;8(29):23–32.

100. Ikeda M, Tanabe H. Two forms of palilalia: a clinico-anatomical study. Behav Neurol. 1992;5(4):241–6.

101. Steinlein OK. Epilepsy-aphasia syndromes. Expert Rev Neurother. 2009;9(6):825–33.

102. Jaraba S, Puig O, Miró J. Refractory status epilepticus due to SMART syndrome. Epilepsy Behav. 2015;11:S1525–5050.

103. Verhoeven J, De Pauw G, Pettinato M, Hirson A, Van Borsel J, Mariën P. Accent attribution in speakers with foreign accent syndrome. J Commun Disord. 2013;46(2):156–68.

104. Moreno-Torres I, Berthier ML, Del Cid Mar M, et al. Foreign accent syndrome: a multimodal evaluation in the search of neuroscience-driven treatments. Neuropsychologia. 2013;51(3):520–37.

105. Luzzi S, Viticchi G, Piccirilli M, et al. Foreign accent syndrome as the initial sign of primary progressive aphasia. J Neurol Neurosurg Psychiatry. 2008;79(1):79–81.

106. Marien P, Verhoeven J, Engelborghs S, Rooker S, Pickut BA, De Deyn PP. A role for the cerebellum in motor speech planning: evidence from foreign accent syndrome. Clin Neurol Neurosurg. 2006;108:518–22.

107. Takayama Y, Sugishita M, Kido T, et al. A case of foreign accent syndrome without aphasia caused by a lesion of the left precentral gyrus. Neurology. 1994;43:1361–3.

108. Luria AR, Tsvetkova LS. Towards the mechanisms of "dynamic aphasia". Acta Neurol Psychiatr Belg. 1967;67(11):1045–57.

109. Robinson GA, Spooner D, Harrison WJ. Frontal dynamic aphasia in progressive supranuclear palsy: distinguishing between generation and fluent sequencing of novel thoughts. Neuropsychologia. 2015;77:62–75. doi:10.1016/j.neuropsychologia.2015.08.001.

110. Mesulam MM. Primary progressive aphasia. Ann Neurol. 2001;49:425–32.

111. Gorno-Tempini ML, Hillis AE, Weintraub S. Classification of primary progressive aphasia and its variants. Neurology. 2011;76(11):1006–14.

112. Ash S, Evans E, O'Shea J, Powers J, Boller A, Weinberg D, et al. Differentiating primary progressive aphasias in a brief sample of connected speech. Neurology. 2013;81:329–36.

113. Goodglass H, Kaplan E, Barresi B. The boston diagnostic aphasia examination (BDAE-3). 3rd ed. San Antonio, TX: Pearson; 2000.

114. Kertesz A. Western aphasia battery revised. Examiner's manual. San Antonio TX: Pearson; 2007.

115. Huber W, Weniger D, Poeck K, Willmes K. The Aachen aphasia test rationale and construct validity. Nervenarzt. 1980;51(8):475–8.

116. Phillips PP, Halpin G. Language impairment evaluation in aphasic patients: developing more efficient measures. Arch Phys Med Rehabil. 1978;59(7):327–30.

117. Berthier ML, Pulvermüller F. Neuroscience insights improve rehabilitation of post stroke aphasia. Nat Rev Neurol. 2011;7:86–97.

118. Berthier ML. Poststroke aphasia: epidemiology, pathophysiology and treatment. Drugs Aging. 2005;22(2):163–82.

119. Naeser MA, Martin PI, Nicholas M, et al. Improved picture naming in chronic aphasia after TMS to part of right Broca's area. Brain Lang. 2005;93:95–105.

120. Robey RR. A meta-analysis of clinical outcomes in the treatment of aphasia. J Speech Lang Hear Res. 1998;41:172–87.

121. Poeck K, Huber W, Willmes K. Outcome of intensive language treatment in aphasia. J Speech Hear Disord. 1989;54(3):471–9.

122. Kendall DL, Rosenbeck JC, Heilman KM, Conway T, Klenberg K, Gonzalez Rothi LJ, et al. Phoneme based rehabilitation of anomia in aphasia. Brain Lang. 2008;105:1–17.

123. Fisher M, Finkelstein S. Pharmacological approaches to stroke recovery. Cerebrovasc Dis. 1999;9:29–32.

124. Hughes JD, Jacobs DH, Heilman KM. Neuropharmacology and the linguistic neuroplasticity. Brain Lang. 2000;71:96–101.

125. Berthier ML, Green C, Lara JP, Higueras C, Barbancho MA, Dávila G, et al. Memantine and constraint-induced aphasia therapy in chronic post-stroke aphasia. Ann Neurol. 2009;65(5):577–85.

126. Wan C, Zheng X, Marchina S, Norton A, Schlaug G. Intensive therapy induces contralateral white matter changes in chronic stroke patients with Broca's aphasia. Brain Lang. 2014;136:1–7.

127. Albert ML, Sparks RW, Helm NA. Melodic intonation therapy for aphasia. Arch Neurol. 1973;29:130–1.

128. Sclaug G et al. From singing to speaking: facilitating recovery from nonfluent aphasia. Future Neurol. 2010;5:657–65.

129. Norton A, Zipse L, Marchina S, Schlaug G. Melodic intonation therapy: shared insights on how it is done and why it might help. Ann N Y Acad Sci. 2009;1169:431–6.

130. Fregni F, Pascual-Leone A. Technology insight: noninvasive brain stimulation in neurology: perspectives on the therapeutic potential of rTMS and tDCS. Nat Clin Pract Neurol. 2007;3:383–93.

131. Saur D, Lange R, Baumgartner A, Schraknepper V, Willmes R, Rijntjes M, et al. Dynamics of language reorganization after stroke. Brain. 2006;129:1371–84.

132. Johansson BB. Current trends in stroke rehabilitation. A review with focus on brain plasticity. Acta Neurol Scand. 2011;123(3):147–59.

133. Fiori V, Coccia M, Marinelli CV, Vecchi V, Bonifazi S, Ceravolo MG, et al. TDC stimulation improves word finding retrieval in healthy and nonfluent aphasic subjects. J Cogn Neurosci. 2010;23(9):2309–23.

134. Fridriksson J, Richardson JD, Baker JM, Rorden C. Transcranial direct current stimulation improves naming reaction time in fluent aphasia: a double blind sham controlled study. Stroke. 2011;42:819–21.

135. Fiori V, Coccia M, Marinelli CV, Vecchi V, Bonifazi S, Ceravolo MG, et al. Transcranial direct current stimulation improves word retrieval in healthy and nonfluent aphasic subjects. J Cogn Neurosci. 2011;23(9):2309–23.

136. Raymer AM, McHose B, Smith KG, Iman L, Ambrose A, Casselton C. Contrasting effects of errorless naming treatment and gestural facilitation for word retrieval in aphasia. Neuropsychol Rehabil. 2012;22(2):235–66.

137. Hanlon Brown RE, Brown JW, Gerstman LJ. Enhancement of naming in nonfluent aphasia through gesture. Brain Lang. 1990;38:298–314.

138. Hadar U, Wenkert-Olenik D, Krauss R, Soroker N. Gesture and processing of speech: neuropsychological evidence. Brain Lang. 1998;62:107–26.

139. Garrison KA et al. The mirror neuron system: a neural substrate for methods in stroke rehabilitation. Neurorehabil Neural Repair. 2010;5:404–12.

140. Bonilha L, Gleichgerrcht E, Nesland T, Rorden C, Fridriksson J. Success of anomia treatment in aphasia is associated with preserved architecture of global and left temporal lobe structural networks. Neurorehabil Neural Repair. 2015;pii:1545968315593808.

141. Altschuler EL, Multari A, Hirstein W, Ramachandran VS. Situational therapy for Wernicke's aphasia. Med Hypotheses. 2006;67(4):713–6.

142. Pulvermüller F, Roth VM. Communicative aphasia treatment as a further development of PACE therapy. Aphasiology. 1991;5:39–50.

143. Barbancho MA, Berthier ML, Navas-Sánchez P, Dávila G, Green-Heredia C, García-Alberca JM, et al. Bilateral brain reorganization with memantine and constraint-induced aphasia therapy in chronic post-stroke aphasia: an ERP study. Brain Lang. 2015;145146:1–10.

144. Stahl B, Henseler I, Turner R, Geyer S, Kotz SA. How to engage the right brain hemisphere in aphasics without even singing: evidence for two paths of speech recovery. Front Hum Neurosci. 2013;7:35. doi:10.3389/fnhum.2013.00035. eCollection 2013.

Acquired Cultural Circuits

11

Reading, writing, calculation as well as literary, musical, visual, and performing arts are subsumed under this topic. However, these evolved at vastly different times in human evolution with the more ancient musical (several 100,000 years ago) and visual arts (~40,000 years ago) preceding the others and writing and reading being the most recent (~3000 years ago).

Dehaene's recycling hypothesis portends that existent primate and hominin brain regions used for one function were gradually exapted to serve another, such as reading, writing, and arithmetic but also the arts, spirituality, religion, and musicality [1]. An important driver of these cultural evolutionary abilities is though to have been the progressive increase in working memory capability to an enhanced working memory capacity or a level 7 as postulated by Coolidge and Wynn [2] (Fig. 11.1). Dehaene's hypothesis, postulates an expansion of the conscious neuronal work space that in turn facilitated cerebral connectivity and enabled new functions for existing circuits.

Petersen et al. first reported the differences in cortical actions in a PET brain, language study detailing the BA 37 activation with reading as opposed to the left superior temporal activation in listening to words [3] (Fig. 11.2). In humans and macaques, the inferior occipitotemporal cortical areas (BA 37) are activated in the process of object identification. Reading also activates this region which supports the premise that reading is not processed by a unique area to humans but one that we have inherited from the primate lineage. Patients with clinical visual agnosia also have lesions in this region. Different objects are processed differently by the two hemispheres (Fig. 11.3). In a magnetoencephalography study, after initial occipital activation at approximately 100 ms, that was bilateral, a segregation followed into one or other hemisphere, left for words and lefts and right for faces. Letters and words activated the left hemisphere BA 37 regions at approximately 50 ms, and hence termed the human letterbox area by Dehaene. Analysis of faces is processed by the right hemisphere BA 37 [4] (Fig. 11.4).

The Neuronal Alphabet and Protoletters

Whereas the detection of lines, shapes and contour occurs within the primary visual cortex (V1), within V2 more complex analysis occurs such as combinations of lines and curves and within the inferior temporal cortex (posteriorly), the combinations of curves. Some inferior temporal neurons respond preferentially to two circles forming resembling a figure of 8, others for T and Y shapes. This neuronal preference for shapes is hypothesized to have been selected during evolution because they appeared more frequently in the natural world. This phenomenon or

M. Hoffmann, *Cognitive, Conative and Behavioral Neurology*,
DOI 10.1007/978-3-319-33181-2_11

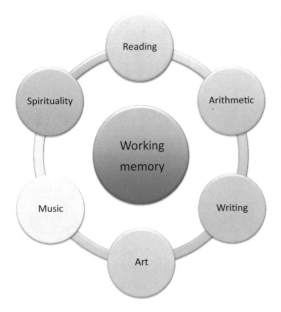

Fig. 11.1 Cultural circuitry: Enhanced working memory was likely the spark that triggered the various cultural evolutions

a type of generic alphabet for shapes phenomenon has been termed protoletters by Dehaene. Common neuronal alphabet shapes such as T, F, Y, O formed part of the building blocks of our alphabet exapted from more primitive (letter) shapes within the inferior temporal cortices and subsequently used for alphabet, words, reading, and writing [1, 5] (Fig. 11.5).

Because reading annexes invades the neuronal circuitry originally destined various identifications of various objects. This suggests that there are consequences of dedicating neuronal space to reading and that due to the concept of the neuronal recycling hypothesis (NRH) there is a price to pay for literacy. The study of Gauthier et al. showed that people who devote a great deal of time to becoming expert at identifying birds or cars for example become less expert at face perception [6].

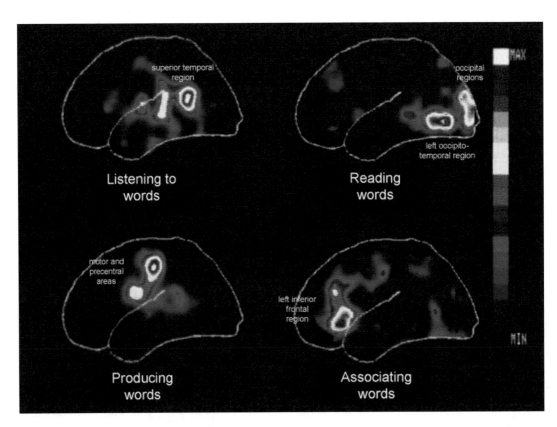

Fig. 11.2 Brain regions involved in various aspects of language. Figure with permission: Petersen SE, Fox PT, Posner MI, Mintun M, Raichle ME. Positron emission computed tomography studies of the cortical anatomy of single word processing. Nature 1988;331:585–589

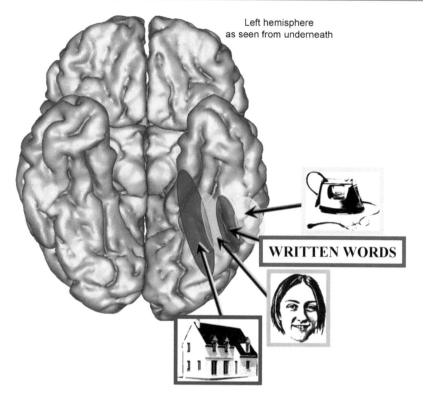

Fig. 11.3 Specialized visual detectors: occipitotemporal regions for recognition of houses and landscapes to faces, to words and tools from medial to lateral. With permission: Dehaene S. The readinginthebrain.com and Ishai A, Ungerleider L, Martin A, Hasby JV. The representation of objects in the human occipital and temporal cortex. Journal of Cognitive Neuroscience 2000;12:35–51 and Puce A, Allison T, Asgari M et al. Differential Sensitivity of Human Visual Cortex to Faces, Letterstrings, and Textures: A Functional Magnetic Resonance Imaging Study. Journal of Neuroscience 1996;16:5205–5216

- Magnetoencephalography records femto-tesla signals and has superb temporal resolution of signals
- There is initial bilateral activation occuring at ~ 100 msec
- Right or left hemisphere activation occurs at ~ 50 msec
- Words are processed in the left hemisphere
- Faces are processed in the right hemisphere

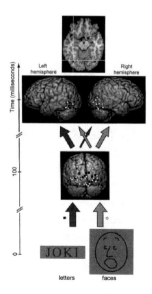

Fig. 11.4 The human letter box area and time course of brain activity of letters and face recognition by magnetoencephalography. Figure with permission: Tarkiainen A, Cornelissen PL, Salmelin R. Dynamics of visual feature analysis and object level processing in face versus letter-string perception. Brain 2002;125:1125–1136

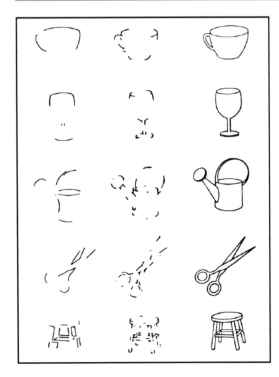

Fig. 11.5 Proto-letters. Complex objects recognition by their particular contours. At joints, contours form shapes; examples: T, L, F, Y. The *left column* figures are difficult to recognize because the junctions are missing that if present enable (*middle*)—enable easier recognition. Figure with permission: Dehaene S. Reading in the Brain. Penguin, New York 2009

The Evolution of Writing: Did Writing Originate in Mythograms and Ideograms? (Fig. 11.6)

Archeological evidence points to a writing origin evolved from visual art forms. A renowned researcher in this field of study, Lero-Gouran postulated that; "originally figurative art was directly linked to language and was much closer to writing than to a work of art. It was a symbolic representation, not a carbon copy of reality. The earliest known paintings do not represent a hunt, dying animal or a family scene, they are a building blocks without any descriptive binder. They are really mythograms closer to ideograms than to pictograms than to descriptive art. Ideography in this form precedes pictography and all Paleolithic art is ideographic" [7].

Thereafter from these pictograms the first cuneiform (wedge shaped) type characters likely originated in the Proto Sinaitic and Phoenicians who adopted hieroglyphs that over time are thought to have evolved into an initial set of letters. The transition from Phoenician to Greek writing forms involved increasing simplification of the original pictograms and involved letter rotation and the development of vowels [1] (Fig. 11.7). Cross modal integration of sound with meaning was enabled by neural circuitry connecting the letter-box to the middle temporal cortex (word meaning) and inferior frontal lobe for auditory articulation.

Numerosity and Calculation

Biparietal horizontal intraparietal sulcus (IPS) (Fig. 11.8) activation has been documented by activation studies whenever numbers are being processed, whether by written or spoken input and across cultures. The IPS activation appears unique to number stimulation and not to letters or colors and activates whether the number stimulation is a series of dots, the word three or the numeral 3 for example [8]. Similar to our color sense for blue or green we have a sense what is 3, 4, or 5. There is a basic numerosity (for magnitude) system located in the IPS of both hemispheres but also a left hemisphere, predominant number system (frontoparietal circuit) that is associated with language and the repository of arithmetic facts. When performing calculations for example both the left hemisphere frontoparietal system and the IPS are activated. With education there is progressively less reliance on the IPS which is initially predominantly activated and with semantic memory formation the frontoparietal region and angular gyrus in the left hemisphere become more activated [9].

Neurophysiology

The various methods of investigation into brain function, including clinical lesion analysis, functional MRI activation studies and microelectrode

Fig. 11.6 The original "Imagenes". Paleolithic art is ideographic and represents a mythogram. Legend: Not a disastrous hunting scene but a mystical representation Lascaux Shaft Scene Painting, Lascaux Cave, Dordogne, France. Prostate man with a bird's head and a bison that has been disemboweled: Trance, death and mystical flight. Figure credit with permission: French Ministry of Culture (Centre national de la préhistoire). Photo credit N. Aujoulat MCC/Centre National de Préhistoire

- Early Sumerian pictographic origins (top row) evolved to progressively more abstract symbols
- Proto-Sinaitic (middle row)
- Phoenician and Greek (bottom rows) evolution to further simplification rotation through 90 and 180 degrees
- Earliest art form from Lascaux Cave, France evolution over time to modern letters – for example the letter "A"
- These ultimately became the letters of our alphabet and each one (eg A). The original core shape of the "A" is something relatively easily recognized by the inferior temporal area

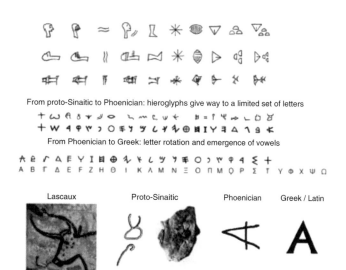

Fig. 11.7 From pictures to early characters. Convention and simplification in the evolution of writing. Figure with permission: Dehaene S. Reading in the Brain. Penguin, New York 2009

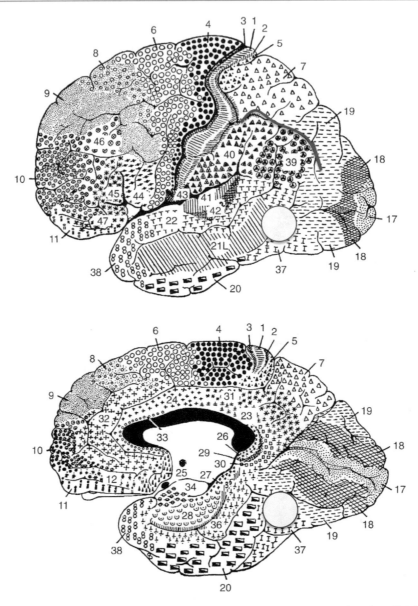

Fig. 11.8 Cortical areas involved in reading and numerosity: The human inferotemporal cortices, BA 37 (*blue*) and intraparietal sulcus (*red*) depicted by the Brodmann areas' map

recordings in monkeys have all pointed to the IPS being activated arithmetic function and to our understanding of numerosity in the brain. For example microelectrode recordings in monkeys have shown that while performing a visual numerical judgment operation, microelectrodes placed in the IPS fired earlier than those placed in the prefrontal cortex. This confirms the basic parietofrontal circuitry and that the activation cascades from the IPS in the posterior parietal cortex to the lateral prefrontal cortex (Fig. 11.9) [10].

The Relationship Between Arithmetic and Language

Pica et al. studied the question of approximate and exact arithmetic in the Amazonian Indigene Munduruku who have limited lexicon for number

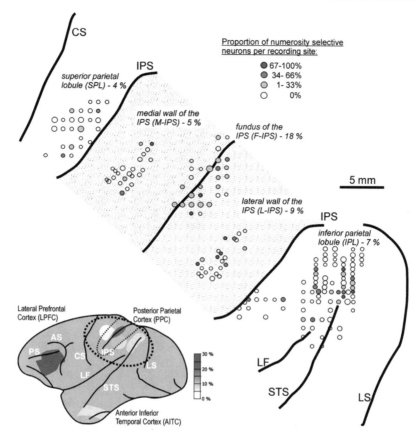

Fig. 11.9 Numerosity in the monkey: A parietofrontal network. Neurons in the intraparietal sulcus (IPS) responded and transmitted numerosity earlier than PFC neurons. Figure with permission: Nieder A, Millker EK. A parieto-frontal network for visual numerical information in the monkey. PNAS 2004;101:7457–7462

words with no words for numbers greater than 5. Yet they are able to make comparisons and add numbers that markedly exceed their naming range—up to 80 dots in this study. Approximate of quantities appears to be an inherent competence amongst people, children as well as some animals. Precise arithmetic however is inaccurate when numbers exceed 5. Taken together these findings support a fractionation of a nonverbal number approximation system and language dependent counting system for may exact arithmetic [11] (Figs. 11.11 and 11.12). The same researchers also discovered that the Munduruku mapped numbers onto a logarithmic scale in a spatial sense that is considered fundamental to mathematics and all people. The linear number line is something that is acquired by education [12]. Both in the Munduruku and in children before the

formal schooling they may for example think of 8 closer to 9 than 1 is to 2, or they may place 10 near the middle of a line ranging from 1 to 100, if asked to represent numbers on a line. This indicates a logarithmic representation. Children devote relatively more space to smaller numbers [12].

The Concept of Subitizing

There seems to be a special status for numerosity pertaining to 1, 2, and 3 which can be readily identified at a mere glance, not involving "counting" which is linked to a distinct circuitry. Subitizing does however require attention, we can briefly store three or four items within a number of different perceptual properties, at most and may be the domain of working memory itself [13].

Developmental Dyscalculia

Refers to problems with arithmetic function and is akin to the dyslexic problem of reading. Here there is a deficiency of numerosity as well as their ability for subitizing of very small numbers, 1, 2, and 3. Reduced gray matter of the left intraparietal sulcus has been documented [13].

Dissociation Between the Basic Arithmetic Operations

The four basic arithmetic functions are represented differently in these brain areas. Activation study analysis has revealed:

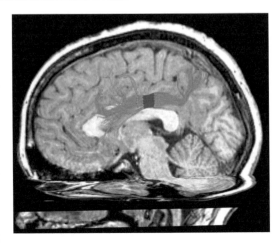

Fig. 11.10 Arithmetic skills measured in children by mental addition task, approximate answer choices and correlated with fractional anisotropy in the left frontoparietal anterior superior longitudinal fasciculus white matter tract. Figure with permission: Tsang JM, Dougherty RF, Deutsch GK, Wandell BA, Ben-Shachar M. Frontoparietal white matter diffusion properties predict mental arithmetic skills in children. PNAS 2009;106:22546–22551

- Bilateral intraparietal activation during subtraction (PET brain and fMRI studies)
- With multiplication there is more posterior and left subangular gyrus activity
- More precise calculations as well as addition functions are more dependent on language and the language circuitry of the left anterior inferior frontal Broca's and angular gyrus regions
- Approximation of quantity approximation is also more dependent on the bilateral IPS and less dependent on language function
- In children, a fMRI study has correlated arithmetic skills with the anterior superior longitudinal fasciculus (aSLF) connecting the parietal and frontal cortices [14] (Fig. 11.10).

Clinical Presentations

Unsurprisingly, given the distributed brain circuit involved with numerosity and the subsequent cultural transformations associated with calculation and mathematics, brain impairment and lesions may cause fractionation of the basic functions. Hence these arithmetic functions may also be differentially affected by lesions such as bland infarct, hemorrhage or tumors. For example impairment with:

- Multiplication is associated with left perisylvian lesions and usually with aphasia

- Subtraction impairment may be associated with left intraparietal sulcus lesions

In a general sense calculations disorders may be primary or due to anarithmetria or secondary where the impairment is explained by an accompanying inattention, visuospatial function, aphasia, or alexia. A proposed classification is suggested:

Dyscalculia, Acalculias

Primary

Transcoding impairment
Asymbolic acalculia
Selective anarithmetria

Secondary

Aphasia
Visuospatial
Alexia,
Inattention

Comparison
Indicate the larger set

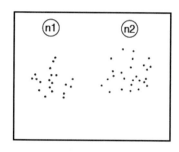

Performance (% correct)

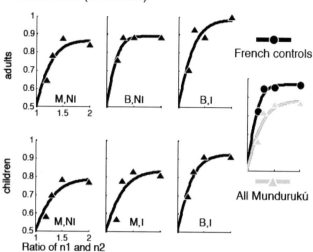

Fig. 11.11 Comparison of magnitude in a traditional Amazonian group versus a French control group. Figure with permission: Pica P, Lemer C, Izard V, Dehaene S. Exact and Approximate Arithmetic in an Amazonian Indigene Group. Science 2004;306:499–503

Approximate addition and comparison
Indicate which is larger: n1+n2 or n3

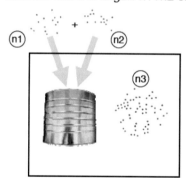

Fig. 11.12 Approximate addition in a traditional Amazonian group versus a French control group. Figure with permission: Pica P, Lemer C, Izard V, Dehaene S. Exact and Approximate Arithmetic in an Amazonian Indigene Group Science 2004;306:499–503

Transcoding Impairment

Transcoding or syntactic processing impairment of numbers is described as correct digit reproduction but misplaced in the order of magnitude, such that 189 may be transcribed as 10,089. The presumed likely pathophysiology is that the number transcoding system includes different code-dependent trajectories for Arabic type transcoding from verbal (spoken) numerals to verbal (written) numerals. These deficits have been described in association with left inferior parietal infarcts [15].

Asymbolic Acalculia

Discrete disability of operation symbol interpretation whereby multiplication, division, addition, or subtraction operations are not discerned, described with left occipitotemporal junction lesions.

Selective Anarithmetria

Refers to a selective calculation impairment or dissociation of impairments. There may be a dissociation between impairment calculations such as simple additions, yet a relative preservation of factual retrieval related to algebraic expression.

Some patients with left parietal damage are adept at subtraction and addition yet unable to perform division and multiplication procedures [benson weir 1972] and [Toghi 1995], yet others with similar left temporoparietal lesions have preserved subtraction but all others, multiplication, addition, division are impaired [Lample 1994]. In line with the bilateral IPS representation of numerosity, the right hemisphere lesions should also affect calculational ability and there may be a specific right hemisphere hub associated with multiplication, perhaps integrated with visuospatial abilities and number comparison ability [16, 17].

In yet another pathological process, that of posterior cortical atrophy, the patient presented with both intact multiplication and addition but deficient subtraction and division [18].

In general left hemisphere lesions are frequently associated with various forms of dyscalculia although there is a degree of independence with respect to the two abilities in the left hemisphere. For example, Rossor's patient with global aphasia had a remarkable preservation of all calculation abilities [19]. Basso maintained that while that calculation is relatively independent, dyscalculia may be diagnosed in an approximately 50 % of those with Broca's and 39 % of those with Wernicke's aphasia [20].

Virtual lesions studies using rTMS in general have corroborated lesion studies and functional activation studies in that left angular gyrus stimulation resulted in an impairment of both finger agnosia and number processing. This is consistent with some of the component deficits of the Gerstmann's syndrome long documented by lesion studies of this area [21].

Pathophysiology

Overall calculation disability has been correlated with left parietal (AG, IPS), frontal, temporoparietal, parieto-occipital, but right hemisphere and subcortical lesions have also been described in association with dyscalculias. In summary:

- Left angular gyrus specifically is concerned with multiplication, exact calculation, and retrieval of calculations that are verbally encoded
- Frontal subcortical circuits connected with language circuits are associated with retrieval of learned calculation procedures such as multiplication.
- Arabic digit identification is dependent of bilateral occipitotemporal areas, the fusiform gyrus in particular, which is shared with letter and object identification
- The IPS, both right and left subserve numerosity and abstract quantification that is not language based [22].

Testing Calculation at the Bedside and by Neuropsychological Batteries

The serial 7 test which is used almost universally in screening tests such as the MMSE and MOCA and tests subtraction, attention and to some degree working memory. Aside from the serial 7 test at a minimum the three other of the four basic arithmetic functions should be tested because of the foregoing evolutionary and clinical fractionation evidence of these functions. A simple procedure that may be adopted includes, in rough order of level of difficulty:

Bedside Testing

Serial 7's

Subtraction	84−13
Addition	56+23
Multiplication	13×4
Division	65/7

Neuropsychological Testing

WAIS IV arithmetic subtest [23]

WRAT 4 (Wide Range Achievement Test, 4th edition) [24]

WIAT III (Wechsler Individual Achievement Test 3rd edition)

The four major components include mathematics, reading, writing, oral language with nine subtests [25]

Johns Hopkins Acalculia Battery [26]

Benton's Battery of Arithmetic tests [27]

Number Processing and Calculation (NPC) Battery

A much more comprehensive test, the NPC is noteworthy in that it assesses all the various calculation abilities such as simple fact retrieval, mental calculation, rule based processing, written calculations amongst all four basic arithmetic operations [28].

Graded Difficulty Arithmetic Test (GDA) [29]

Acalculia and Dyscalculia Treatments

Although the study and rehabilitation of acalculias is associated with a relatively sparse literature, a recent review by Lochy et al. is a comprehensive review provides guidance for therapists challenged with these deficits [30]. Dehaene has developed a more practical web based program, the Number Race Game (http://thenumber-race.com/nr/home.php), that is particularly suitable for children but also applicable to patients with acquired dyscalculia such as after stroke. This has shown improvement in children's performance on a variety of tasks such as subtraction and subitizing [31, 32].

Alexias

Alexias; the acquired inability to read despite intact elementary visual function such as acuity, object and color identification. First described by Dejerine in 1891, the first subtypes were alexia with agraphia, also termed acquired illiteracy, due to a left parietal lobe lesion and alexia without agraphia (pure alexia) due to a left occipital lobe lesion resulting in a right homonymous hemianopia and being able to write but not read [33]. Observations of these patients may include the ability to perform what is termed letter by letter reading with very slow and laborious reading ability of short words at best. There was in addition to the left occipital lesion another lesion at the posterior corpus callosum preventing transfer from the left visual field to the left hemisphere language circuitry (Fig. 11.9). In general alexias may be classified into peripheral and central subtypes.

Alexia Subtype Classification

Central Deep alexia (paralexia)
  Paralexia (deep dyslexia)
  Peripheral
  Alexia with agraphia
  Alexia with agraphia
  Frontal alexia (third alexia)
  Hemialexia
  Spatial alexia
  Increased function
  Hyperlexia
  Congenital
  Dyslexia
  Central alexias are also referred to as deep dyslexias or paralexias. In the latter entity there is a remarkable tendency to enunciate semantically related words such as "boat" instead of "ship" [34]. There is also increased difficulty in reading abstract words, non words and one theory of the disorder pertains to the right hemisphere being involved in processing words with deep dyslexia. The causative lesions are mostly left perisylvian and associated with Broca's and sometimes global aphasias [35]. A paralexia deficit has also been described in relation to a left thalamic lesion [36].

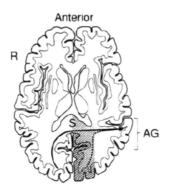

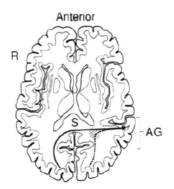

Fig. 11.13 Alexia without agraphia: the relatively large left hemisphere occipital lesion extends sufficiently forward to prevent the right hemisphere visual tracts from communicating with the left angular gyrus. Figure with permission: Quint DJ, Gilmore JL. Alexia without agraphia. Neuroradiology 1992;34:210–214

Alexia Without Agraphia

Being able to write but not read. A left occipital lesion results in the elementary neurological deficit of a right homonymous hemianopia and a posterior callosal lesion prevents the transfer of information from intact R occipital region to left hemisphere language circuitry responsible for the graphic decoding (Figs. 11.13 and 11.14)

Alexia with Agraphia

Also termed acquire illiteracy, this may occur with, and sometimes without, aphasia in association with left parietal lobe lesions [37].

Frontal Alexia

Sometimes also referred to as the "third alexia," this infrequently used term refers to the association of reading difficulty that occurs in conjunction with Broca's aphasia and may be related to syntax and recursion comprehension difficulty [38].

Hemialexia

Only the right half the word read due corpus callosal lesions interrupting fiber tracts crossing from the right hemisphere with impaired left hemifield vision.

Spatial (Neglect) alexia

Not strictly an alexia but associated with visual neglect syndrome is generally attributed to left visual field neglect in association with right hemisphere lesions. More complex words such as "workplace" may be read instead as "place" [39].

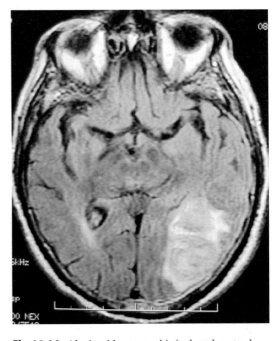

Fig. 11.14 Alexia without agraphia in the subacute phase of an approximately 30 cc intracerebral hemorrhage. With resolution of associated edema the deficit improved to a left homonymous hemianopia

Hyperlexia

This phenomenon has been reported several times in various isolated case reports. The condition is characterized by a compulsion to read aloud verbage in their immediate environment including labels, warning, traffic signs [40]. It may well be the same condition described as forced hyperphasia, subsumed under the entity of field dependent behavior and loss of autonomy from the environment (frontal systems chapter). There may also be an association of hypergraphia [41].

Dyslexia

A reading disorder in the context of otherwise preserved and normal cognition and intelligence may be due to a visual perceptual or language related dysfunction. The condition usually presents during childhood at the time of formal education when a delay in speech onset, problems with distinguishing right from left and letter reversal may be noted. Although mirror writing occurs commonly in children across cultures, for letters such as p and q, b and d for example, functional imaging studies have revealed abnormalities. These include left inferotemporal activity in the so called letterbox area, left middle frontal cortical (Exner's area) and right temporoparietal activation. Ectopias and excess gray matter in the middle temporal cortex has also been documented [42, 43].

Agraphias

Writing straddles the abilities of both language and praxis and hence may be recognized with either of these syndromes as well as others. Agraphia is invariable with left hemisphere language (if dominant) disorders with the expressive impairment in Broca's aphasia for example evident in both vocal and written output forms. However a number of other agraphic syndromes may be recognized:

Agraphia Classification

Primary agraphia

Pure isolated agraphia
Central (aphasic) agraphia (nonfluent, phonological, lexical, semantic)
Deep agraphia

Secondary agraphia

Aphasic agraphia
Apractic agraphia

Visuospatial agraphia
Phonological
Micrographia
Peripheral agraphia

Hypergraphia
Dystypia

Pure or Isolated Agraphia

There is an alexia but no associated apraxia or aphasia. This has been associated with superior left parietal lesion as well as left posterior frontal lesions. Sigmund Exner first proposed in 1881 that lesions of the inferior medial frontal gyrus lesions may cause writing impairments. Over one hundred years passed before lesion studies and autopsy reports provided supportive confirmation. A cortical stimulation and MR functional imaging confirmed the BA 6 to be critical to the orthography to motor programs that exemplify handwriting [44]. The recent reports of two patients b Keller and Meister with significant dysgraphia in lesions located by MR imaging to this region is supportive of phoneme–graheme processing of this region [45]. An autopsy study amyotrophic lateral sclerosis patients with obvious agraphia, but no aphasia revealed TDP-43-pathology with the degeneration and immunohistochemical pathological abnormalities were most profound in the inferior left middle frontal gyrus, or Exner's area (Fig. 11.15) [46].

Central (Aphasic) Agraphia

Both anterior aphasic nonfluent and posterior aphasic with lexical and semantic agraphias may be evident In addition phonological agraphias are noted by their inability to write non-words or words that are not easily pronounced.

Aphasic agraphia of the left hand may be seen with anterior corpus callosum and aphasic disorder with left hemisphere damage. Similar to the sympathetic dyspraxia with Broca's dysphasia where there is an inability in transferring language information the right hemisphere.

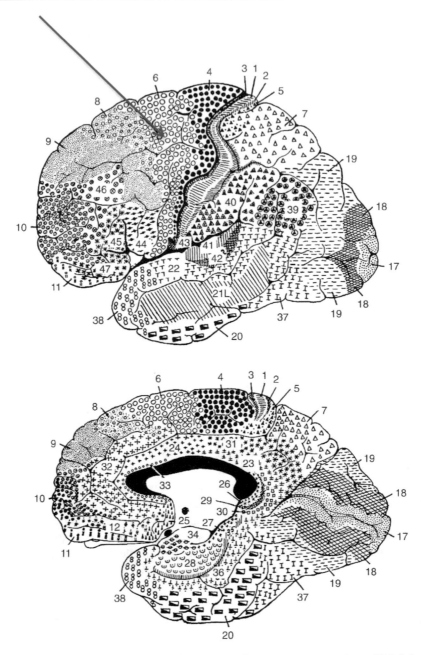

Fig. 11.15 Exner's area in the posterior superior frontal gyrus, the upper premotor regions of BA 6 depicted by the Brodmann areas' map

Paretic Agraphia

Peripheral nerve, neuromuscular junction, myopathic process is responsible

Movement Disorder Associated Dysgraphia

Due to tremor, tics, chorea, dystonia, and varieties of focal dystonias such as Writer's cramp

Micrographia

Most commonly due to Parkinsonian syndromes, less often due to corticospinal tract lesions [47]

Hypergraphia

Frequently regarded as a right hemisphere syndrome, this may take the form of very extensive or compulsive writing. The former is a feature of the Geschwind Gastaut syndrome, or mania. Compulsive writing may be found in association with schizophrenia and bilateral frontal lobe injury perhaps a subtype of utilization behavior [48–50].

Dystypia

Isolated typing impairments due to left frontal strokes, specifically without accompanying aphasia or apraxia and in one case even without agraphia have been reported. The relevant lesion was in the second left frontal gyrus and adjacent frontal operculum [51, 52].

Testing

Bedside Testing

Write name, address, and signature
Copy words
Write letters, words, sentences, and numbers to command
Write a few sentences or a short paragraph relating to something familiar such as what does their occupation entail
Check fluency, speed of writing, letter or word size, grammar, and sentence complexity

Neuropsychological Testing

Boston Diagnostic Aphasia subtest
Western Aphasia Batteries subtest (see under Aphasia chapter)

Art and the Brain

The arts refer to the three most commonly appreciated forms, the visual, music, and performing arts (dance), but culinary arts and those related to the perfume industry for example represent sophistication in the areas of taste and smell. Art was long thought to be the last bastion of humans as other animals had language, culture, toolmaking, and counting abilities. However, with the discovery of the elaborate artistic display of the male Bower Bird in Australia, this too is no longer the sole attribute of humans [53].

The first indisputable works of art date to the Franco-Cantabrian cave art dated to about 40,000 years ago (Fig. 11.16). These were regarded as initial attempts conveying abstract thinking about the afterlife and early spirituality. An example is the Lascaux painting with the bird-man, disemboweled bison, bird on a stick all of which were previously interpreted as a hunting scene with adverse outcome but more recently it has renewed interpretations. It is now regarded as a metaphorical, representation of trance, demise, transformation in an afterlife, and mystical flight. The art represents a communication that is imagistic which preceded modern language. Visual art solidified groups and so enhanced survival and social unification [54, 55]. In his book 'The Art Instinct", Dutton postulated that; "There was not enough time for human hereditary to cope with the extent of the new contingent possibilities revealed by high intelligence. The arts filled the gap, allowing humans to develop more flexible and sophisticated responses to new situations. Art is a vestigial fitness indicator and a way of knowing another mind in social interchange" [56].

Since then art through the ages has changed in its progressively more realistic representation of nature, more recently photography and the nineteenth century artists by using figural and emotional primitives. Klimt for example portrayed the unconscious aspect of humans painting mostly women, Kokoschka achieved this through facial and hand gestures and Schiele by depicting whole body postures that included himself.

Fig. 11.16 Franco-Cantabrian cave art: Two bulls on the right wall of The Hall of the Bulls, Lascaux cave, Dordogne, France. Figure credit with permission: French Ministry of Culture (Centre national de la préhistoire). Photo credit N. Aujoulat MCC/Centre National de Préhistoire

Kandel has contended that a flurry of neurochemical and neural circuits are activated when viewing a picture such as "Judith and Holofernes" painted by Klimt. For example he proposed that:

> The luminous gold surface, smooth body contours and appearance and the harmonious colors trigger pleasure circuits that initiate dopamine release. The smooth skin and curvaceous appearance trigger endorphins, the decapitated head of Holofernes she is hold release norepinephrine. The brushwork techniques might stimulate serotonin release and acetylcholine release may come about due to requirement of storing the image complexities for analysis. All together result in a complexity of emotional states triggered by this visual art [57].

Ramachandran attempted to solidify the neurophysiological aspects pertaining to art and aesthetics with his nine laws of aesthetics:

The Laws of Aesthetics

1. Grouping
2. Peak shift
3. Contrast
4. Isolation
5. Peekaboo or perceptual problem solving
6. Abhorrence of coincidence
7. Orderliness
8. Symmetry
9. Metaphor [58]

Neuroesthetics has been investigated with functional imaging which has revealed the neurobiological components which include the putamen, globus pallidus, supplementary motor area, cerebellar vermis, amgydala, and medial orbitofrontal cortex. In formulating their brain-based theory of beauty, Ishizu and Zeki using functional MRI concluded that, "beauty" as an independent modality, is associated with activation in the medial orbitofrontal cortex, that is not restricted to the visual or musical inputs [59] (Fig. 11.17).

How does art help with neuroscientific insights and monitoring in neurological disease? Clinical observations by Mendez have provided remarkably images of an artist with frontotemporal lobe

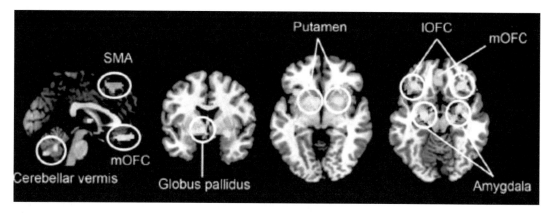

Fig. 11.17 Neuroesthetics: The brain "beauty" circuitry. Figure with permission: Ishizu T, Zeki S. The brain's specialized systems for aesthetic and perceptual judgment. Eur J Neurosci 2013;37:1413–1420

dementia (FTD) who charted his facial imagery over the years as he declined cognitively, emotionally, and behaviorally due to the FTD (Fig. 11.18) [60]. Schott provided a catalogue of visual art forms painted by people with autism, stroke, Parkinson's disease, epilepsy, and migraine [61] (Fig. 11.19). Artistic talent may also emerge with brain disease such as frontotemporal lobe dementia [62].

Music and the Brain

Music has been called the universal sense. The neural circuitry of musicality is primordial with respect to the neural circuits of the mind [63]. Musicality antedated not only the newer cultural competencies but also language. We may trace music origins back at least to our origins of bipedalism. There is an important evolutionary link between pedalism and rhythm. Not only is rhythm essential for walking, running, and complex coordination but also for music. Recent clinical studies supporting rhythmic auditory stimulation therapy as a therapy for people with stroke, traumatic brain injury, and Parkinson's disease attests to this [64–66]. Mithen and Blacking proposed that a musicolanguage preceded modern language as well as modern musical abilities (Fig. 11.20) [67, 68].

Neurobiology and Neurophysiology

From a neurobiological point of view the temporal lobe regions concerned with auditory input and processing are connected to all frontal regions of the brain. The core belt and parabelt areas of the superior temporal lobe transmit via a dorsal stream to the parietal and frontal regions areas and ventrally to the inferior temporal lobe integrating memory, motor, and executive function areas (Fig. 11.21) [69]. The auditory cortical areas are connected to every major prefrontal cortical region particularly the medial PFC and frontopolar cortex (BA 10) is spatially unconstrained and auditory cortical regions connect and impinge on every area of the PFC. In a neurophysiological sense, musicality is thought to facilitate cooperation amongst individuals and promote cohesive group behavior. The Neanderthals for example were regarded as a musical species with modern humans with relatively constrained musical abilities attributed to the evolution of language [70].

The ACC has been implicated in motivational processes, perception of music and speech, movement, monitoring of performances, autonomic function, and physiological arousal. The musicality neural circuitry allows entrainment of multiple brain regions (Fig. 11.22). This allows improved perception, attention, both epi-

Fig. 11.18 Artist with frontotemporal lobe dementia. Progressively more bizarre and intimidating caricatures with evolution of the dementia. Image in (**a**) drawn prior to his illness, (**b**) and (**c**) about 2 years after dementia, and (**d**) about 3 years after dementia diagnosis. Figure with permission: Mendez MF, Perryman KM. Disrupted facial empathy in drawings from artists with frontotemporal dementia. Neurocase 2003;9:44–50

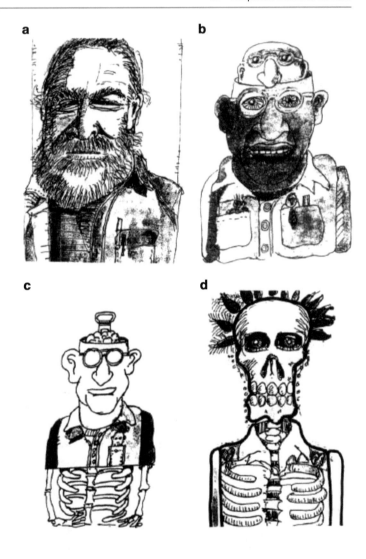

sodic and working memory, movement modulation, emotional modulation and promoting sociality and social bonding [71]. Barbas et al. have proposed that auditory neural circuitry is fundamental to organized thinking presumably because of the strong connection of the auditory temporal cortex to the frontopolar metacognitive area and ACC that are concerned with internal cognitive processing [72]. Musicality is particularly effective at inducing emotions and Plutchik has suggested that music is able to induce a panoply of emotive states of varying intensities. A major contention is that the basic emotions (rage, joy, fear, anger), are not always experienced in their most dramatic or intense states but may be present

in more subtle or milder forms. He proposed the cone model of basic emotions with a vertical intensity dimension and a horizontal distribution of different emotional subtypes. Juslin proposed that music might be able to trigger low intensity emotional responses or milder variations of the basic emotions (Fig. 11.23) [73, 74].

Pathology

There can be music without language and language without musicality (amusia). The disturbances associated with music may involve the basic temporal lobe auditory unimodal and

Fig. 11.19 Color and happiness: Color–emotion synesthesia. A young woman with sudden artistic ability due to a stroke with post stroke pain during which she painted "cold" colors (**d**, *bottom right image*) with warm colors (**a**, **b**, **c**, *red* and *orange*) when happy. Figure with permission: Thomas-Anterion C, Creac'h C, Dionet E, Borg C, Extier C, Faillenot I et al. De novo artistic activity following insular-SII ischemia. Pain 2010;150: 121–7 and Schott GD. Pictures as a neurological tool: lessons from enhanced and emergent artistry in brain disease. Brain 2012;135:1947–1963

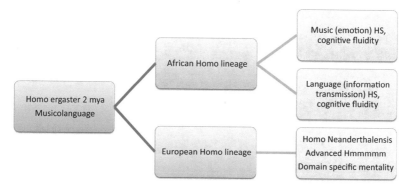

Fig. 11.20 Evolution of music and language. Adapted from Mithen S. The Singing Neanderthals. Weidenfeld and Nicolson London 2005 and Blacking J. How Musical is Man? Seattle University of Washington Press 1973

heteromodal association cortices, familiarity of music with left inferior frontal localization and reading, writing and production of music, which has major hubs in the left hemisphere as well as the right (Fig. 11.24 and 11.25) [75] In addition to deficits in musical appreciation or ability there may an increased ability or a sudden savant like musicophilia [76]. A paradoxical functional type

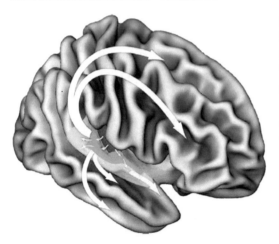

Fig. 11.21 Human auditory processing begins in the superior temporal lobe core areas: belt (*red*) and parabelt (*green*, *yellow*) regions and dorsal stream to parietal and frontal and ventral stream to the inferior hippocampal memory areas. Figure with permission: Zatorre RJ, Salimpoor VN. From perception to pleasure: Music and its neural substrates. PNAS 2013;110:10430–10437

facilitation, such as may occur with frontotemporal lobe disorders of increased musical ability similar to the visual arts abilities described. These are usually "released" by left hemisphere damage [77].

Amusia Classification

Clinical classification
 Receptive amusia
 Expressive amusia
 Instrumental amusia
 Musical alexia
 Musical agraphia
 Musical amnesia
 Musical variable disorders
 Timbre disorders
 Pitch and melody disorders
 Rhythm disorders
 Harmony disorders [78, 79]

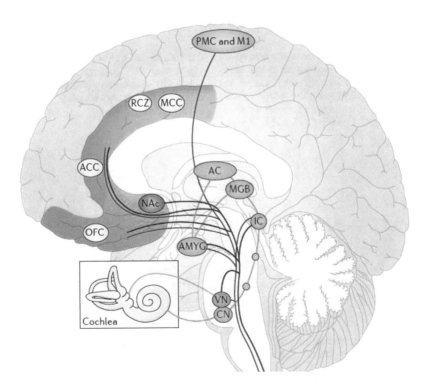

Fig. 11.22 Musicality and the brain: Principal circuits mediating autonomic and neuromuscular effects of music. *OFC* orbitofrontal cortex, *ACC* anterior cingulate cortex, *RCZ* rostral cingulate zone, *MCC* middle cingulate zone, *NAc* nucleus accumbens, *AMYG* amygdala, *VN* vestibular nuclei, *CN* cochlear nuclei, *IC* inferior colliculus, *M1* primary motor cortex, *MGB* medial geniculate body, *AC* auditory cortex. Figure with permission: Koelsch S. Brain correlates of music-evoked emotions Nature Reviews Neuroscience 2014;15:170–180

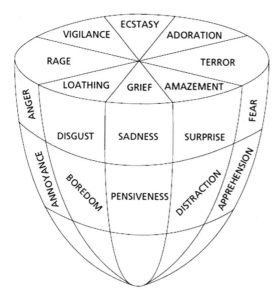

Fig. 11.23 Plutchik's cone model of emotion: The concept of partial emotions or less intense emotions that may be triggered by music. Figure credit: Juslin PN. What does music express? Basic emotions and beyond. Frontiers in Psychology 2013:4:596 and adapted from Plutchik R. The Psychology and Biology of Emotion. Harper Collins, 1994, New York

Receptive amusia

Musical perception is impaired such that familiar songs or tunes sound strange. Often correlated with right superior temporal and or insula region lesions.

Expressive amusia

An acquired impairment in singing, humming tunes or whistling

Instrumental amusia

Loss of ability of playing an instrument previously learnt

Musical alexia

The inability to read music such as reported by a personal account by MacDonald. He sustained a stroke which involved the right angular and supra-marginal gyrus due to embolic stroke [80]

Musical agraphia

Inability to write musical notes which may involve both writing rhythm, pitch and copying musical notes [81].

Musical amnesia

Memory loss for recalling particular tunes, melodies, or songs. Memory for music may however

be spared even with otherwise severe memory disorders including transient global amnesia [82]

Timbre disorders

Described as the frequency, intensity, and duration of quality of a sound. Seen with left more than right hemisphere disorders

Pitch and melody disorders

Right hemisphere associated

Rhythm disorders

Temporal segmentation of sound seen with left hemisphere disorders

Harmony disorders

Right, sometimes bilateral hemisphere disorders [83]

Testing musical competence

Bedside testing

This can be accomplished by simple inquiry as to loss of function of playing an instrument, singing or being able to appreciate tunes and songs previously known or liked.

Neuropsychological musical testing

Montreal Battery of evaluation of amusia [84]

Gordons Musical aptitude profile [85]

Seashore's tests for musical ability [86]

Music therapy

In general, music therapy can augment both psychological and physiological well-being due to influencing attention, emotion cognition, and behavior. Stroke patients may improve wit respect to fine and gross motor abilities and cognitive domains of attention and verbal memory post piano playing therapy. People with Parkinson's hypokinesia can improve with musical arousal due to entrainment of the motor system to the musical beat for example. One of the mechanisms may include augmentation of working memory [87]. Working memory and brain health and music that has been supported by ERP studies for both auditory and visual working memory. Attention and executive function [88–90].

Other possible mechanisms may involve the important role of dopamine that occurs in response to music. Music is a relatively abstract stimulus which can induce euphoria and desire for something. A PET brain scan study with (11) C] raclopride) showed that dopamine released in the striatal region, nucleus accumbens, peaked

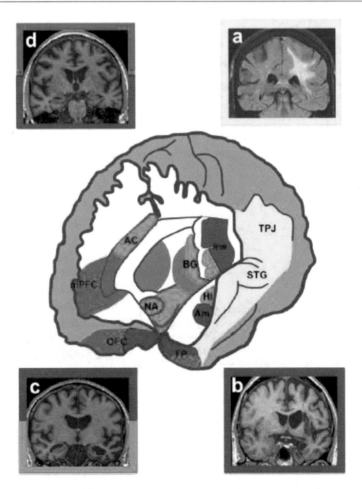

Fig. 11.24 Neuropathological processes affecting music processing: (*a*) temporoparietal tumor triggering musical hallucinations, (*b*) infarction of insula and amygdala with impaired emotive response to music, (*c*) hyperfunction or musicophilia secondary to left temporal semantic dementia, (*d*) frontotemporal lobe dementia and abnormal emotional coding to music. *Yellow* perceptual analysis and imagery, *green* biological motivation and reward encoding, autonomic responses, *red* expectancies, associations and affective evaluation, *blue* mental state processing and behavioral evaluation. Figure with permission: Clark CN, Downey LE, Warren JD. Brain disorders and the biological role of music. SCAN 2014;doi:10.1039/scan

during emotional arousal while music was played caudate activity during anticipation [91]. With melodic intonation therapy for Broca's type aphasia for example it may be that decreasing hyperactivity in the right homologous Broca's region (with transcranial magnetic stimulation) with inhibitory effects can improve the function of the left hemisphere Broca's region [92].

Spirituality, Mysticism, and Religion

Creativity, higher order consciousness, mysticism, and spirituality may be considered apical neural activities unique to humans [93, 94]. Neuro-archeological, neuropsychological, and cognitive

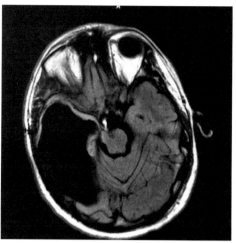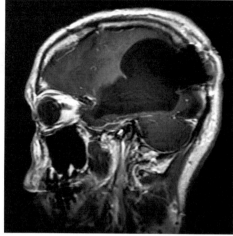

Fig. 11.25 Right hemisphere traumatic brain injury in a young man involving the right temporal and parietal areas, with receptive and expressive amusia and minimal sensorimotor deficit

neuroscience research has inferred that the human brain and mind circuitry is wired to be spiritual and mystical. The differing cultural interpretations of such experiences are the avenues whereby the various formal religions ensue, recognized today [95–97]. Insightful interpretations of the Franco-Cantabrian cave art have been considered the earliest (known) attempts at communicating abstract thinking, pondering the afterlife and spirituality.

Musical rituals in conjunction with statues, relics, and scriptures represent some of the fundamental components of the neural circuitry of early religious deliberation.

Day portends that religious material culture provided a kind of "cognitive scaffolding" that allowed people to communicate the more complex and perplexing concepts concerning supernatural agents, forces, or powers [98].

Regardless, clinical lesion studies have yielded a number of case reports that both augment and destroy spiritual or religious circuitry in the human mind. Both left and right inferior posterior parietal damages have been associated with an increase in self-transcendence [99]. Excessive or sustained analytical thinking may promote religious disbelief. This has been postulated to occur in otherwise normal people, with overusing their analytical (left) brain may sometimes translate

into diminution or abandonment of their religious belief [100]. Ecstatic seizures or mystical seizures are described in association with some temporal lobe epilepsy syndromes [101]. There are also reports of "non-believers" having profound religious experiences in the course of a seizure [102]. Both religious and atheistic states therefore form part of the human neurobiology.

References

1. Dehaene S. Reading in the brain. New York, NY: Penguin; 2009.
2. Coolidge FL, Wynn T. Working memory, its executive functions, and the emergence of modern thinking. Cambr Archaeol J. 2005;15:5–26.
3. Petersen SE, Fox PT, Posner MI, Mintun M, Raichle ME. Positron emission computed tomography studies of the cortical anatomy of single word processing. Nature. 1988;331:585–9.
4. Tarkiainen A, Cornelissen PL, Salmelin R. Dynamics of visual feature analysis and object level processing in face versus letter-string perception. Brain. 2002;125:1125–36.
5. Tanaka K. Columns for complex visual object features in the inferotemporal cortex: clustering of cells with similar but slightly different stimulus selectivities. Cereb Cortex. 2003;13(1):90–9.
6. Gauthier I, Skudlarski P, Gore JC. Expertise for cars and birds recruits brain areas involved in face recognition. Nat Neurosci. 2000;3:191–7.

7. Leroi-Gourhan A. The dawn of European art: an introduction to palaeolithic cave painting. Cambridge: Cambridge University Press; 1982.

8. Damarla SR, Just MA. Decoding the representation of numerical values from brain activation patterns. Hum Brain Mapp. 2013;34(10):2624–34.

9. Dehaene S, Piazza M, Pinel P, Cohen L. Three parietal circuits for number processing. Cogn Neuropsychol. 2003;20:487–506.

10. Nieder A, Millker EK. A parieto-frontal network for visual numerical information in the monkey. Proc Natl Acad Sci U S A. 2004;101:7457–62.

11. Pica P, Lemer C, Izard V, Dehaene S. Exact and approximate arithmetic in an Amazonian indigene group. Science. 2004;306:499–503.

12. Dehaene S, Izard V, Spelke E, Pica P. Log or linear? Distinct intuitions of the number scale in Western and Amazonian indigene cultures. Science. 2008;320:1217–20.

13. Dehaene S. The number sense. Oxford: Oxford University Press; 2011.

14. Tsang JM, Dougherty RF, Deutsch GK, Wandell BA, Ben-Shachar M. Frontoparietal white matter diffusion properties predict mental arithmetic skills in children. Proc Natl Acad Sci U S A. 2009;106:22546–51.

15. Macoir J, Audet T, Breton MF. Code-dependent pathways for number transcoding: evidence from a case of selective impairment in written verbal numeral to Arabic transcoding. Cortex. 1999;35(5):629–45.

16. Cochon F, Cohen L, van de Moortele PF, Dehaene S. Differential contributions of the left and right inferior parietal lobules to number processing. J Cogn Neurosci. 1999;11:617–30.

17. Grana A, Hofer R, Semenza C. Acalculia from a right hemisphere lesion dealing with "where" in multiplication procedures. Neuropsychologia. 2006;44(14):2972–86.

18. Delazer M, Karner E, Zamarian L, Donnemiller E, Benke T. Number processing in posterior cortical atrophy – a neuropsychological case study. Neuropsychologia. 2005;44:36–51.

19. Rossor N, Warrington E, Cipolotti L. The isolation of calculation skills. J Neurol. 1995;242:78–81.

20. Basso A, Burgio F, Caporali A. Acalculia, aphasia and spatial disorders in left and right brain damaged patients. Cortex. 2000;36:265–80.

21. Rusconi E, Walsh V, Butterworth B. Dexterity with numbers: rTMS over left angular gyrus disrupts finger gnosis and number processing. Neuropsychologica. 2005;43:1609–24.

22. Kahn HJ, Whitaker HA. Acalculia: a historical review of localization. Brain Cogn. 1991;17:102–15. and [Ardila A, Roselli M. Acalculia and dyscalculia. Neuropsychol Rev 2002;12;179–31].

23. Wechsler D. Wechsler Memory Scale IV. San Antonio, TX: Pearson; 2009.

24. Wilkinson GS, Robertson GJ. Wide Range Achievement Test 4 professional manual. Lutz, FL: Psychological Assessment Resources; 2006.

25. Wechsler D. Wechsler Individual Achievement Test. 3rd ed. San Antonio, TX: The Psychological Corporation; 2011.

26. McCloskey M, Aliminosa D, Sokol SM. Facts, rule and prodecures in normal calculation: evidence from multiple single patient studies of impaired arithmetic fact retrieval. Brain Cogn. 1991;17:154–203.

27. Benton AL, Hutcheon JF, Seymour E. Arithmetic ability, finger-localization capacity and right-left discrimination in normal and defective children. Am J Orthopsychiatry. 1951;21:756–66.

28. Delazer M, Girelli L, Grana A, Domahs F. Number processing and calculation – normative data from healthy adults. Clin Neuropsychol. 2003;17:331–50.

29. Jackson M, Warrington EK. Arithmetic skills in patiens with unilateral cerebral lesions. Cortex. 1986;22:611–20.

30. Lochy A, Domahs F, Delazer M. Rehabilitation of acquired calculation and number processing disorders. In: Campbell JID, editor. Handbook of mathematical cognition. New York, NY: Psychology Press; 2005.

31. Siegler JC, Ramani GB. Playing linear numerical board games, promotes low income children's numerical development. Dev Sci. 2008;2:655–61.

32. Siegler JC, Ramani GB. Playing linear board games, but not circular ones, improves low income preschooler's numerical understanding. J Educ Psychol. 2009;101:545–60.

33. Dejerine J. Sur en case de cecite verbal avec agraphie, suivi d'autopsie. Comp Rendu Seances Soc Biol. 1891;3:197–201.

34. Marshall JC, Newcombe F. Patterns of paralexia: a pyscholinguistic approach. J Psycholinguist Res. 1973;2:175–99.

35. Béland R, Mimouni Z. Deep dyslexia in the two languages of an Arabic/French bilingual patient. Cognition. 2001;82(2):77–126.

36. Hoffmann M. Thalamic semantic paralexia. Neurol Int. 2012;4(1), e6. doi:10.4081/ni.2012.e6.

37. Benson DF. The alexias: a guide to the neurologic basis of reading. In: Kirschner HS, Freemon FR, editors. Neurology of aphasia. Amsterdam: Swets Publishing Co; 1982.

38. Benson DF. The third alexia. Arch Neurol. 1977;34:327–31.

39. Kinsbourne M, Warrington EK. A variety of reading disability associated with right hemisphere lesions. J Neurol Neurosurg Psychiatry. 1962;25:339–44.

40. Suzuki T, Itoh S, Hayashi M, Kouno M, Takeda K. Hyperlexia and ambient echolalia in a case of cerebral infarction of the left anterior cingulate cortex and corpus callosum. Neurocase. 2009;15(5):384–9.

41. Kerbeshian BL. Hyperlexia and a variant of hyper-graphia. J Percept Mot Skills. 1985;60(3):940–2.

42. Cohen L, Henry C, Dehaene S, Martinaud O, Lehéricy S, Lemer C, et al. The pathophysiology of letter-by-letter reading. Neuropsychologia. 2004;42(13):1768–80.

43. Henry C, Gaillard R, Volle E, Chiras J, Ferrieux S, Dehaene S, et al. Brain activations during letter-by-letter reading: a follow-up study. Neuropsychologia. 2005;43(14):1983–9.

44. Roux FE, Dufor O, Giussani C, Wamain Y, Draper L, Longcamp M, et al. The graphemic/motor frontal area Exner's area revisited. Ann Neurol. 2009;66:537–54.

45. Keller C, Meister IG. Agraphia caused by an infarc-tion in Exner's area. J Clin Neurosci. 2014;21(1): 172–3.

46. Ishihara K, Ichikawa H, Suzuki Y, Shiota J, Nakano I, Kawamura M. Is lesion of Exner's area linked to progressive agraphia in amyotrophic lateral sclerosis with dementia? An autopsy case report. Behav Neurol. 2010;23:153–8.

47. Greenblatt SH. Neurosurgery and the anatomy of reading: a practical review. Neurosurgery. 1977;1(1): 6–15. Review.

48. Van Vugt P, Paquir P, Keels L, Cras P. Increased writ-ing activity in neurological conditions: a review and clinical study. J Neurol Neurosurg Psychiatry. 1996;61:510–29.

49. Frisoni GB, Scuratti A, Bianchetti A, Trabucchi M. Hypergraphia and brain damage. J Neurol Neurosurg Psychiatry. 1993;56(5):576–7.

50. Yamadori A, Mori E, Tabuchi M, Kudo Y, Mitani Y. Hypergraphia: a right hemisphere syndrome. J Neurol Neurosurg Psychiatry. 1986;49(10): 1160–4.

51. Cook FA, Makin SD, Wardlaw J, Dennis MS. Dystypia in acute stroke not attributable to aphasia or neglect. BMJ Case Rep. 2013;pii:bcr2013200257. doi:10.1136/bcr-2013-200257.

52. Otsuki M, Soma Y, Arihiro S, Watanabe Y, Moriwaki H, Naritomi H. Dystypia: isolated typing impair-ment without aphasia, apraxia or visuospatial impairment. Eur Neurol. 2002;47(3):136–40.

53. Diamond J. Animal art: variation in bower decorat-ing style among male bowerbirds Amblyornis inor-natus. Proc Natl Acad Sci U S A. 1986;83(9): 3042–6.

54. Whitley DS. Cave paintings and the human spirit. The origin of creativity and belief. New York, NY: Prometheus Books; 2009.

55. Curtis G. The cave painters. Probing the mysteries of the world's first artists. New York, NY: Anchor Books; 2006.

56. Dutton D. The art instinct. Sydney: Bloomsbury Press; 2009.

57. Kandel ER. The age of insight. New York, NY: Random House Publishing House; 2012.

58. Ramachandran VS. The tell tale brain. New York, NY: WW Norton & Company; 2011.

59. Ishizu T, Zeki S. The brain's specialized systems for aesthetic and perceptual judgment. Eur J Neurosci. 2013;37:1413–20.

60. Mendez MF, Perryman KM. Disrupted facial empa-thy in drawings from artists with frontotemporal dementia. Neurocase. 2003;9:44–50.

61. Schott GD. Pictures as a neurological tool: lessons from enhanced and emergent artistry in brain dis-ease. Brain. 2012;135:1947–63.

62. Miller B, Cummings J, Mishkin F, Boone K, Prince F, Ponton M, et al. Emergence of artistic talent in fron-totemporal dementia. Neurology. 1998;51:978–82.

63. Horowitz SS. The universal sense: how hearing shapes the mind. New York, NY: Bloomsbury; 2012.

64. Thaut MH, McIntosh GC, Rice RR. Rhythmic facili-tation of gait training in hemiparetic stroke rehabili-tation. J Neurol Sci. 1997;151:207–12.

65. Thaut MH, McIntosh KW, Hoernberg C. Auditory rhythmicity enhances movement and speech motor control in patients with Parkinson's disease. Funct Neurol. 2001;16:163–7.

66. Hurst CP, Rice RR, Mctintosh GC, Thaut MH. Rhythmic auditory stimulation in gait training for patients with traumatic brain injury. J Music Ther. 1998;35:228–41.

67. Mithen S. The singing Neanderthals. London: Weidenfeld and Nicolson; 2005.

68. Blacking J. How musical is man? Seattle, WA: Seattle University of Washington Press; 1973.

69. Zatorre RJ, Salimpoor VN. From perception to plea-sure: music and its neural substrates. Proc Natl Acad Sci U S A. 2013;110:10430–7.

70. Freeman W. A neurobiological role for music in social bonding. The origins of Music. Cambridge: MIT Press; 2000.

71. Koelsch S. Brain correlates of music-evoked emo-tions. Nat Rev Neurosci. 2014;15:170–80.

72. Barbas H, Bunce JG, Medalla M. In: Stuss DT, Knight RT, editors. Principles of frontal lobe func-tion. 2nd ed. Oxford: Oxford University Press; 2013.

73. Plutchik R. The psychology and biology of emotion. New York, NY: Harper Collins; 1994.

74. Juslin PN. What does music express? Basic emo-tions and beyond. Front Psychol. 2013;4:596.

75. Clark CN, Downey LE, Warren JD. Brain disorders and the biological role of music. Soc Cogn Affect Neurosci. 2015;10(3):444–52. doi:10.1039/scan.

76. Sacks O. Musicophilia. New York, NY: Vintage Books. Random House Inc; 2008.

77. Miller BL, Boone K, Cummings JL, Read SL, Mishkin F. Functional correlates of musical and

visual ability in frontotemporal dementia. Br J Psychiatry. 2000;176:458–63.

78. Stewart L, Von Kriegstein K, Warren JD, Griffiths TD. Music and the brain: disorders of musical listening. Brain. 2006;129:2533–53.

79. Zatorre RJ, Evans AC, Meyer E. Neural mechanisms underlying melodic perception and memory for pitch. J Neurosci. 1994;14:1908–19.

80. McDonald I. Musical alexia with recovery: a personal account. Brain. 2006;129:2554–61.

81. Midorikawa A, Kawamura M. A case of musical agraphia. Neuroreport. 2000;11(13):3053–7.

82. Brust JCM. Music and language: musical alexia and agraphia. Brain. 1980;103:367–92.

83. Matthews BR. The musical brain. In: Goldenberg G, Miller BL, editors. Handbook of clinical neurology. Neuropsychologhy and behavioral neurology, vol. 3. Edinburgh: Elsevier; 2008.

84. Peretz I, Champod AS, Hyde K. Varieties of musical disorders. Montreal battery of evaluation of amusia. Ann N Y Acad Sci. 2003;999:58–75.

85. Gordon E. Musical aptitude profile. Chicago, IL: GIA; 1965.

86. Seashore CE. Seashore measure of musical talents. New York, NY: Psychological Corporation; 1960.

87. Koelsch S. Towards a neural basis of music related emotions. Trends Cogn Sci. 2010;14:131–7.

88. George EM, Coch D. Music training and working memory: an ERP study. Neuropsychologica. 2011;49:1083–94.

89. Burunat I, Alluri V, Toiviainen P, Numminen J, Brattico E. Dynamics of brain activity underlying working memory for music in a naturalistic condition. Cortex. 2014;57:254–69.

90. Schulze K, Koelsch S. Working memory for speech and music. Ann N Y Acad Sci. 2012;1252:229–36.

91. Salimpoor VN, Benovoy M, Larcher K, Dagher A, Zatorre RJ. Anatomically distinct dopamine release during anticipation and experience of peak emotion to music. Nat Neurosci Rev. 2011;14(2):257–62.

92. Naeser MA, Martin PI, Nicholas M, Baker EH, Seekins H, Kobayashi M, et al. Improved picture naming in chronic aphasia after TMS to part of right Broca's area: an open-protocol study. Brain Lang. 2005;93(1):95–105.

93. Mellars P. Cognitive changes and the emergence of modern humans in Europe. Camb Archeol J. 1991;1:63–76.

94. Renfrew C, Morely I, editors. Becoming human Innovation in prehistoric material and spiritual culture. Cambridge: Cambridge University Press; 2009.

95. Niels G. Genetic and environmental influences on religious interests, attitudes and values: a study of twins reared apart and reared together. Psychol Sci. 1990;1:138–41.

96. Newberg A, d'Aquili E. Why god won't go away: brain science and biology of believing. New York, NY: Balantine Books; 2001.

97. Renfrew C. Situating the creative explosion: universal or local? In: Renfrew C, Morely I, editors. Becoming human. Innovation in prehistoric material and spiritual culture. Cambridge: Cambridge University Press; 2009.

98. Day M. Religion, off-line cognition and the extended mind. J Cogn Cult. 2004;4:101–21.

99. Urgesi C, Aglioti SM, Skrap M, Fabbro F. The spiritual brain: selective cortical lesions modulate human self-transcendence. Neuron. 2010;65:309–19.

100. Gervais WM, Norenzayan A. Analytic thinking promotes disbelief. Science. 2012;336:493–6.

101. Devinsky O, Lai G. Spirituality and religion in epilepsy. Epilepsy Behav. 2008;12:636–43.

102. Dewhurst K, Beard AW. Sudden religious conversion in temporal lobe epilepsy. Br J Psychiatry. 1970;117:497–507.

Definitions: Frontal Syndromes and Frontal Network Syndromes

Clinically, frontal lobe syndromes, frontal network syndromes, frontal systems syndromes, executive dysfunction, and metacognition have all been used to describe disorders of frontal lobes and their extended networks although they are not all synonymous. Anatomically they refer to those parts of the brain rostral to the central sulcus. However, because the frontal lobes network with every other part of the brain, strictly speaking, frontal network syndromes constitute the most accurate neurobiological depiction. The term, frontal network syndromes (FNS) emphasizes the universal connectivity of the frontal lobes with all other brain regions. For example, the stroke literature is replete with FNS that have been reported with discreet lesions outside the anatomical boundary of the frontal lobe, such as subcortical gray matter, subcortical white matter, with isolated lesions of the brainstem, cerebellum, temporal and parietal lobes [1–8]. For the purposes of simplification, five primary, core or elementary syndromes and numerous secondary syndromes may be delineated. Impairment in working memory, executive function, conation, inhibition, and emotional control may be regarded as the elementary deficits of FNS. In addition a number of secondary manifestations may be identified such as a wide array of behavioral abnormalities such as loss of social norms, imitation behavior, compulsions, and obsessions [9, 10]. Fig. 12.1.

Evolutionary Aspects and Relevance to Clinical Syndromes

To begin to understand the most complex object in the universe, the human brain and in particular the frontal lobes, it is most illuminating to study the evolution of our mind and thereby gain a better understanding of the clinical syndromes we are faced with today. In the words of Theodosius Dobzhansky, "nothing in biology makes sense except in the light of evolution" [11]. Life on earth evolved approximately 3.7 billion years ago and thereafter continuously shaped by extraterrestrial and geological events, punctuated by a number of key events. The inclusion of prokaryotes into eukaryotic cells furnished cells with a powerhouse, the mitochondria. Some time after "Snowball Earth," when glaciers reached the equatorial regions about 620–590 million years ago (mya), with the Cambrian explosion of organism diversity, vertebrates (bony fish, amphibians, reptiles, birds, and mammals) formed (~520 mya) [12]. Formation of the vertebrate skeleton allowed rapid movement, an advanced nervous system and high degree of encephalization even though 98 % of animal species are invertebrates

Fig. 12.1 Clinical cognitive disciplines may interact in multiple ways

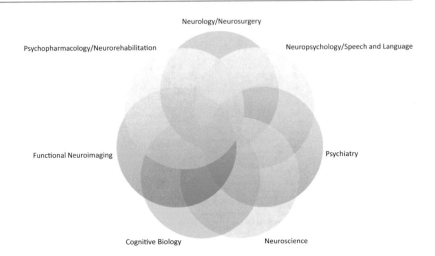

versus 2% being vertebrates. Myelination enabled a vastly improved neural transmission, speeding up neural transmission by a factor of 10 (~9 m per second in unmyelinated fiber versus 50–100 m per second in myelinated fiber), with increased temporal precision, faster communication between the brain and body parts, and ability to react more rapidly to prey and predator [13, 14]. With warming conditions, fish evolved lungs, fins transformed into rudimentary limbs, and occupation of the terrestrial environments occurred about 365 mya. Early mammalian evolution (~200 mya) followed and the subsequent proliferation occurred after the non avian dinosaur extinction ~65 mya. Mammals developed advantageous thermoregulation, thanks to fur and the advantage of mammalian glands and the six-layer cortex that also progressively enlarged in comparison to fishes, amphibians, and reptiles [12] (Fig. 12.2). Primates evolved about 85 mya and about 6 mya the "East Side Story" event (African Rift Valley formation leading to a hot and dry East Africa) precipitated bipedalism, increase in brain size, tool making [15]. However, the impetus of hominin encephalization are multifarious and included climatic drying and cooling that led to savannah expansion, marked climatic fluctuations, complex topography, the risk of predation, which was in turn countered to some degree by sociality, which in turn led to improved communication and ultimately language and the nutritional and metabolic requirements of increased brain size [16].

The emergence of dopamine as a key neurotransmitter was critical in cooling our bodies and brains in a thermally stressed environment and later exapted for executive function [17]. Around this time our frugivorous diet (since ~60 mya) was supplemented with meat and with the advent of Marine Isotope Stage 6 (180–120 mya) may have also served as an important event that highlighted the key dietary changes to seafood that may have played a factor in advancing our cerebral connectivity that ultimately made us modern humans [6, 7]. Shell-fish (scallops oysters, prawns) are rich in both iodine and essential fatty acids, both of which have been correlated with boosting dopamine activity and intellectual development [18–20]. Morphological brain changes as well as connectivity changes were key features in our development.

Brain Volume

As a starting point, using the so-called "missing link" hominid, *Australopithecus africanus* (brain volume approximately 450 ml) there was an increase in size to approximately 1500 ml in Neanderthals over a 3 million year period and subsequently a slight decrease again in modern humans *Homo sapiens sapiens* to 1350 ml [21]. During this time, there was a reduction in the size of the visual striate area (BA 17) with a relative increase in the posterior parietal cortices and frontal lobe reorganization at the network,

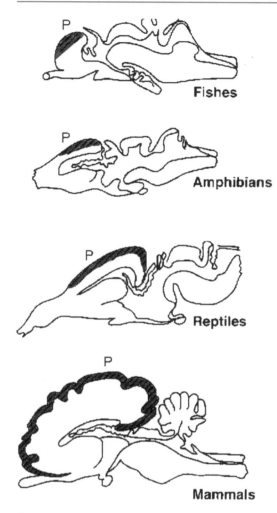

P

Fishes

P

Amphibians

P

Reptiles

P

Mammals

Fig. 12.2 Key innovation of mammals: six-layer neocortex and comparison to fish, amphibian, and reptile cortices. *P* pallium refers to both paleocortex or neocortex. Figure with permission from: Fuster JM. The Prefrontal Cortex. Philadelphia Lippincott – Raven 1997 and von Bonin G. Essays on the cerebral cortex of man. Charles C Thomas, Springfield Illinois 1948

neurotransmitter and receptor levels [22]. Brain size and reorganization probably occurred simultaneously. There is endocast data evidence that in *Australopithecus sediba* dating back about 2 mya, reorganization of the brain occurred before an increase in overall size. This was deduced from the fact that although brain size was relatively small as with other Australopithecines, the orbitofrontal morphology from endocasts was similar to that of typical of later Homo species [23]

Frontal Lobe Size

The size of frontal lobes in various mammalian species is frequently cited as steadily progressing allometrically from the so-called lower forms (rats, mice) to dogs and cats with primates and humans having the biggest proportionally (Fig. 12.3) [24]. Frontal lobe size is not proportionally bigger in humans compared to apes, but prefrontal cortex is, specifically frontopolar cortex or BA 10 and BA 13 is smaller (Fig. 12.4) [25]. The frontal lobes comprise 37–39 % of the cerebral cortex macroscopically and connect to all other parts of the brain, often in a reciprocal manner [26]. The frontal lobe in humans is as large as would be calculated for an ape of human brain size overall, not larger as is often reported [27]. However, what sets us apart from other mammals is not so much brain size, including hemispheric asymmetry with cerebral torque (right frontal and left occipital petalias), but reorganization of our brains in terms of connectivity, neuropil reorganization, neurotransmitter changes, and receptor modification [28].

The Prefrontal Cortex Evolved in Phases

Early mammals were primarily nocturnal characterized by visual, auditory, and somatosensory cortical development but with agranular (relatively devoid of layer four of the six-layer cortex) prefrontal cortex. With the evolution of primates in the angiosperm environment and the move to a diurnal life style, foveal vision was required for higher acuity and depth perception. In addition trichromatic color vision developed that accompanied brighter light accommodation as well as for discernment of ripe fruits. The frontally directed fields of vision and larger binocular fields enabled stereopsis that facilitated survival in the arboreal habitat that required leaping from branch to branch and picking out colored fruit, insects, flowers, nectar, seeds, and younger leaves also referred to as the fine branch niche environment [29]. Accompanying tactile foveal evolution with

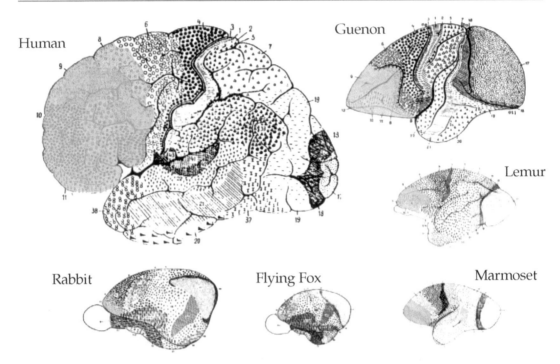

Fig. 12.3 Prefrontal lobe size in mammals, primates and humans. In humans granular prefrontal cortex constitutes 29 % of the cortical area, compared to 11.1 % in guenons, 8.9 % in marmosets, 8.3 % in lemurs, 2.3 % in flying foxes, and 2.2 % in rabbits. Figure with permission: Elston GN. Cortex, cognition and the cell: New insights into the pyramidal neuron and prefrontal function. Cerebral Cortex 2003;13:1124–1138

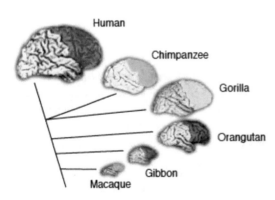

Fig. 12.4 Frontal lobe size is not proportionally bigger in humans compared to apes, but prefrontal cortex is, specifically frontopolar cortex or BA 10 and BA 13 is smaller. Figure with permission: Passingham RE. The frontal cortex: does size matter? Nature Neuroscience 2002;5:190–192 and Semendeferi K, Lu A, Schenker N, Damasio H. Humans and great apes share a large frontal cortex. Nature Neurosci 2002;5:272–276

Meissner's corpuscles that occurred in conjunction with specialization of the primary motor cortex (BA4) in the earlier anthropoids with the newly formed region, the caudal part of the primary motor cortex, issuing most projections to the spinal motor neurons. Cutaneous inputs to this area developed allowing fine manipulation of objects (younger leaves and fruit) and their selection and the development of the somatosensory Meissner's corpuscle receptors. The density of these corpuscles has been correlated amongst nine anthropoid species with increased fruit consumption [30].

In the earlier primates, evolution of the ventral premotor cortex (VPMC) was associated with development of corticospinal projections to the facial nucleus and upper cervical cord segments. These projections enabled improved facial motility and expression and control of the mouth, head and upper limb reaching movements respectively. The dorsal premotor cortex (DPMC) as well as the pre-supplementary motor area (pre-SMA) is involved in lower limb control [31].

The first granular prefrontal cortical areas existed in early primates ~65 mya. Subsequent primate evolution included the divergences between the haplorhines and strepsirrhines ~55 mya, the anthropoids and tarsiers ~45 mya, the old world (catarrhines) and new world monkeys (platyrrhines) ~34 mya, the apes and catarrhines ~23 mya, and finally the hominins and chimpanzee ancestor ~7 mya. During these events there appeared additional prefrontal granular cortical areas: area 46 (midlateral PFC), 9 (dorsomedial PFC), 12/44/45/47 (ventral PFC), and 10 (polar PFC) [32] (Figs. 12.5, 12.6, and

12.7). However, the approximate sequence of development included an approximate caudal to rostral increase in the internal granular layer or layer 4. This is associated with an increased density of cells occupying this level and when above a certain minimum it signifies granular cortex. This development is associated with increased connectivity and increases from monkeys to humans as depicted by Mackey and Petrides [33] Fig. 12.8.

Fig. 12.5 Primate brain and frontal cortex evolution lobe and initiating events: Granular layer 4, of the six-layer mammalian cortex expanded

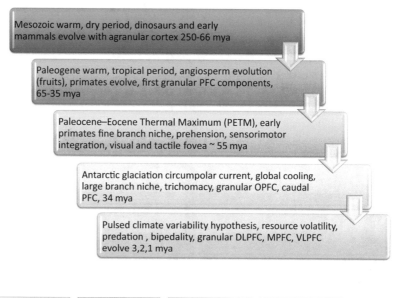

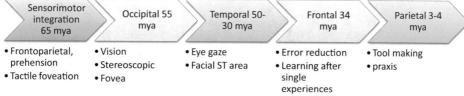

Fig. 12.6 Primate brain evolution: A mosaic pattern with increasing granular frontal cortical areas. *mya* million years ago

Fig. 12.7 Anthropoid frontal lobe evolution sequence since ~34 mya. Legend: *Numbers* denote Brodmann's areas, *DLPFC* dorsolateral prefrontal, *VLPFC* ventromedial prefrontal

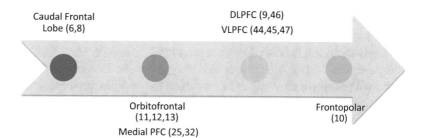

Fig. 12.8 Granular cell layer (IV) and pyramidal neurons in the prefrontal cortex of macaques (*dark blue*) and humans (*light blue*). The granule cell layer IV density is seen to increase from posteriorly to anterior dimension depicted in (**a**). Layer Va pyramidal neuron density increases from medial to laterally noted in (**b**). Figure with permission: Mackey S, Petrides M. Quantitative demonstration of comparable architectonic areas within the ventromedial and lateral orbital frontal cortex in the human and the macaque monkey brains. European Journal of Neuroscience 2010;32:1940–1950

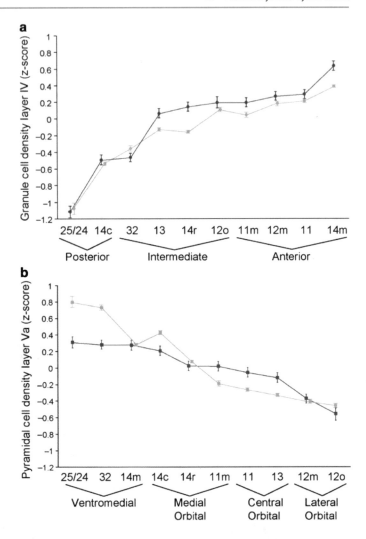

The extensive studies of Passingham and Wise have made major contributions to the likely course of evolutionary developments of the frontal cortical subcomponents and following represents a brief synopsis of their postulates. Evidence points to the first granular PF areas having evolved in the region of the caudal PF cortex (the frontal eye fields and area 8) and orbital prefrontal cortical region (areas 11, 13, 14). Other new granular PF areas appeared subsequently including the dorsal PFC, ventral PFC and polar PFC, all of which provided the primates with adaptive advantages in responses to their most major adversities, chiefly climatic induced food and resource challenges as well as primate society relationships. Overall these frontal regions reduced errors from food scarcity and predatory risk. The significant global cooling and extensive aridity and food at this time has been attributed to Antarctic glaciation with inconstant food availability required diversification of feeding and anthropoid brain expansion, mostly of the granular PFCs can be traced to these events (Fig. 12.9) [29].

The various prefrontal subcomponents (Fig. 12.10) each have distinctive attributes. For example the OPFC designates value to items whereas the caudal PFC and frontal eye field regions promote searching for such objects by virtue of eye orientation and importantly engaging attention. This is due to the caudal PFC/FEF component having connections with the ventral

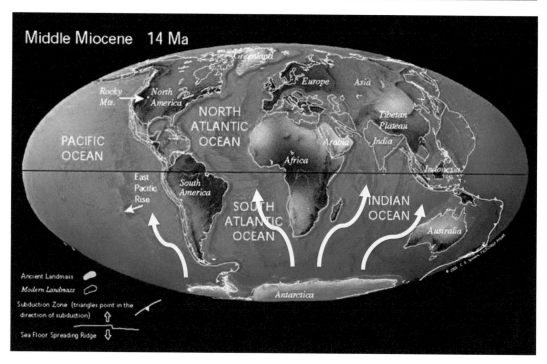

Fig. 12.9 Global cooling consequent to the relatively rapid development of the Antarctic ice sheet 34 mya This led to the thermal isolation of Antarctica and the circumpolar ocean current, in turn due to the southern ocean gateway openings. Cold dense water (*blue*) travels northwards from Antarctica through the sea troughs and basins due to sea ice formation that helps form Antarctic bottom water. Figure with permission: Scotese CR. Atlas of Earth History, Volume 1, Paleogeography, PALEOMAP Project, Arlington, Texas, 2001. References; Goldner A, Herold N, Huber M. Antarctic glaciation caused ocean circulation changes at the Eocene–Oligocene transition. Nature 2014; 511:574–577 and Bradley RS. Paleoclimatology. Reconstructing Climates of the Quarternary (3rd edition). Elsevier, Amsterdam, 2015

Fig. 12.10 Schematic representation of the evolutionary development of frontal lobe regions: (1) caudal frontal (*purple*), (2) ventrolateral frontal (*yellow*), (3) dorsolateral frontal (*pink*), (4) orbitofrontal (*orange*), (5) frontopolar (*green*), and 6 medial frontal (*blue*)

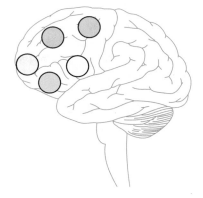
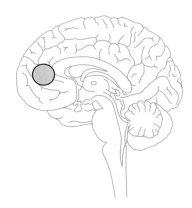

and dorsal radiations emanating from the posterior cortical visual areas and connections to the brainstem eye movement nuclei, 3, 4, and 6 by way of the superior colliculus and basal nuclei. The medial prefrontal cortex (MPFC) facilitates an action choice based on prior behavioral experiences or outcomes mediated by connections with the hippocampus for eliciting previous events or topography, the amygdala and the medial premotor cortex for ultimate action. Whereas the MPFC allows the choice and action in the absence of external sensory stimuli, the

Fig. 12.11 Frontopolar
Cortex: comparative
analysis. Tsujimoto S,
Genovesio A, Weiss
SP. Frontal pole cortex:
encoding ends at the end
of the endbrain. Trends
in Cognitive Sciences
2011;15:169–176

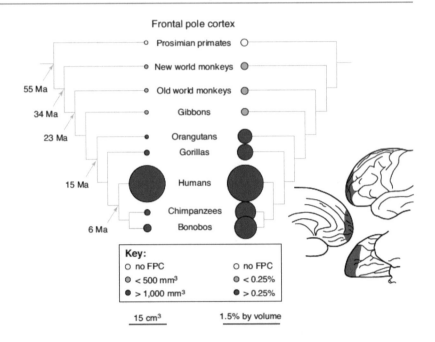

OPFC differs in that the action follows the prompts from external stimuli. The extensive connections of the dorsolateral prefrontal cortex (DLPFC) to the OPFC, premotor cortex, posterior parietal cortex, and hippocampus enables integration of the processes information from these areas and ultimate action output through the premotor cortex. Acoustic (superior temporal) and visual (inferior temporal) information feed into the ventrolateral prefrontal cortex (VLPFC) allowing advantageous decision making based on integration of the data from both visual and acoustic sources. Perhaps because it is part of the mirror neuron circuitry and imitation capability this linking between decisions and consequences may occur even after an isolated event. Such an attribute reduces errors in the complex topography of the East African Rift Valley environment [29].

The frontopolar cortex is particularly enlarged in humans compared to all other extant apes and is the largest architectonic PFC subregion comprising ~500 million neurons, and an area of 28,000 mm³. The neuronal density is relatively decreased and dendritic spines and density higher compared to other areas and comprises subregions 10p frontal pole, 10r and

10m medially. The FPC develops relatively late with maturation of dendritic spines developing up to adulthood 23–30 years [34–36]. The FPC region allows what is termed metacognition or self-reflection. The principal components proposed by Christoff et al. included episodic memory, multitasking, relational integration, self-referential evaluation, and introspection [37]. From an evolutionary perspective, the FPC in humans is relatively much larger in size amongst primates (Fig. 12.11) [38].

Overall this resulted in a key advance for evolving hominoids by allowing error reduction, being able to learn by imitation as well as the important attribute of mental time travel and imaging an action before engaging. There is archeological evidence from the Bodo skull from Ethiopia dated at ~600 kya and the Kabwe skull from Zambia dated at ~300 kya (both *Homo heidelbergensis* species) that allow evaluation of these archaic human skulls from modern humans in terms of the differing large brow ridges, but the slope of the inner frontal brain case is not significantly different. The implication is that the frontal lobes had reached their modern size and approximate shape at least by ~600–300 kya or relatively early in our evolution and considerably before the

Mammals primates, anthropoids	~mya
Early mammals agranular frontal cortex	200-65
Early primates with first parts of granular frontal cortex, caudal, granular OPFC	65
Visual area expansion	55
Split between haplorrhines (fovea) and strepsorrhines (no fovea)	55
Split between anthropoids and tarsiers	45
Catarrhines/platyrrhine split –foveation visual, tactile, trichomacy, stereopsis	34
Granular PFC evolves, dorsal, ventral, polar (erratic resources)	34
Apes and catarrhines separate ways	23
Humans and chimpanzees diverged	
Parietal lobe enlargement (klinorhynchy)	3
Humanoid evolution	~kya
Earliest evidence of modern frontal lobe size (Kabwe and Bodo skulls)	600
Language (recursive) develops	50
Cave art, musicality, spirituality	40
Earliest evidence of executive function –Lionman	33
Earliest Temple, Gobekli Tepe (before agriculture)	11

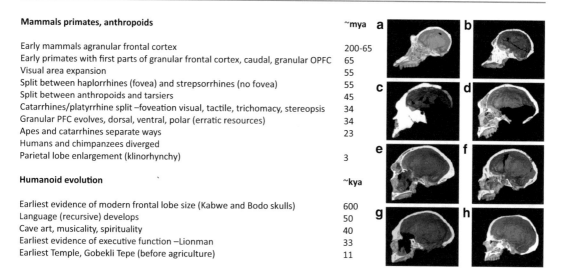

Fig. 12.12 Frontal lobe evolution timeline. The archaic human skulls differ from those of early modern humans in having large brow ridges, but the slope of the inner frontal brain case does not differ, suggesting that the shape of the frontal lobes reached a modern state in archaic humans ~300–600 kya as evidenced by the Kabwe and Bodo skulls

(**a** Chimp, **b** Australopithecus, **c** Bodo, **d** Kabwe, **e** Homo heidelbergensis, **f** proto Neanderthal, g Neanderthal, **h** *Homo sapiens*. Figure with permission: Bookstein F, Schaefter K, Prossinger H et al. Comparing Frontal Cranial Profiles in Archaic and Modern Homo by Morphometric Analysis. The Anatomical Record (New Anat) 1999;257:217–22

so-called cultural evolution that occurred 70–50 kya [39]. A summary table of the frontal lobe evolution timeline is noted in Fig. 12.12.

Prefrontal Cortex Size Is Mainly Due to White Matter Increase

Data from anthropoid primates, revealed a scaling parameter of approximately 4/3 between white and gray matter of the frontal and the non-frontal lobe. This hyperscaling is largely due to white matter increases. The frontal lobe hyperscales with the rest of the brain, which in turn suggests that the hyperscaling of the neocortex is attributed to the frontal lobe white matter changes. Hence, it has been deduced that frontal white matter is the primary component in the explanation of increased brain size [40]. The dramatic white matter connectivity changes is evident in the depiction of the progression indices (how many times larger a particular brain component is in a species and compared to the same brain structure in the in a hypothetical Tenrecinae of the same body size). The progression indices of white matter in the forebrain was compared to the primary visual area

for basal insectivores (Tenrecinae, Madagascar hedgehogs), prosimians, simians, and humans (Fig. 12.13) [41, 42]. Prefrontal gray matter and prefrontal white matter is also significantly different in the 11 primate species examined by Schoenemann [43] (Fig. 12.14).

Histological Architectural Changes of the Neuropil

Frontal Lobes

Axons, dendrites and space between the neurons and glial cells constitute the neuropil which is decreased in BA 10 in humans relative to other primates. In Broca's area (BA 44, 45), the cortical architectural units or mini-columns, are wider in humans relative to primates. BA 10 constitutes the frontopolar and BA 13 the posterior frontorbital region. BA 10 is twice as large in terms of overall brain volume compared to any of the other great apes (1.2 % in humans versus 0.46–0.74 % in great apes). Interestingly, BA 13 is relatively reduced in humans [44, 45].

Spindle cells or von Economo cells appear in layer Vb in both the fronto-insular cortex as well as

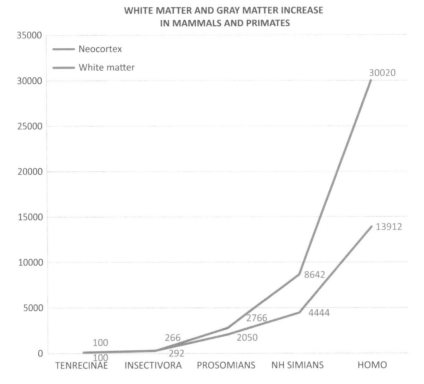

Fig. 12.13 The hallmark of human brain evolution was the profound allometric volume increase of white matter proliferation more so than the neocortical volume increase in *Homo sapiens*. Comparison in the figure is made to Primates, nonhuman Simians, Prosimians, Insectivora, and the stem mammals here represented by the extant Tenrecinae the Madagascar hedgehog. The Tenrecinae are regarded as representing stem Insectivora with relatively primitive brain morphology. The size of the various brain components is referred to the Tenrecinae base of 100 on the *Y* axis . Figure compiled from the data of Stephan H, Baron G, Frahm H. Comparative Brain Research in Mammals, vol 1, Insectivora, Springer, New York 1991

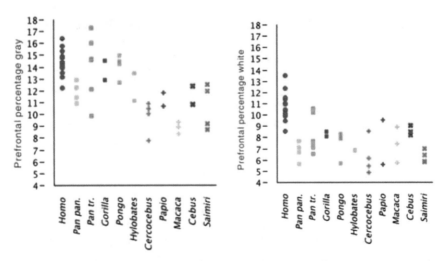

Fig. 12.14 The percentage of prefrontal gray matter volume (*left figure*) and the percentage of volume of prefrontal white matter volume (*right figure*) in humans and 11 primate species. Human total prefrontal gray matter averaged 2.1 ml (3%) smaller but white matter averaged 16.5 ml (41%) larger than predicted from comparisons to 11 primate species. Figure with permission: Schoenemann PT, Sheehan MJ, Glotzer LD. Prefrontal white matter is disproportionately larger in humans than in other primates. Nature Neuroscience 2005;8:242–252

the anterior cingulate cortex, only in humans and African (not Asian) great apes and are approximately 30 % more numerous in the right hemisphere of these species. As they arose in our last common ancestor about 10 mya, they probably played a role in social and emotional processing which arose many millions of years before language. In view of this, they may constitute one of the neurobiological deficits of autism. As an example of convergent evolution, other intelligent species such as the cetaceans have spindle cells [46].

The neuronal density of the important BA 10 and 13 in humans are about half (human BA 10 ~32,000 and BA 13 ~30,000 neurons per cubic millimeters) that of the great apes (chimpanzee BA 10 ~60,000 and BA 13 ~43,000) often much less than half (Orangutan BA 10 ~78,000 neurons per cubic millimeter). The increase in neuropil particularly of BA 10 is likely related to the connectivity of this region with other tertiary association cortex and the other hemisphere [47].

The Temporal Lobe

Surprisingly this cortical region is larger in size in humans than would be predicted from for an ape of human brain size. There is a relative increase in the size of white matter and the ratio of gyral to core white matter in the temporal lobes is larger than would be predicated from other hominoids. This relatively enlarged gyral white matter compared to core white matter is interpreted as reflecting greatly increased interconnectivity subserved by short association fibers [48].

The Amgydala

Of the component amygdaloid nuclei (lateral, basal, and accessory nuclei), the lateral nucleus is relatively larger in humans than would be expected in an ape of human brain size. This has been attributed to the increased interconnectivity with the temporal lobe's unimodal and polymodal sensory information [49].

Overall Brain Reorganization and Mosaic Systems

In human evolution, there has been a differential expansion and reorganization not only of the temporal lobes and amygdaloid complex, but also of the inferior parietal lobes. Specific networks evolved in a coordinated manner that has been termed mosaic evolution. This implies that evolution may have acted on neural systems rather than discrete anatomical structures [50]. Areas that are critical to social behavior which include the amygdaloid nuclei and limbic component of the frontal cortex are both volumetrically larger and revealed reorganization in their networks. This is in direct contrast to the traditional view that limbic structures are conserved whereas the frontal lobes had enlarged. Within the frontal lobes themselves, however, many organizational and network changes have of course taken place. Structures and networks that are implicated in social and emotional processing include the orbitofrontal cortex, the amygdala, fronto-insular cortex, and temporal polar cortex with the latter also important in language processing [51]. These represent a so-called mosaic reorganization that has been a feature of human evolution [52].

Neurotransmitter Systems: Evolutionary Aspects

The quick acting excitatory (glutamate) and inhibitory (GABA) neurotransmitters (NT) act via ion channels with charged ions enabling a relatively quick response in terms of microseconds and seconds. The neuromodulators (NM) such as serotonin (5-HT), dopamine (DA), norepinephrine (NE), acetylcholine (Ach), histamine (H) act differently in that they promulgate longer lasting and more diffuse actions via the G-protein cascade system The concept of connectional and behavioral fingerprints emphasize the brain connectivity as a whole on both an anatomical and physiological basis. The connectional finger print of a particular region not only constrains what that area may do but also provides insights into what particular function can be deduced [53].

Overall, there is regional heterogeneity of neurotransmitters in the human brain. Different NTs subserve different higher cortical functions (HCF) and in neurodegenerative disease NT deficits occur in varying combinations. The downside of NT modulatory systems for intellectual advances, probably made humans more susceptible a number of neuropsychiatric diseases (schizophrenia, autism, bipolar disease) and neurodegenerative diseases, unique to humans [54].

Dopamine (DA)

Within the frontal subcortical circuits, DA is the principal NT. The reasons for this have been proposed as part of a very plausible and well researched hypothesis by Previc and their open loop systems, DA and Ach became the predominant NTs in the left hemisphere and NE and 5HT in the right hemisphere [17].

DA is considered to have been one of the key factors in the emergence of human intelligence. After the geological events that led to the "East Side Story" with East African becoming relatively dry and arid, heat management and combating the deleterious effects of hyperthermia (including the so-called heat stroke) was a critical factor in survival of the mammals and the newly emerged bipedalist, *Australopithecus africanus* [15]. The function of DA in lowering body temperature presumably enabled early hominoids to better tolerate the hyperthermia of chase hunting and catching prey that succumbed to chase myopathy [55, 56]. Thereafter in an evolutionary sense, DA expansion occurred, due to increased calcium metabolism from prolonged aerobic activity as well as the increased tyrosine (a dopamine precursor) consequent to increasing meat supplementation about 2 mya [57]. The clinical sequelae of blocking dopamine (by drugs such as haloperidol, risperidone, quetiapine) as malignant hyperthermia and neuroleptic malignant syndromes may be therefore interpreted in an evolutionary perspective. DA became exapted as the most important NT in our evolving brains, eventually concerned with most of the core frontal functions; working memory, cognitive flexibility, motor planning, abstract representation, temporal sequencing, and generativity [53].

DA exerts a modulatory effect (affects signal to noise ratio) on the PFC G-protein linked receptors on dendritic shafts and spines of glutaminergic pyramidal neurons and dendrites of

GABA-ergic neurons [58]. These neurobiological features enable DA to regulate working memory, reasoning and language. Humans and great apes feature DA input to all cortical areas, in contradistinction to the paucity of DA-ergic innervation of rodents [59]. This was determined by measuring cortical DA innervation (axon density) using tyrosine hydroxylase immunereactivity. In addition there is a regional DA-ergic distribution, most intense in layer I and V–VI of the association cortices [60]. Furthermore, compared to great apes, humans have a generalized increased DA-ergic input to the prefrontal cortical regions. The dopaminergic hypothesis in human evolution purports that the expansion of human DA-ergic in particular was the most important factor in human tool making, exploration, cultural and scientific developments [53]. This theory also proposes that the drawbacks are the propensity for hyperdopaminergic syndromes such as schizophrenia, bipolar disease, autism, attention deficit hyperactivity disorders and neurodegenerative diseases [54].

Serotonin

In humans and great apes, compared to other mammals, there is an overall increase in the cortical output 5-HT efferents [37]. The 14 different serotonergic G-protein related receptors and one ion channel receptor (5-HT 3) enable the modulation of several different functions simultaneously, including memory, learning and inhibition. This occurs via receptors on pyramidal cells and dendritic shafts and via inter-neurons, which allows signal modulation from local circuits with reference to extrinsic stimuli [61]. From a clinical point of view, serotonin in the OFC circuitry has been linked to self-control, emotional, processing, and inhibition regulation [62].

Acetylcholine

In the neuromodulatory axons of humans and great apes exist varicose type axons that are likely to have a role in cortical plasticity [63], associated with advanced traits such as superior learning capability, social learning, advanced tool manufacture, and self-awareness. These effects are mediated via five muscarinic receptors (M1–M5), all of which are G-protein linked. Nicotinic receptors are all ligand gated ion channels. Both transmit mediating excitatory and inhibitory effects on GABA interneurons and pyramidal cells. The neurophysiological effects translate into cognitive flexibility, learning, and working memory [64].

The Mosaic Cognitive Evolution

Hominoids are able to imitate behavior and imitation ability of primates and hominoids was crucial to the cultural evolution. The imitation ability has been termed appropriately as an "all purpose learning mechanism" by Subiaul [65]. The imitation circuitry likely evolved through anatomical, chemical and organizational changes. The circuit has been termed "mosaic" in that it involves the PFC, parietal, temporal and cerebellar regions. The core features of memory and attention as well as more specific cognitive domains such as language and tool use are presumed to be based on the mosaic pattern, itself based on the imitation behavior circuitry that may have a visual, auditory and tactile dimension [65]. A summary of these changes is depicted in Table 12.1.

Table 12.1 Neuroarcheologically based neurobiological and changes in human brain evolution including brain size, reorganization, and NT changes

Frontal lobe size as expected within hominoid evolution
Frontoparietal sensory motor integration including mirror neuron circuitry
Lunate sulcus moves more posteriorly with reduction in primary visual cortex
Petalias left occipital, right frontal (cerebral torque)
Neuropil less dense
BA 10 increased
BA 13 decreased
Temporal lobe increased in size
Amgydala basolateral nucleus increase in size

The Brain Is a Connectome Consisting of Both Neurochemical Tracts Macroscopic Hardwired Tracts

Neurochemical Tracts

The fast acting excitatory (glutaminergic) and inhibitory (GABA) amino acid neurotransmitters are modulated by a number of widely projecting slower acting (most G-protein linked) neurotransmitters. This type of chemical architecture is useful in coordinating many neurons and neuronal circuits in response to a stimulus or threat. There are currently eight chemical or neuromodulatory tracts (DA, 5-HT, NE, Ach, H, oxytocin/vasopressin, and orexin) (Fig. 12.15) with their nuclei of origins in the brainstem, basal forebrain or hypothalamus, and extensive cortical ramifications.

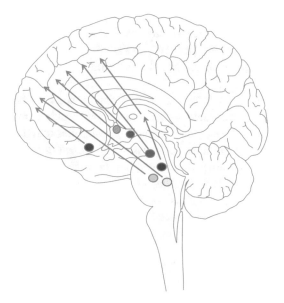

Fig. 12.15 Neurotransmitter or connectional fingerprints are a function of the ascending neurotransmitter systems: Acetylcholine (*purple*), noradrenaline (*yellow*), serotonin (*green*), dopamine (*blue*), histamine (*pink*), orexin (*red*)

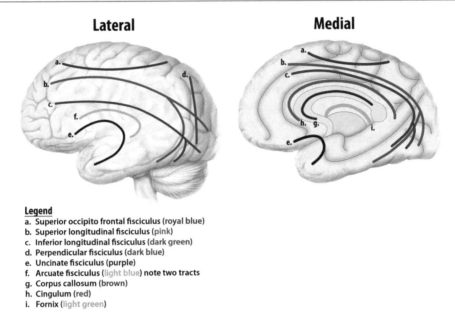

Fig. 12.16 The major fiber tracts of the brain, lateral and medial

Network Neuroanatomy and Neurophysiological Functional Systems

The Major Cerebral Fasciculi as They Pertain to the Frontal Lobes and Network Systems

The main functions of the frontal lobes may be viewed as motor action and the temporal integration of behavior. Frontal lobe evolution may be conceived of as a progressive refinement of pyramidal pathway responses (motor, speech, behavior) by incorporating cognitive and emotional processes. Optimal decision-making requires a flexible system that can incorporate a wide range sensory input, at the same time prioritizing and choosing the most effective response in a changing environment. This has resulted in a complex circuitry and is achieved through the major cerebral network systems; superior frontal occipital fasciculus, superior longitudinal fasciculus, inferior longitudinal fasciculus, cingulum, and uncinate fasciculus as well as seven frontal subcortical circuits [66, 67]. The major fasciculi from an anatomical point of view are depicted in Fig. 12.16, and the neurophysiological frontal functional sys-

tems in Fig. 12.17. These can also be imaged by diffusion tensor imaging with directional specificity that is color-coded.

1. U fibers (orange)
2. Superior occipitofrontal fasciculus (royal blue)
3. Superior longitudinal fasciculus (pink)
4. Inferior longitudinal fasciculus (dark green)
5. Perpendicular fasciculus (dark blue)
6. Uncinate fasciculus (purple)
7. Arcuate fasciculus (light blue) — note two tracts
8. Corpus callosum (brown)
9. Cingulum (red)
10. Fornix (light green)

Neurophysiological Frontal Functional Systems

Frontal lobe presentations can be enigmatic, bewildering, covert, and silent. Standard clinical neuroscience texts typically list two dozen or more different presentations [68, 69]. Understanding frontal network presentations in the context of their evolutionary heritage aids insight and understanding. They may be grouped as follows:

1. Frontoparietal group comprising the mirror neuron, working memory, and arcuate

Fig. 12.17 Frontal lobe neurophysiological functional network systems: an evolutionary syndrome approach

Principal Circuits

1. Frontoparietal group
2. Frontal subcortical circuits
3. Uncinate fasciculus group
4. Fronto-ponto-cerebellar circuit
5. Ascending neurotransmitter
6. Evolutionary repurposed circuits

Cognitive domains/deficits

language, working memory, praxis, mirror neurons
executive, disinhibtion, abulia spectrum
social and emotional syndromes
cognitive deficits, cognitive efficiency
attention, arousal, learning, sleep
arithmetic, reading, writing

fasciculus. Representative syndromes include the field dependent behavior disorders, dyexecutive, dysmemory, dysnomia, and multitasking disorders.

2. The seven frontal subcortical circuits with index syndromes including disinhibition, dysexecutive, abulia spectrum of disorders, and obsessive-compulsive disorders.

3. The uncinate fasciculus group with syndromes including social, emotional, and episodic memory disorders.

4. The frontopontocerebellar circuitry associated with the cerebellar cognitive affective syndrome complex.

5. The ascending neurotransmitter based systems that mediate state dependent conditions such as arousal, learning, attention disorders.

6. Evolutionary repurposed circuitry that mediate reading and arithmetic as well as hodological hyperconnectivity and hypoconnectivity disorders.

The Frontoparietal Working Memory (WM) Group

Archeological evidence exists for modern WM capacity and development to a hypothetical seven stages. Wynn and Coolidge speculated that "enhanced WM capacity powered the appearance of the modern mind." The technical evidence of stone tools is considered more a function of procedural and long-term memory function. However, reliable weapons (projectile tools, spears, hafting, bows and arrows, traps and snares), foraging, and using external storage devices (Blombos cave engravings from ~70 kya) all necessitated a response inhibition of the central executive that required delayed gratification. Within the last 50 kya, with the development of enhanced working memory (EWM) (stage 6–7), a far greater number of abstract conceptualizations, off-line thoughts and options translated into improved behavioral flexibility as well as creativity.

EWM underlies attention, intelligence, language, memory processes, inhibition, and theory of mind (TOM). This in turn enabled the cultural evolution [70].

From direct cell recordings it was established that neurons with DLPFC (BA 46/9) fire during the delay period fitting the cellular basis of maintenance of information, ultimately for preparation for action. The mid DLPFC is critical for the monitoring component (tracking of relevant stimuli) of the information pertaining to working memory, termed epoptic process [71]. On the other hand, maintenance of information in WM depends on posterior cortical association areas that include the superior temporal gyrus for auditory representations and the rostral inferotemporal cortex for object representations. Maintenance of information in WM differs from the active retrieval of memory representations, which is a function of the midventrolateral PFC (VLPFC) [72, 73]. In addition the posterior parietal region is also is concerned with the manipulation of information such as mental rotation and mental arithmetic. Petrides has postulated that the involvement of the MDLPFC in manipulation tasks is necessary for close monitoring of information while this is being manipulated. Notable in this regard is the finding of the BA 46 axons targeting especially layers I–III of components of the intraparietal sulcus supporting this modulatory role of the MDLPFC on parietal function—the epoptic process [74, 75] (Fig. 12.18).

Working memory hubs

1. Epoptic process: Mid dorsolateral prefrontal cortex monitors information
2. Manipulation of information: Posterior parietal region around the intraparietal sulcus
3. Both regions interact with the superior temporal region for sensory information (auditory, visual verbal, multimodal) resides

Principal fiber tracts sub-serving working memory

A. Superior longitudinal fasciculus
B. Middle longitudinal fasciculus
C. Extreme capsule

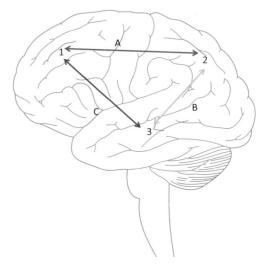

Fig. 12.18 The most fundamental and cardinal function of the frontal lobes is working memory: The extensive evolutionary evolved frontoparietal circuits, were exapted for working memory

Fig. 12.19 Frontal subcortical circuits. Parallel circuits each with same components with the exception of the medial prefrontal cortex via nucleus accumbens instead of caudate nucleus

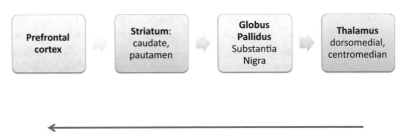

The Seven Frontal Subcortical Networks (FSC)

The FSC share a similar basic anatomy, namely; cortex-caudate-globus pallidus-thalamus-cortex (Fig. 12.19). Confusion of terminology may exist because of the rather cumbersome terminology related to subcortical structures, best delineated by a diagrammatic representation (Fig. 12.20). The neurotransmitters and neuropeptides integrated in this circuitry are mainly excitatory (glutamate) and inhibitory (GABA) being the principal ones but many others involved in a neuromodulatory capacity including the monoamines, enkephalin, neurotensin, substance-P, dynorphin, adenosine, and neuropeptide-Y [9, 10].

The FSC connectivity

1. Direct and indirect pathways
2. Connections with the other circuits (cortico-cortical)
3. Connections to areas outside the FSCs

Direct pathway: Glutamate release occurs from frontal cortical regions to the striatum, mostly caudate nucleus, less often to putamen and nucleus accumbens. This releases GABA at internal segment of GP and SN, GABA from GPi (globus pallidus interna) to thalamus diminishes and in turn the thalamus increases glutaminergic excitation of cortical regions. The striatal neurons within this pathway that project to the GPi are termed striosomes with D1 receptors. The net result is a thalamic disinhibition [76–78].

The indirect pathway balances the direct pathway. Here the striatal efferents are termed matrix efferents wit D2 receptors and project from the globus pallidus externa (as opposed to interna) with a net thalamic inhibition. Some of the GPe neurons are cholinergic but most are GABA-ergic with two different types of GABA-ergic cell types within the GPe termed GP-TI and GP-TA, which mediate cross talk between the afferent and efferent circuits of the GPe as well as the direct and indirect pathways [79].

Fig. 12.20 Basal
ganglia anatomy

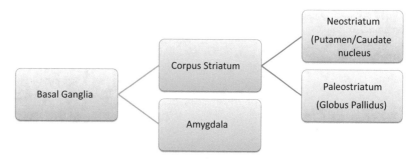

Dorsal striatum: Caudate and putamen
Ventral striatum: nucleus accumbens
Additional terminology: Lentiform nucleus – Globus Pallidus and Putamen

Connections to regions outside the FSCs include (a) DLPFC (BA 46) to parietal (BA 7), (b) anterior cingulate to temporal lobe (hippocampus, amygdala, entorhinal cortex), (c) medial OFC to temporal lobe, and (d) lateral OFC to heteromodal sensory cortex [9].

There is a complex interplay of the neurotransmitters, receptors and circuit plasticity such as enhancement of striatal dopamine release by cholinergic agonists. The importance of recognizing and understanding this circuitry is in the relationship they have to clinical syndromes clinicians appreciate with regards to frontal pathology. These may be conveniently conceived of as three principal syndromes of:

1. Predominant dorsolateral prefrontal cortex (DLPFC); temporal organization of information; executive function, working memory, multitasking [80].
2. Anterior cingulate circuitry; motivation for behavior. Impairments here lead to akinetic mutism and the abulic spectrum of disorders [81, 82].
3. Orbitofrontal (OFC) circuitry; the cortical component of the limbic system where the emotional and other limbic components are integrated into behavioral output. Hence, disinhibitory syndromes may occur in response to lesions of the medial as well as the lateral OFC [9].
   (a) The medial OFC circuitry mediates empathic and socially appropriate behavior. Personality change in the context of frequently normal cognitive (DLPFC)

function is usual and a manifold of presentations are encountered principally. The various forms of echopraxia or imitation behavior, utilization behavior and environmental dependency syndromes (field dependent behaviors) are sometimes the overriding clinical manifestations of these lesions. From a neurophysiological point of view the medial OFC attaches emotional valence to events, which in turn determines the strength of the episodic memory. Impaired autonomic and endocrine processing is associated with lesions of the medial OFC [83].

(b) Lateral OFC lesions have been correlated with OCD, depression, irritability, mood disorders, and field dependent behaviors. Obsessive-compulsive disorders spectrum group of disorders include OCD spectrum itself that comprise obsessions (intrusive urges thoughts, images), compulsions (repetitive, ritualistic type of activities of a physical or cognitive nature), Tourette's syndrome, kleptomania, risk-seeking behavior, pathological gambling, body dysmorphic disorder, and the immune disorder of pediatric autoimmune neuropsychiatric disorders (PANDAS) [84].

In addition to these three principal behavioral FSCs two others have recently been added. The inferotemporal subcortical circuit may be associated with deficits in visual discrimination, visual scanning, visual hallucinations and psychosis [10]. In addition a circuit between the posterior

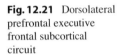
Fig. 12.21 Dorsolateral prefrontal executive frontal subcortical circuit

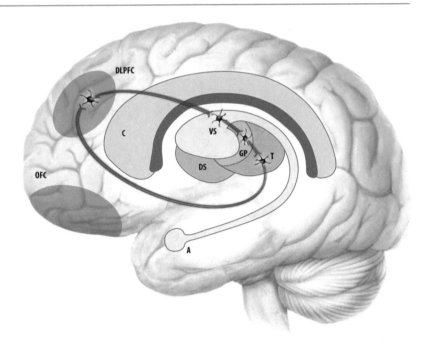

parietal region (BA 7) and the prefrontal region (BA 46) also contributes to processing of visual stimuli of significance and accordingly visuospatial processing [10, 79].

Individual lesions or disease processes usually affect more than one to differing degrees and in various combinations. To facilitate the diagnostic components and sometimes treatment options, it is useful to consider the FSCs and their clinical counterparts in terms of neurophysiological core components and correlation to the clinical syndromes (Fig. 12.21).

Uncinate Fasciculus (UF) Group

The UF is a long-range association fiber tract, relatively enlarged amongst humans compared to other primates and connecting the OPFC and anterior temporal lobe (Fig. 12.22). It has a particular propensity to injury in traumatic brain injury and is also involved in psychiatric syndromes because of its relatively unique late maturation in the third decade [85].

The three main types of disorders associated with UF impairment include episodic memory, language and social and emotional conduct disorders.

Neuropsychiatric clinical disorders associated with the UF include anxiety states, schizophrenia, antisocial personality disorder. Less frequent syndromes include epilepsy, uncinate fits (dreamy states, olfactory and gustatory hallucinations), emotional and sexual arousal, involuntary movements of mouth and face, Kluver–Bucy syndrome, and delusional misidentification syndromes (DMIS) such as Capgras and reduplicative paramnesia. Hirstein and Ramachandran [86] proposed that these syndromes result from severed connections between the limbic areas and face processing areas that emotionally color facial percepts. For example with Capgras syndrome although the face that is seen is accurately identified as that of a familiar person or family it is devoid of feeling and the person reaches the conclusion that it must be someone else. UF dysfunction therefore likely plays a role in DMIS [86]. The UF links memory, social and emotional processing and allows

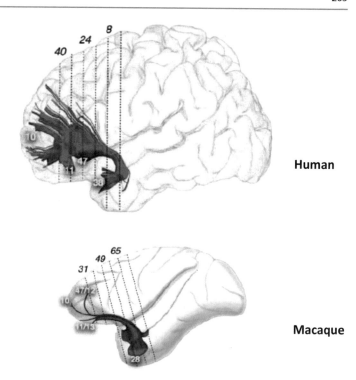

Fig. 12.22 Uncinate fasciculus evolution: comparative human (*top figure*) and macaque (*bottom figure*) analyses. Figure with permission from: Thiebaut de Schotten M, Dell'Acqua F, Valabregue R, Catani M. Monkey to human comparative anatomy of the frontal lobe association tracts. Cortex 2012; 48: 82–96

Human

Macaque

mnemonic associations on a temporal basis such as the face, voice, and emotional disposition to a particular person [87].

Frontopontocerebellar Circuit

Modern human brain size was substantiated by ~200 kya, yet modern cerebellum size only at ~28 kya. A recent escalation in cortico-cerebellar connectivity occurred, particularly the prefrontal cortex occurred in the late Pleistocene period. This is mostly likely to have been associated improved cognitive efficiency such as working memory [88]. Two syndromes, the cerebellar cognitive affective syndrome and cerebellar cognitive dysmetria are examples of disruption of this circuitry most commonly by stroke for example whereby pathophysiologically there is a corticocerebellar crossed diaschisis [89]. Recently diffusion tension imaging has been employed to image this extensive tract that mediates frontal syndromes, particularly executive dysfunction as well as behavioral abnormalities form a remote posterior fossa insult (Fig. 12.23) [90].

Ascending Neurotransmitter

Both deficiency and excess of one or more neurotransmitters may be encountered. Attention, working memory, and executive function are dependent on the monoaminergic systems and deficiencies in these lead to impairments in these functions. Parkinson's (dopamine deficiency state) is a prototypical disorder with dopamine deficiency is associated with motor (nigrostriatal, mesocortical) and cognitive dysfunction (mesolimbic). However, neurotransmitter perturbations usually of excess are features of the serotonin toxidrome, the neuroleptic malignant syndrome, the cholinergic toxidrome, and malignant hyperpyrexia (see below).

Evolutionary Repurposed Circuits

Relatively recent cultural inventions such as reading, mathematics and art were fashioned on existing primate brain circuitry. Dehaene's neuronal recycling hypothesis posits that the increase in the conscious neuronal workspace enabled this

Schematic fronto-ponto-cerebellar circuit,
schematic and by MRI-DTI tractography

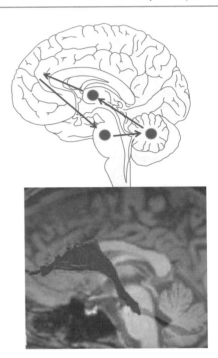

Fronto-ponto-cerebellar circuit imagedby
MRI-DTI tractography

Figure with permission: Kamali A, Kramer LA, Frye RE, Butler
IJ, Hasan KM. Diffusion Tensor Tractography of the Human
Brain Cortico-Ponto-Cerebellar Pathways: A Quantitative
Preliminary Study. Journal of Magnetic Resonance Imaging
2010;32:809-817

Fig. 12.23   Modern human cerebellum evolution occurred relatively late (~28 kya)

- Numerosity in the monkey has been shown to
  require a parieto-frontal network

- Neurons in the intraparietal sulcus responded and
  conveyed numerosity earlier than PFC neurons

- The anterior part of superior longitudinal fasciculus
  (aSLF) connects parietal and frontal cortex

- In children, fractional anisotropy in the left aSLF
  correlates (positively) with arithmetic skill

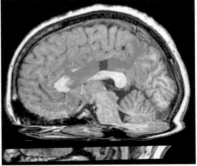

Fig. 12.24   Evolutionary repurposed circuitry use of
frontoparietal networks: arithmetic, reading skills. Figure
with permission: Tsang JM, Dougherty RF, Deutsch GK,
Wandell BA, Ben-Shachar M. Frontoparietal white matter
diffusion properties predict mental arithmetic skills in
children. PNAS 2009;106:22546–22551

repurposing [91]. For example, in humans and
macaques, the temporal regions responsible for
object identification are located within the inferior occipitotemporal region (BA 37). Reading is
handled by this brain region (termed the brain's
letterbox) that has been used for visual object
identification for ~10 million years [92]. Monkey
microelectrode recordings have revealed that

neurons in both the intraparietal sulcus and prefrontal cortex respond and convey numerosity,
suggesting that numerosity has been correlated
with a parietofrontal network [93]. The horizontal IPS activates only when numbers are seen, not
colors or letters, and arithmetic skills have been
correlated with the anterior part of superior
longitudinal fasciculus [94] (Fig. 12.24).

Clinical Aspects: Recognizing Primary and Secondary Syndromes of FNS and Their Usefulness in Treatment Strategies

The complexities associated with prefrontal cortex function have yielded a number of different hypothetical explanations, summarized and reviewed by Wood and Grafman table) [95].

Current Theories of Prefrontal Function

Working memory is regarded as the core function of prefrontal function by many investigators. From a practical clinical point of view, a vast panoply of symptoms have been associated with frontal lobe lesions but may conveniently grouped under primary and secondary domains:

Primary (Fig. 12.25)

Core frontal systems include:

Initiation
Disinhibiton
Working memory
Attention
Monitoring
Emotional control

Secondary (Fig. 12.25)

Phenotypic presentations of these primary processes may include a myriad of secondary syndromes that may be grouped into:

Motor
Eye movement disorders
Alien hand syndrome
Obsessive-compulsive disorders

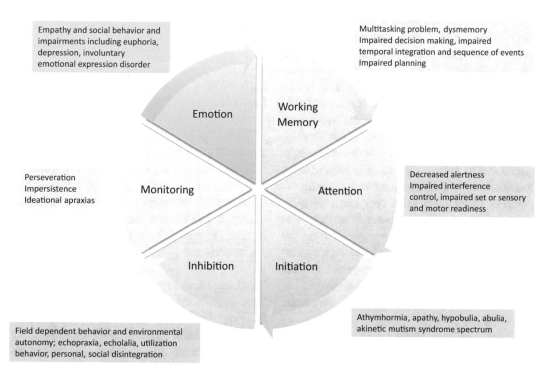

Fig. 12.25 Core frontal systems and the associated primary memory) and secondary syndromes

Fig. 12.26 Disorders of conation: the athymhormia or abulia spectrum of disorders

Behavioral
Various disinhibition syndromes
Field dependent behaviors

Cognitive
Dysexecutive
Multitasking problems
Motor sequencing difficulty
Motor language disorders

Disconnection (hodological effects)
Emergent visual artistry
Emergent musical artistry
Cerebellar cognitive affective syndrome

Initiation
The akinetic mutism, abulia, hypobulia, apathy spectrum of disorders denoted degrees of impairment in conation from no movement and muteness to an overall generalized slowing of motor and speech behavior. Conation, from the Latin: conatus—the natural tendency to strive towards or directed effort, is an important component of both neurological and psychiatric disease syndromes Both Kraeplin and Bleuler referred to the very significant negative symptoms of schizophrenia, a type of avolitional disorder that correlates with the functional decline. Neuropharmacological agents primarily target the positive symptoms, the delusions and hallucinations, however, and came to dominate schizophrenia therapeutics during the twentieth century. Unsurprisingly, the antipsychotics, whether haloperidol or more recent later generation agents such as clozapine or risperidone, have not translated into significant functional improvements [96].

There is an extensive list of subcomponents of the avolitional syndrome that include apathy, asociality, anhedonia, affective flattening, alogia, attentional impairment, hypophonia,

emotional withdrawal, impaired abstract thinking, and incoherence. Social cognition, viewed as a decoupling of TOM, comprises social knowledge, social perception, emotional processing, and attribution bias (attributing cause to someone or agents forming persecutory delusions) [97]. A recent review regards the avolitional component as the core feature in schizophrenia and at the same a de emphasis for the concept of anhedonia, with asociality and alogia being clinical or phenotypic exhibitions of the core avolitional deficit [98]. In addition, loss of creativity, curiosity, initiative, and emotion may also be present. Alexithymia refers to a personal inability to describe and identify emotions within oneself (Fig. 12.26).

Monitoring
Perseveration
Impersistence

Disinhibition
Impulsivity

Field dependent behavior forms and manifestations (Fig. 12.27) [99].

Loss of judgment, loss of insight, impairment comportment, inappropriate social behavior, loss of empathy, irritability, aggression, irascible, excessive jocularity, irresponsible behavior, restlessness, hyperactivity, hypersexuality, hyperorality, incontinence.

Working Memory
Verbal and nonverbal working memory, multitasking, abstract thought, planning ahead, temporal sequence of events. Examples of executive functions include learning new information both verbal and visuospatial, searching memory systems, activation of past memories,

More elementary forms

Imitation behavior
Utilization behavior
Echolalia
Environmental dependency syndrome

More complex forms

Forced hyperphasia
Zelig-like syndrome
Echoing approval
Oral spelling behavior
Exaggerated startle responses
Command-automatism and echopraxia to television
Forced collectionism
Excessive television watching
Response to next patient stimulation
Hypermetamorphosis
Tourette's syndrome
Obsessive compulsive disorder, echoplasia

Fig. 12.27 Classification of field dependent behavior syndromes; elementary and more complex forms. From: Hoffmann M. The panoply of field dependent behavior in 1436 stroke patients. The Mirror Neuron System uncoupled and the consequences of loss of personal autonomy. Neurocase 2014;20(5):556–68

temporal organization of behavior, attention, generation of motor activity that includes speech, writing, or limb movement. These involve at least four of the five sub-processes; task setting, initiation of the task, monitoring, error detection, and behavioral self-regulation [100]. These are therefore modulated by up to four of the FSCs with task setting correlating with left DLPFC and monitoring/error detection correlating with right DLPFC activity. Initiation of a new task involves both left and right superior medial frontal FSCs and behavioral regulation disrupted after medial OFC circuit lesions [101]. Cerebral lesions distinct from the FSCs may be associated with FNS such as parietal lesions. Even memory analysis can at times be correlated with FSC topography. For example, caudate lesion patients may have poor recall but relative preservation of recognition. Thalamic lesion patients on the other hand may have impairment of recall and recognition [102].

Attention

Under this rubric are included alertness and arousal. Differentiating attention and working memory: attention allows certain stimuli, be they sensory or cognitive, to be given preference over other competing ones. Working memory refers to the keeping a limited amount of information usually for a few seconds to allow manipulation or use of that information for another task. Attention may be usefully categorized as follows:

1. Focused attention—respond discretely to specific visual, auditory, and tactile stimuli.
2. Sustained attention—maintain consistent behavioral response during continuous and repetitive activity.
3. Selective attention—maintain behavioral or cognitive set in context of distracting or competing stimuli.
4. Alternating attention—mental flexibility that allows individuals to shift their focus of attention and move between tasks.
5. Divided attention—highest level of attention with ability to respond simultaneously to multiple tasks [101, 103].

Language

Motor language disorders include:
Broca's aphasia (usually left hemisphere)
Expressive aprosodia (usually right hemisphere)
Transcortical motor aphasia
Aphemia (nonfluent speech, normal comprehension and writing, initial muteness)
Central aphasia

Emotional Control

Clinical lesions studies have implicated in particular the orbitofrontal cortex as part of the neural network for emotional responses [104]. Patients with orbitoprefrontal and medial frontal regions were significantly impaired in both cognitive and affective empathy as compared to parietal patients and healthy controls and those with damage restricted to the prefrontal cortex, no matter which side, resulted in impaired empathy and lesions involving the right parietal lesions [105, 106]. Subsequently from a registry analysis a much more widely distributed lesion site network impairs EI, in keeping with the extensive contemporarily appreciated neurobiological emotional network proposed by Pessoa et al. [107]. Many different brain lesions may affect EI, including frontal, temporal, subcortical and even subtentorial stroke syndromes with the strongest relationship (EI scores) pertained to the frontal and temporal regions [108].

These are presented under headings and categories that we currently and traditionally see them. However, they are all frontal network syndromes that just happen to be treated by differing brain related clinicians for historical reasons. Frontal network syndrome phenotypes may comprise a mixture of the primary and secondary syndromes in various combinations. Clinical syndromes are also treated by different disciplines, with considerable overlap by Psychiatry, Neurology, neuropsychology, Speech and Language, and Physical Medicine and Rehabilitation.

Other Disinhibitory Control Syndromes
Alien Hand and Anarchic Hand Syndrome

Usually the hand and arm are noted to behave autonomously whereby inadvertent object grabbing, inadvertent hitting of bed partner during sleep, the one hand interferes with the function of the other while performing a task or they may make opposing movements. Of note is that because these actions are not voluntary there may be an unawareness of these actions. Alien hand subtypes have been identified, including:

Subtypes
Frontal
Parietal
Corpus callosal
Ictal

The Anterior or Frontal Variant Alien Hand Syndromes

These are characterized by grasping or reaching actions, sometimes also with difficulty in releasing the object that was grasped. The relevant lesion is usually a unilateral, anteromedial frontal.

The Posterior Parietal Type

There is a tendency to an incoordinate type of withdrawal from contact (akin to the parietal avoidance syndrome) in contradistinction to the grasping action typical of the frontal variant. The posterior alien hand may also levitate, termed the avoidance reaction by Denny Brown [109]. The posterior type is due to is due to either parietal lobe or occipital lobe damage.

Corpus Callosal Type

In this subtype there is the typical intermanual conflict between the two hands whereby the two hands perform approximately opposite actions. This has also been termed diagnostic ideomotor apraxia.

Ictal Alien Hand

This is an intermittent presentation only with seizural activity due to mesial frontal or corpus callosal injury

Pathophysiologically there is a primary motor cortex isolation from the premotor cortical area as suggested by fMRI studies. Management has been recommended to occupy alien hand as much as possible or with the frontal variant in particular to avoid grasping and grabbing by placing an object in their hand [110–112].

A simplified clinical and neuranatomical and functional connectivity classification of the principal frontal network syndromes based on the frontal subcortical network framework is presented (Fig. 12.28):

Clinical syndromes	Frontal cortical hubs
Dysexecutive disorders	Dorsolateral prefrontal
Disinhibitory disorders	Orbitofrontal
Abulic spectrum of disorders	Mesial frontal, anterior cingulate

Hodological Perspectives, Hyperfunction and Improved Behavior

There are other frontal network presentations that are not readily classifiable under the above system. As a brain lesion may cause both hypo or

Fig. 12.28 Clinico-anatomically based frontal systems presentations: dysexecutive dorsolateral prefrontal cortex (*1*), disinhibitory orbitofrontal cortex (*2*), and abulic, mesial, anterior cingulate region (*3*) syndromes

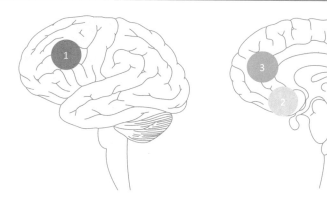

hyperfunction of a circuit or a lesion may cause hypo or hyperfunction remotely (diaschisis) because of the hodological nature of brain function or connectomics. One theory termed the paradoxical functional facilitation proposes that one brain area reverses inhibition in other areas or results in compensatory augmentation, resulting in counter intuitive paradoxical improvement in certain functions function. According to this theory, increased originality requires inhibition of the left hemisphere and an intact right hemisphere [113]. Examples of such syndromes include:

- Emergent visual artistic ability in the setting of neurodegenerative disease especially with those affecting the left temporal lobe (progressive aphasia syndromes) such as has been reported in association with frontotemporal lobe disorders, stroke, Alzheimer's disease, Parkinson's, epilepsy, and migraine [114]
- Emergent literary artist ability especially with impairment of the right temporal lobe [115]
- Delusional misidentification syndromes, seen particularly with right frontal stroke [116]
- Increased humor, particularly after right frontal lesion such as stroke [117]
- Loss of visual imagery in dreaming [118]
- Savant syndromes that may include the sudden, acquired prodigious, sudden, splinter, or talented subtypes [119]

Diagnostic Tests: Beside and Metric

The time pressured nature of clinical practice, limited interview time, emergency room evaluation and patient cooperation all place constraints on the nature of tests and how much testing can be performed. Hence, it is useful to consider available tests in a time orientated hierarchical manner. The overall decision on how to deploy more time consuming neuropsychological tests is detailed in the recommendations by the AAN Neuropsychological testing guidelines [120]. An emerging viewpoint is that subcortical processes, mostly cerebrovascular, may be the most frequent cause of cognitive disorders, particularly in the mild cognitive impairment domain. The cognitive component in question is frontal network systems particularly executive function, attention and working memory [121, 122].

Rapid (Bedside) Diagnosis

1. Montreal Cognitive Assessment (MOCA) administration time approximately 10–15 min [123]
2. Frontal Assessment Battery (FAB)—administration time approximately 15–20 min) [124]
3. Executive Interview bedside test (EXIT)—administration time approximately 15–20 min [125]
4. Comprehensive cognitive neurological test in stroke (Coconuts). Frontal Network System component—administration time approximately 20–30 min [126]
5. Metacognitive test—administration time approximately 20–30 min [127]

Computerized Screening Tests

Neurotrax Mindstreams [128]
CANTAB [129]

Cognistat [130]
CNS Vital Signs [131]

Metric Tests

Global tests

1. Delis Kaplan Executive Function System (DKEF) [132]
2. Wechsler Adult Intelligence Scale (WAIS-IV) components [133]

Clinical syndrome orientated, questionnaire based

3. Frontal Systems Behavior Scale (FRSBE) [134]
4. Behavior Rating Inventory of Executive Function (BRIEF) [135]
5. Frontal Behavioral Inventory (FBI) [136]

Working memory/executive function tests

6. Trail Making Tests (Comprehensive, Trails A and B, Color Trails) [137]
7. Letter and category fluency list generation [138]
8. Wisconsin Card Sorting Test [139]
9. Tower of London Test [140]
10. Working memory tests (verbal and non verbal)

Emotional intelligence tests

11. Emotional Intelligence Quotient (EQ)—Baron [141]
12. Mayer-Salovey-Caruso Emotional Intelligence Test (MSCEIT) [142]

Tests of disinhibition/inhibition

13. Stroop Neuropsychological Screening Test [143]
14. Iowa Gambling Test [144]

Autism

15. Autism Diagnostic Interview—Revised [145]
16. Autism Spectrum Quotient [146]

ADHD

17. Brown Attention Deficit Disorder Scales [147]

Depression

18. CES-D (Center for Epidemiological Studies Depression Scale) [148]
19. Beck Depression Scale [149]
20. Hamilton Depression Inventory [150]

Behavioral Neurological Tests

21. Faux Pas Test [151]
22. Reading the Mind in the Eyes Test [152]
23. Hotel Task [153]
24. Multiple Errands Test [154]
25. Ambiguous Figures Test [155]

Creativity tests

26. Torrance Test of Creative Thinking [156]

Other tests that predominantly assess frontal network systems

27. Visual Search and Attention Test [157]
28. Rey Complex Figure Test [158]

Clinical tests—qualitative without normative data

29. The Executive Control Battery [159]

A novel approach

30. The metacognitive battery incorporating neurological, neuropsychological, and neuropsychiatric syndromes [127].

Aphasia tests useful in motor aphasia, dysnomia and aphasias in general

31. Western Aphasia Battery [160]
32. Boston Diagnostic Aphasia Evaluation [161]

Elementary neurological

Clinical neurological assessment of:
Olfaction (The Smell Identification Test, Sensonics [162]
Gait (modified Gait Abnormality Rating Scale) [163]
Incontinence
Primitive reflexes (grasp reflex, palmo-mental reflex, sucking reflex)
Volitional eye movements [164]

Neuropathological States

Due to the expansive frontal subcortical circuits and their open connections, it may be readily appreciated that the frontal lobes connect with all other regions of the brain including the cerebellum and brainstem [165]. Clinical lesion studies have repeatedly shown that no matter where the brain lesion is, whether subcortical gray matter, subcortical white matter, cerebellum, brainstem, or even parietal and occipital lobe, a greater or lesser degree of frontal systems syndrome is present [1–8]. Even transient ischemia has been associated with a transient frontal network syndrome [166].

An appreciation of the tropism of the various neuropathological states is important. FTD is relatively confined to the frontal and anterior temporal lobes, similarly herpes simplex encephalitis. However, cognitive vascular disorders (CVD), disorders of white matter such as the leukodystrophies and CADASIL, vasculitis TBI, MS, most of the toxic metabolic encephalopathic all tend to affect the frontal subcortical networks in a more diffuse fashion and hence present with inattention, dysexecutive syndrome, and dysmemory as the hallmark signature syndromes, considered the most common presentations of frontal network syndromes [167]. Many of these will have varying degrees of neurospsychiatric syndrome admixtures most commonly depression and anxiety and perhaps less often disinhibtory behavior, irritability/aggression, obsessive-compulsive disorders, and adult onset ADHD.

Syndromes affecting the modulatory systems of the brain may also present with cognitive alterations. The serotonin toxidrome may present with barely perceptible symptoms to coma with cognitive, somatic, and autonomic manifestations [167]. The cognitive features may include hypomania, agitation, hallucinations, autonomic hyperthermia, hypertension, tachycardia, diarrhea, and somatic features; myoclonus, tremor, and hyperreflexia [168]. Pathophysiologically there is an increase in cerebral serotonin or overstimulation of the 5HT 2A receptors, often due to MAOIs in combination with SSRIs, SNRIs, TCAs, appetite suppressants, or opioids [169]. The diagnosis remains a clinical one. Similarly with the neuroleptic malignant syndrome, cognitive changes, autonomic instability, tremors, muscle cramps, tremors, and elevated creatinine phosphokinase are noted. The fever is caused by a hypothalamic dopamine receptor blockade and the muscular effects due to blockade of the D2 receptor and the pathophysiology due to low dopamine or dopamine receptor blockade with sympatho-adrenal hyperactivity [170]. With the malignant hyperpyrexia syndrome, exposure to anesthetic drugs such as succinylcholine, a neuromuscular blocking agent, halothane or desflurane, an abnormal muscular activity is induced with hyperpyrexia and circulatory collapse that can be fatal. The inheritance is autosomal dominant, usually for the ryanodine receptor (gene RYR1) located on the sarcoplasmic reticulum and opens in response to increased in intracellular calcium which is exaggerated in this condition. There is also a relationship with central core myopathy [171]. The cholinergic toxidrome may take the form of anticholinergic or cholinergic toxidrome. Both have a similar presentation including cognitive autonomic and muscular symptoms and signs but with some important differences such as tachycardia, mydriasis urinary retention, hyperthermia with the former and diarrhea, urination, hypothermia miosis, and bradycardia with the latter. Both also present with psychosis, seizures, hallucinations, delirium, and myoclonus [172]. PAIDS (paroxysmal autonomic instability and dystonia syndrome) is usually

Table 12.2 Presentations of frontal network syndromes: clinical and radiological

(a) Lesion studies (multimodality MRI or CT imaging)
1. Symptom related; most conditions present with the triad of inattention, executive dysfunction, and dysmemory. All of these may be on the basis of a working memory disorder. However, depending on the lesion site there may be an extensive list of behavioral symptoms
2. Syndrome related basic clinical (abulia, disinhibition, dysexecutive)
3. Syndrome pathophysiologically related. Examples include frontal stroke, herpes simplex encephalitis, leukoaraiosis, watershed infarction such as "Man in the Barrel Syndrome" or tumor related such as the Foster Kennedy syndrome
1. Anatomically lobar: motor, premotor prefrontal dorsolateral, prefrontal mediobasal, and prefrontal orbitofrontal
2. Anatomically network; frontal subcortical circuits
3. Anatomically long range network—brainstem, cerebellar, occipital lesions associated with FNS
(b) No radiological abnormality: neurotransmitter syndromes
Serotonin syndrome
Neuroleptic malignant syndrome
Malignant hyperpyrexia
Cholinergic and anticholinergic toxidromes
Paroxysmal autonomic instability and dystonia syndrome (PAIDS) (appendix 1)
(c) Synaptopathies (for example, limbic encephalitis)
Disorders with antibodies against synaptic proteins such as NMDA, AMPA, and GABA-B receptors. Present with seizures, encephalopathies and yet are treatable [174] (appendix 1)
(d) Networktopathies and participatory networks (f-MRI)
The default mode network, salience network and attentional network may be evaluated by f-MRI (abnormal in AD, FTD, TBI, MS, depression for example) [175]
Functional MRI task related activity seen for example with the Stroop, Word List Generation tests and Wisconsin Card Sorting Test activating particular networks [176]

seen after significant traumatic brain injury and also presents with a combination of cognitive alterations, autonomic abnormalities, and in this instance dystonias rather than muscular rigidity and tremor [173].

Frontal syndromes or frontal network syndromes may come to attention by way of the patients symptoms, a clinical syndrome elicited clinically or primarily by neuroimaging findings bearing in mind that this part of the brain has sometimes been termed clinically silent (Table 12.2 and Fig. 12.29).

Frontotemporal Lobe Degeneration

Frontotemporal lobe degeneration as a generic term rather than dementia is recommended because many remain in a category of MCI for a long time before frank dementia supervenes. Because of the protean manifestations and long duration of decline, sometimes for decades,

diagnostic difficulty is the rule. Eventually neuroimaging may reveal focal degeneration of frontal insular and temporal lobes although the basal ganglia and spinal motor neurons may also be involved and the presentations may accordingly be initially be one of the parkinsonian dementia syndromes (CBD, PSP even DSDB) and MND. Hence, it is useful to consider FTLD in the context of other primary dementias [177] (Tables 12.3 and 12.4).

Epidemiology

FTD is more common than in AD in those <60 years and the prevalence is about the same as AD in 60–70 group and there are reports of FTD presenting in the third decade. The most common subtype is bv FTD and a rapidly progressive FTD is seen in association with motor neuron disease with a mean survival of approximately 2 years. By comparison the survival of bv FTD is between

Fig. 12.29 Frontal network syndromes may be caused by focal or diffuse processes with differing pathophysiologies. *FTD* frontotemporal degeneration/dementias, *TBI* traumatic brain injury, *MS* multiple sclerosis, *CVD* cognitive vascular disorder, *NMS* neuroleptic malignant syndrome, *SS* serotonin syndrome

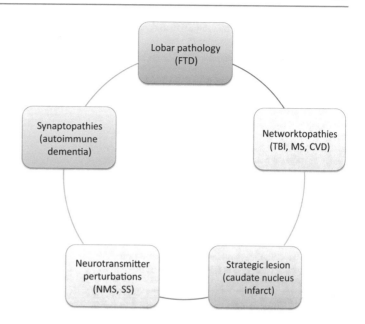

6 and 8 years with the language variant subtypes currently demonstrating the longest survival times of approximately have longest survival 8–12 years [177, 181, 182].

Clinical Presentations

People with FTLD have a particularly protean clinical presentation from subclinical to requiring institutionalization. The symptoms and signs initially are subtle, often undiagnosed for years and frequently misdiagnosed as bipolar disease, mania, obsessive-compulsive disorder, personality disorder, and depression. Dysexecutive function is not specific to FTD and is also seen with other dementias including AD; however, the emotional impairment is. Memory complaints are unusual. When the pathology is left sided, language impairment is a major clue and all present with word finding problems. However, in logopenic progressive aphasia (LPA) the person retains the so-called "islands of normal speech" without dysarthria [182].

Hodological effects or hyperfunction or hypofunction syndromes may occur with FTLD. For example artistic and literary art excellence may be observed with left temporal and right temporal lobe pathology respectively.

Clinical Diagnostic Criteria

The criteria of Neary et al. [183] have been augmented by the Revised International Consensus Criteria of Racovsky et al. which are presented in abbreviated form with the major features including [184] (Fig. 12.30):

- progressive deterioration of behavior and or cognition by observation
- early behavioral disinhibition
- early apathy
- early loss of sympathy or empathy
- early perseverative, stereotyped or compulsive/ritualistic behavior
- hyperorality and dietary changes
- executive deficits with relative sparing of memory and visuospatial functions

Pathology of FTD

TAR DNA binding protein 43 (TDP-43) was identified as the major ubiquitinated protein, with positive inclusions in neurons and glia, occurring in FTD and approximately half of FTD patients have TDP 43 inclusions due to mutations in the progranulin gene and found in

Table 12.3 The more common clinical disorders presenting with neurological and/or psychiatric FNS

I. Neurological
(a) Neurodegenerative
Frontotemporal disorders (FTLD)
Alzheimer's disease (AD)
Cognitive vascular disorders (CVD)
Frontal variant of AD
Cortico-basal-ganglionic disorders (CBD)
(b) Cerebrovascular and cognitive vascular disorders
Bland infarcts
Strategic infarct
Subcortical infarct
Watershed infarct
Frontal, sometimes bilateral as with common origin of both anterior cerebral arteries of the anterior communicating artery
Leukoaraiosis
Brainstem infarct
Cerebellar infarct
Strategic infarct such as caudate nucleus, basal ganglia, thalamus
Frontal lobe amyloid angiopathy
Hemorrhage
Amyloid angiopathy
Micro-hemorrhage
Subcortical hypertensive related
(c) Tumors
Frontal lobe meningioma (Foster Kennedy syndrome)
(d) Traumatic brain injury
Diffuse axonal injury
Chronic subacute encephalopathy
(e) Multiple sclerosis
(f) Parkinson's, Huntington's
(g) Frontal lobe epilepsies
(h) Normal pressure hydrocephalus
(i) Neurotoxicology—alcohol, chemical
II. Neuropsychiatric
Schizophrenia
Mania and hypomania
Depression
Anxiety
Obsessive-compulsive
Tourette's
Attention deficit hyperactivity disorder (ADHD)
Autism
William's syndrome
Pervasive developmental disorders

Table 12.4 Additional neuropathological states and conditions in which FNS is invariably part of the neurological syndrome [178–180]

Subcortical gray matter
HIV dementia
Wilson's disease
Huntington's
Neuro-acanthocytosis syndrome
Prionopathies (Creutzfeldt Jakob, GSS FFI, BSE)
Fahr's syndrome—calcification of the BG
Pantothenate kinase 2 associated neurodegeneration (PANK2)
Adult neuronal ceroid lipofuscinosis
Subcortical white matter
Leukodystrophy disorders (Metachromatic, Krabbe's, adrenal, orthochromatic)
Fabry's disease
Vanishing white matter disease
Mixed cortical and subcortical pathology
Vasculitides
Meningitis/encephalitis
CADASIL
Alexander's disease
Canavan disease
Cerebrotendinous xanthomatosis
Polycystic lipomembranous osteodysplasia with sclerosing leukoencephalopathy (PLOSL) or Nasu–Hakola disease
Mitochondrial diseases (MELAS, MERFF, Kearns Sayre)

the frontal and temporal neurons and dentate gyrus. Most the remainder are due to tau inclusions caused by mutations in the tau gene. Other pathological-clinical associations of TDP 43 include the FTD-MND association, SV is mostly linked to TDP 43, hippocampal sclerosis (frequently seen in FTD) and is usually associated with TDP 43. The bv-FTD may be associated with tau or TDP 43. PNFA is more likely due to tau pathology. About one quarter of Progressive Lewy Body Disease patients may also show TDP 43 inclusions [185–189].

Overall in FTD the pathology is therefore either a tauopathy or TDP-43 proteinopathy with a small number associated with other pathologies (Table 12.5). Tau is a microtubule associated protein that is integral to microtubule assembly and in stabilizing microtubules [190–192].

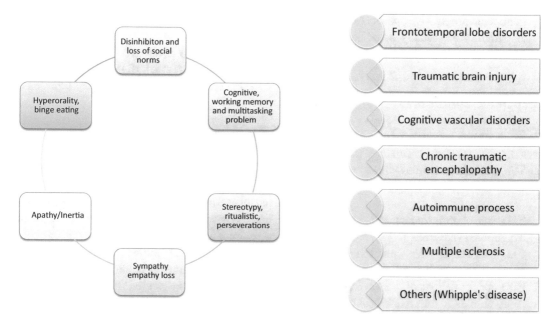

Fig. 12.30 Frontotemporal disorders (FTD) phenotypes: New International Consensus Criteria after Rascovsky et al. for diagnosis of FTD (*left*) and pathophysiological conditions presenting with FTD syndromes (*right*).

Rascovsky K, Hodges JR, Knopman D, Mendez MF, Kramer JH et al. Sensitivity of revised diagnostic criteria for the behavioral variant of frontotemporal dementia. Brain 2011:134;2456–2477

Table 12.5 Dementias have clinical pathological and molecular components

Domain	Pathology	Clinical	Subtypes
Linguistic	Tauopathies	Picks, SD, PPA	3
Comportmental	Tauopathies	FTD behavioral	1
Amnestic	Amyloidopathies	AD	4
Movement Dis.	Synucleinopathies	PD, DLB, PSP, CBD	5

Histologically, von Economo neurons a relatively unique spindle shaped cell found in mammals of such as cetaceans, hominoids, elephants are found the anterior cingular cortex and fronto-insular regions and thought to be important in social networks. These cells selectively degenerate in the early stages of bv-FTD. Neurochemically, the presentations of FTD may be due to a significant reduction of serotonin receptors in the cingulate, oribitofrontal, and insular regions [193].

Genetics

Overall 10 % of FTD cases are associated with an autosomal dominant pattern of inheritance with the rest sporadic. Both tau and progranulin mutations have been linked to chromosome 17, and other genetic loci linked to FTD occur on chromosome 3 and 9. Both tau and progranulin mutations cause inclusions in neurons and glia. In general, bv-FTD has the strongest inheritance, SV less, and PNFA intermediate [194].

Clinical Subtypes

1. Behavioral variant
2. Progressive non fluent aphasia (left perisylvian degeneration associated with tau pathology)
3. Semantic variant (anterior temporal degeneration associated with TDP 43 proteinopathy but about 10 % have AD)

4. Logopenic progressive aphasia (angular gyrus of parietal lobe usually associated with AD pathology)

Associated Syndromes

Other neurological deficits may be elicited including peripheral neuropathy, parkinsonism, apraxia, gaze abnormalities which would support the existence of one of the following syndromes:

1. FTD and Parkinsonism and cortico-basal-ganglionic syndrome.
2. FTD and ALS—more common with bulbar onset alS than limb onset alS.
3. FTD and PSP. Present, vertical gaze impairment, with early falls axial rigidity.
4. CBD: asymmetric movement disorder with alien hand syndrome and cortical sensory loss. Rather than just a primary disorder, CBD may be caused by secondary conditions such as FTD, AD and Creutzfeld Jakob disease. For example CBD, PSP and half of people with FTD have MAPT (microtubule associated protein tau) [192].

Investigations

Clinical Cognitive/Behavioral

It is important to employ specific frontal behavioral batteries such as FBI and FRSB as neuropsychological cognitive testing may be normal with a distinct discrepancy in the relative absence of amnesia and visuospatial impairment.

Neuroimaging

Structural (MRI/CT) may reveal frontal or temporal atrophy and functional (PET, SPECT) revealing frontotemporal abnormality much earlier with frontal and/or anterior temporal lobe hypometabolism or hypoperfusion respectively. A study using MR perfusion scanning revealed similar results and may prove to be more practical [195].

Laboratory

CSF tau and ratio of tau to A-beta 42 are significantly lower in FTD than AD with a sensitivity of 79–90 % and specificity of 65–97 % [196].

Treatment

In brief, pharmacological management may be employed using serotonergic agents for decreasing obsessions disinhibition, overeating, repetitive behaviors, as exemplified by the trial of trazodone has been shown in a randomized control trial. Atypical antipsychotics for agitated behavior. Avoidance of cholinesterase inhibitors as these may cause agitation [197]. Investigational treatments include both tau and PGRN based approaches. For example, inhibition of tau kinases to prevent tau hyperphosphorylation may be accomplished using lithium chloride or valproate. Prevention of polyubiquitination to decrease tangle maturation by using HSP-90 and so antagonizing tau fibrillation or stabilizing microtubules using paclitaxel or anti-inflammatory agents are other possible avenues. Low progranulin levels might be treated with replacing progranulin [198].

Cerebrovascular/Cognitive Vascular Disorders: An Example of a Networktopathy with One or More FSC Affected

The spectrum of cognitive vascular disorders cerebrovascular disorders include:
   Cognitive impairment—no stroke

Brain at risk stage (risk factors only; hypertension, diabetes, dyslipidemia)
Transient ischemic attacks and cerebral infarct with transient symptoms

   Cognitive impairment—subcortical infarct

Strategic infarct—caudate, thalamus, basal ganglia
Leukoaraiosis (Fig. 12.31)

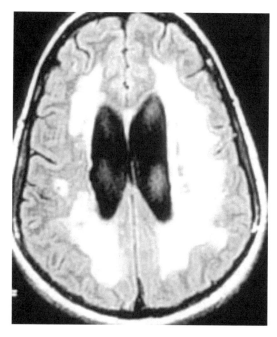

Fig. 12.31 Extensive subcortical leukoaraiosis: a common cause of frontal network syndromes

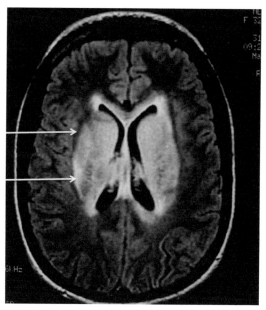

Fig. 12.33 Athymhormia spectrum disorders (akinetic mutism, abulia, hypobulia, apathy). Akinetic mutism secondary to deep, central, tegmentothalamic ischemia (reversible with good recovery) due to deep venous system thrombosis post partum (*arrows*)

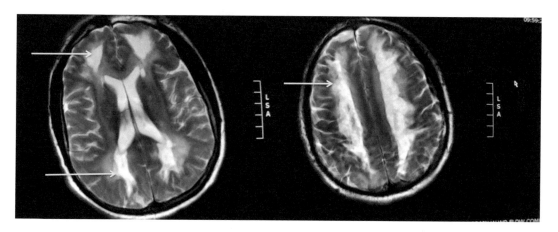

Fig. 12.32 Frontal and posterior watershed infarction (*arrows*)

Watershed infarction (Fig. 12.32)
Deep venous system tegmentothalamic lesions (Fig. 12.33)

Cognitive impairment—cortical infarct

Left angular gyrus, right temporal lobe, frontal

Cognitive impairment—subtentorial stroke

Brainstem and cerebellum

Multiple infarcts

Vascular dementia

Most of the various cerebrovascular syndrome such as small vessel disease, leukoaraiosis, vasculitis as well as multiple sclerosis and traumatic brain injury affect the brain more diffusely. Consequently the prototypical deficits involve attention, memory, and executive functioning. This is important to consider as many of these patients have a considerably reduced attention span (and some may be irritable or irascible at the same time due to their disease process) and this greatly impacts the mode of testing. Extensive neuropsychological testing in such patients is not usually practical and one of the shorter batteries may suffice. However, decline in memory is strongly associated with AD and decline in executive function strongly associated with CVD [199].

The origin of cognitive impairment in the context of vascular disease and whether neurodegenerative disease might be implicated, the following are considered:

• The pattern and severity of cognitive impairments
• The size and location of infarcts including symptomatic infarcts, silent infarcts, and leukoaraiosis
• The severity of atrophy patterns in particular hippocampal atrophy and pattern of atrophy of frontal versus parietal
• Do the vascular lesions adequately explain the cognitive impairment
• If the vascular lesions do not adequately explain the cognitive impairment, then neurodegenerative disease is likely present as well

In studies of neuropsychological series of vascular cognitive impairment and AD, the former had executive dysfunction and less impairment in verbal episodic memory. There were no differences in language, constructional abilities, attention [200, 201].

The current definition of dementia is memory centered and requires ADL impairment and does not emphasize the predominant executive dysfunction of the CVD subtypes. For minor cognitive deficits due to vascular disease the term VCI-ND been recommended (cognitive deficits that do not meet criteria for dementia but impair minor chores) [202].

Neuroimaging has been helpful in differentiating AD and CVD. More WM changes or leukoaraiosis and less medial temporal lobe atrophy are seen in CVD when compared to AD although overlap occurs. Two types may be discerned; when the WM lesions are distinct and separated from ventricles this tends to be more specific for vascular dementia. When the WM hyperintensity is periventricular this tends to support a neurodegenerative rather than vascular process [203]. A formidable number of conditions can present with white matter hypertensities on MRI brain scan (Table 12.6).

Leukoaraiosis may affect one, some or all of the FSC and quantification will likely become more important in the future and apart from neuroimaging software capabilities the Junque classification system (Table 12.7) is a very useful manner of quantifying and measuring over time.

These findings have treatment implications with the milder forms more amenable to cholinergic therapy and the more severe forms might perform better with dopaminergic therapy.

Neuropathology of vascular lesions

• Cerebral infarcts (Fig. 12.34)
• Lacunes
• Microinfarcts—up to 5 mm
• Widening of perivascular spaces
• Incomplete infarction
• Leukoencephalopathy—associated with SVD.
• Laminar necrosis selective involvement third and fifth layers.
• Granular atrophy patches of gray matter between two or three arterial territories
• Lobar hemorrhages—linked to A-ß angiopathy and AD (Fig. 12.35)
• Small hemorrhages—association HTN at cortico-subcortical junction (slit hemorrhages) and BG. The latter are called type II lacunes [205]

Neurobiology of Cognitive Vascular Disorder

Leukoaraoisis (LA) may interrupt the neurotransmitter modulatory systems such as aminergic or corticostriatal and thalamocortical networks.

Table 12.6 Neuropathologic subtypes

Histopathology	Subtypes	Clinical	Chromosome
Tauopathy	PiD	bvFTD, PNFA	17 (MAPT)
PSP	PSP		
CBD	CBD, PNFA		
AGD	bvFTD, MND		
MST			
TDP-43	Type 1	bvFTD	
	Type 2	SD, MND	9 (IFT74)
	Type 3	bvFTD, SD, PNFA	17 (PGRN)
	Type 4	bvFTD, myopathy, Paget's	9 (VCP)
FTLD-UPS	FTD 3	bvFTD (CHMP2B)	3
BIBD	–	FTD-MND	
FTLD-IF	–	–	–
FTLD-ni	–	–	–

Modified from Josephs KA [204]

PiD Pick's disease, *PSP* progressive supranuclear palsy, *CBD* cortico-basal-ganglionic degeneration, *AGD* argyrophilic grain disease, *MND* motor neuron disease, *PNFA* progressive nonfluent aphasia, *SD* semantic disease/variant, *MST* sporadic multisystem tauopathy, *BIBD* basophilic inclusion body disease, *FTLD IF* FTLD with intermediate filament inclusions, *FTLD ni* FTLD with no inclusions, *FUS* fused in sarcoma, *NIF* neuronal intermediate filaments, *VCP* valosin containing protein, *CMBP 2B* charged multivesicular body protein 2B, *MAPT* microtubule associated protein tau, *TDP-43* TAR DNA binding protein 43, *IFT74* intraflagellar transport protein 74

Table 12.7 Conditions that can present with white matter hyperintensities on MRI brain scan

- Cerebrovascular (hypertension, atrial fibrillation, diabetes mellitus, homocysteine)
- Alzheimer's
- APOE 4 status
- Trauma
- Migraine
- AIDS dementia
- Psychiatric (bipolar disease, schizophrenia)
- Autism
- CADASIL
- Wilson's, Hallervorden–Spatz
- Dystonia
- Neuroacanthocytosis
- Fragile X associated tremor and ataxia
- Susac's syndrome
- Myotonic dystrophy type 1 and 2
- Hypoglycemic encephalopathy
- Leucodystrophies (Metachromatic, Krabbe)
- Multiple sclerosis
- Autoimmune vasculitic (SLE, Sjogren's)

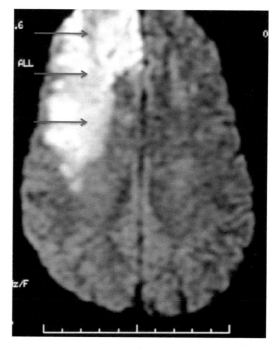

Fig. 12.34 Large right frontal infarct (*arrows*) associated with minimal long tract signs (NIHSS = 2), minor neuropsychological test abnormalities and marked behavioral disturbance of imitation and utilization behavior

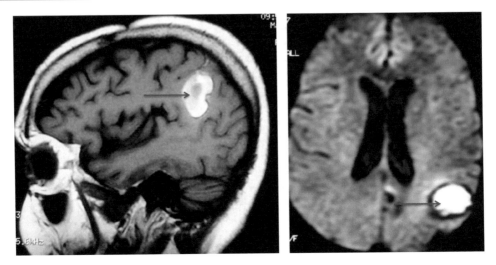

Fig. 12.35 Discreet hemorrhage (*arrows*) in left angular gyrus region presenting with Gerstmann's syndrome

Patients with moderate MRI LA may respond better to Aricept especially with Junque LA scores of ≥10 compared to those with minimal MRI LA. There may be at least moderate MRI LA in context of a dysexecutive syndrome, which may be a marker for the relative preservation of cholinergic neurons. Delayed recognition memory measure is relatively preserved in subcortical vascular disease compared to AD. Perhaps important in this context, is that the cerebral vasculature may be observed real time by fundoscopy. The following were predictive of lacunar stroke; narrower central retinal arterial artery equivalent, wider central retinal vein equivalent, focal arteriolar narrowing, and arteriovenous nicking [206].

Overall mixed a mixed dementia is the most common type, one that has been aptly termed a vascular tsunami (Hakim). Population based autopsy studies indicate that less than 50 % of patients have pure AD and many people diagnosed with AD may in fact have VCD. The neuropathology of AD and VCI may coexist and influence each other and eight of ten of the traditional vascular risk factors also pertain to AD such as hypertension, hyperlipidemia, hyperhomocysteinemia, APOE2 and APOE 4 [207]. Importantly treatment of vascular risk factors is associated with slower decline in AD with no CVD [208, 209].

Frontal Variant of Alzheimer Disease (AD)

AD is normally regarded as disease process afflicting the posterior brain, namely parietotemporal regions with a clinical correlate of dysmemory, visuospatial impairment, geographical disorientation and only much later behavioral abnormalities that are characteristic of FNS. The usual AD variants include a primary progressive aphasia, visuospatial and posterior cortical atrophy syndrome subtypes or variants. Reports indicate that about 14–17 % of AD patients have non-memory presentations as an atypical subtype [210]. In addition executive function is not considered a component of AD at least in mild to moderate disease [211]. However, a subpopulation of AD patients may present with a FNS early on including all three principal frontal syndromes of abulia, dysexecutive syndrome, and disinhibition. The neurobiology may be a disturbance of the frontal subcortical systems but the underlying etiology remains to be determined. One study found a tenfold greater neurofibrillary tangle pathology in BA 8 of the frontal lobes [212]. Other explanations may include white matter disease, coexistent Lewy body disease and other cerebrovascular pathologies that influence the clinical presentation of the AD as a frontal variant [213, 214].

Multiple Sclerosis (MS)

As a pathophysiological process with a few or numerous subcortical plaques, MS is a process that may impair one or man of the FSC's. Recent studies cite a data indicating that 43–72 % of patients with MS are considered to have cognitive impairment [215]. The correlation with MS plaques as measured by standard MRI is not always present. MS plaques frequently lie adjacent to the ventricles and so tend to interrupt the long association fibers in the cerebrum such as the fronto-occipital fasciculus, the superior and inferior longitudinal fasciculi.

A recent study underscoring the important neurobiology of the frontal subcortical tracts employed frontal neuropsychological as well as behavioral neurological metric tests in comparison to diffusion tensor imaging (DTI) of the frontal subcortical systems. Tests sensitive to the orbitofrontal and cingulate regions of the frontal lobes were used including the Hotel Task Test, Iowa Gambling Test, Faux Pas Test, Multiple Errands Test, and Reading the Mind in the Eyes Test. DTI evaluations were performed in the frontomedial, frontolateral, orbitofrontal anterior cingulate and a significant difference found with lower FA values in the FM and FL in the MS patients compared to controls. A significant correlation was also found with loss of fiber integrity in the frontolateral regions and an impairment on the Hotel Task Test and Multiple Errands Test [216].

When cognitive impairment in MS is evaluated, the symptoms elicited are principally inattention, abulia (apathy), dysexecutive syndrome, dysmemory (working memory), and disinhibition (inappropriate jocularity). Elicitation of these symptoms is important in that each of these is potentially treatable. Stimulant therapy with dopaminergics pertain to the first four and sodium valproate or carbamazepine is often effective for treating disinhibtion syndromes.

Traumatic Brain Injury (TBI) and Concussion

TBI is associated with a complex syndrome characterized by cognitive (memory, attention and executive problems), elementary neurological symptoms (headache dizziness, vertigo, imbalance), neuropsychiatric impairment (anxiety, depression, irritability, irascibility, mania, disinhibition, impulsivity). Symptoms may be relatively mild to severe and may be present with normal neuroimaging and even anatomical pathology leading to frequent misdiagnosis and under appreciation of the severity of the syndrome (Fig. 12.36). Key to understanding the complexity of these syndromes is the current understanding of the pathophysiology (Table 12.8).

The dramatic advances in neuroimaging and alterations in the biochemistry and vascular system might best be described as a networktopathy with neurotransmitter and vascular perturbations that often escape anatomical imaging. The realization that vasospasm is a frequent accompaniment during concussion and TBI has lead to transcranial Doppler study initiatives as a novel way of non invasive monitoring [218]. Default mode network imaging may prove to be the most sensitive diagnostic tool yet in diagnosis [219]

Alcohol Excess

The frontal lobes are particularly susceptible to the effects of alcohol, as revealed by recent magnetic resonance spectroscopy studies [220]. The hippocampal CA1 and CA3 regions are also affected, particularly in animal models [221]. In addition all of the following alcohol related cerebral conditions can affect cognitive functioning in particular frontal systems of working memory, attention and executive function.

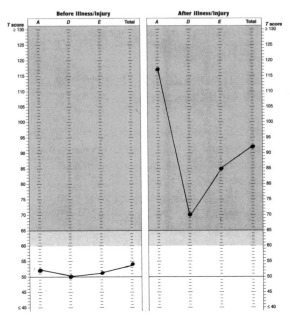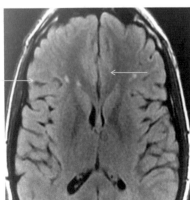

Fig. 12.36 Traumatic brain injury: clinico-radiological discrepancy with abnormal apathy, disinhibition and executive function scores by Frontal Systems Behavioral Examination (*left image*) normal neuropsychological tests and minimal subcortical shear injury evident on anatomical brain imaging (*right image, arrows*)

Table 12.8 Junque leukoaraiosis grading [217]

Evaluate five areas in each hemisphere	
• Centrum semi-ovale in frontal region	
• Centrum semi-ovale parietal region	
• White matter surrounding frontal horn	
• White matter surrounding corpus of the lateral ventricle	
• White matter surrounding the atrium and occipital horn of lateral ventricle	
Numerical scores 1–4 of T2 hyperintensity	
• No changes	0
• <25 %	1
• 25–50 %	2
• 50–75 %	3
• >75 %	4
Total score	0–40

In general the leukoaraiosis (LA) on MRI can be interpreted as follows:

1. Mild MRI LA is correlated with markedly lower scores on episodic memory compared to working memory and is a neuropsychological feature associated with AD

2. Moderate MRI LA correlates with both amnesia and executive dysfunction

3. Severe MRI LA correlates with significantly lower scores on working memory and executive dysfunction [208]

- Alcohol intoxication—coma, pathologic intoxication, alcoholic blackouts/memory loss
- Abstinence or withdrawal syndromes
- Nutritional disease (Wernicke–Korsakoff)
- Cerebrovascular infarcts and hemorrhages
- Cardiomyopathy
- Cerebellar degeneration
- Marchiafava–Bignami disease (Fig. 12.37)
- Central pontine myelinolysis
- Alcoholic dementia
- Cerebral atrophy
- Neurologic conditions secondary to liver cirrhosis, portal shunts
- Traumatic brain lesions during intoxication

Normal Pressure Hydrocephalus

Previously characterized as a subcortical dementia, this syndrome is still diagnosed by the clinical triad of cognitive, gait impairment and urinary incontinence. The predominant cognitive presentation is a frontal syndrome characterized by dysmemory, dysexecutive syndrome, inattention and speed of information processing slowing.

Fig. 12.37
Marchiafava–Bignami
syndrome may present
with florid frontal
syndromes, as well as
paranoia, mania and
various degrees of
cognitive impairment as
well as dementia. The
pathological hallmark is
transcallosal
abnormality evidenced
by the hyperintensity on
MRI scan (*arrow*)

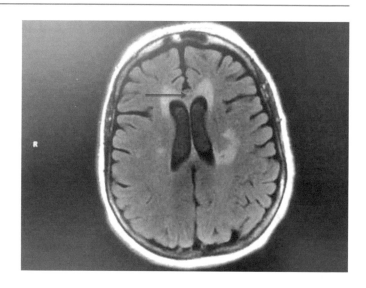

The frontal lobe syndrome or FNS is much more profound than that seen with AD and the memory disturbance is relatively mild with NPH, being much more severe with AD. The pathophysiology includes damage to FSC, corpus callosum, thalamus, basal ganglia, and hippocampus. The cognitive impairments respond to appropriate shunt procedures but usually less so than the improvement in gait impairment [222]. Recently the importance of specific assessment of executive function assessment has been proposed as this can be differentially improved by shunting relative to other cognitive impairments [223]. Neuroimaging with PET (or SPECT which measures perfusion rather than metabolism) has been shown to be the best diagnostic tool and a frontal hypometabolism (in NPH) as opposed to a posterior parieto-occipital hypometabolism (in AD) been shown to differentiate NPH from AD better than cisternography [224].

Autoimmune Disorders and Limbic Encephalitis as Examples of Synaptopathies

A reappraisal of autoimmune conditions associated with cognitive impairment and at times frank dementia has led to the concept of immunotherapy responsive dementias and encephalopathies. In addition to being treatable and reversible causes of dementia, they may account for up to 20 % of the so-called young dementia patients (<45 years) [174]. The cognitive profile includes the usual combination of FSC impairments in speed of information processing, dysmemory, and behavioral abnormalities typically of a fluctuating nature as well as agitation, hallucinations, and seizures [225].

The finding that 1 alpha dendrotoxin antibodies against VGKC noted in patients with Morvan's disease, limbic encephalitis, and neuromyotonia completely changed the understanding from limbic encephalitis being a rare paraneoplastic condition with poor outcome and associated with anti-Hu, anti-Ma2, or CV2/CRMPS antibodies. Autoimmune dementia syndromes (AID) in general are disorders with antibodies against synaptic proteins such as NMDA, AMPA, and GABA-B receptors and are treatable (Table 12.9). The identification of a number of neural specific autoantibodies such as voltage gated potassium channel (VGKC) antibodies have increased the number of phenotypic presentations of AID. The diagnosis is made by establishing cognitive impairment or an encephalopathic state with clinical, radiologic, or serologic evidence of autoimmune etiology. Other causes of dementia require exclusion and a beneficial response to an immunotherapy trial [228, 229].

In addition to the specific antibody tests, cerebrospinal fluid analysis is useful with the following

Table 12.9 Pathophysiology of concussion and traumatic brain injury [226, 227]

1. Excessive or indiscriminate release of excitatory neurotransmitters
Increased glutamate binding to NMDA receptors causes efflux of potassium out of the cell and influx of calcium, alteration of the neuronal membrane potential, Na-K pump is upregulated and consequently requires more ATP
2. An uncoupling of glucose metabolism and cerebral blood flow occurs
A glucose hypermetabolism ensues and there is a simultaneous diminished cerebral blood flow, which may be reduced as much as 50 % of normal
Even more important from a clinical point of view, the cerebral glucose may be reduced for up to 4 weeks (measured by PET brain scan in humans)
3. Calcium accumulation occurs
Intracellular Ca^{++} accumulation causes mitochondrial impairment, cell death by phosphokinases, protein kinases, NO synthase, endonucleases, calpains, and plasmalogenase culminating in free radical accumulation and apoptosis
4. Chronic alterations in neurotransmission
Glutaminergic, cholinergic, and adrenergic alterations account for the memory and cognitive deficits seen after concussion and TBI. The neurochemical findings include: LTP may be persistently impaired after TBI, loss of cholinergic input from the basal forebrain, and impaired GABA inhibitory function of the hippocampal dentate granule cells occurs which predisposes the injured brain to seizures.
5. Axonal disconnection occurs
Diffuse axonal injury may occur due to mechanical stretching or calcium influx with subsequent microtubule breakdown. Axonal bulbs may result due to intra-axonal cytoskeletal injury, accumulation of organelles at the site of damage axonal damage with localized axonal swellings appearing (axonal bulbs). Secondary axonotomy (constrictions) with axonal disconnection which may occur many weeks after TBI

Table 12.10 Different classification systems and current status of antibodies implicated: Autoimmune dementia may be idiopathic or secondary to cancer (paraneoplastic) [230]

Eponymous
Morvan syndrome
Syndromic
Progressive encephalomyelopathy with rigidity and myoclonus
Serologically
VGKC antibody associated encephalopathy
Pathologically
Nonvasculitic autoimmune meningoencephalitis
Antibodies
VGKC
NMDA receptor antibody
AMPA receptor
GABA$_B$ receptor
GAD 65
ANNA-1 (anti Hu)
ANNA-2 (anti Ri)
ANNA-3
AGNA (SOX-1)
PCA-2
CRMP-5 (anti CV2)
Amphiphysin
Ma/Ta proteins
NMO-IgG

VGKC voltage gated potassium channel, *NMDA* N-methyl-D-aspartate, *AMPA* alpha amino 3 hydroxy 5 methyl isoxazolepropionic acid, *GABA* gamma amino butyric acid, *GAD* glutamic acid decarboxylase 65, *ANNA* antineuronal nuclear antibody, *AGNA* antiglial nuclear antibody, *PCA* Purkinje cell cytoplasmic antibody, *CRMP 5* collapsing response mediator protein 5, *NMO* neuromyelitis optic IgG antibodies (Modified from Mckeon et al.) [228]

parameters regarded as support for AID: raised protein especially over 100 mg/dl, pleocytosis, oligoclonal bands, and IgG index elevation. Cancer screening is important with computerized chest, abdomen, and pelvis required in all patients, mammography in women and prostate specific antigen and testicular ultrasound in men.

Treatment options for AID include intravenous methylprednisolone (1000 mg×3–5 days then weekly for 6–8 weeks), IVIG if seropositive for GAD 65 or IA2 autoantibodies with plasma exchange if cannot tolerate steroids or IVIG. Long-term therapy is indicated in responders that may require a steroid sparing inhibitor such as azathioprine, mycophenolate, cyclophosphamide, or methotrexate [140, 229] (Table 12.10).

Cerebral Vasculitides, Infectious Disorders, Autoimmune Disorders and Chronic Inflammatory Disorders

These conditions share several features in that they involve the brain diffusely often with small or microscopic lesions, have relatively covert onset and often subtle signs and syndromes, and may often be overlooked as consequence. The FSCs are affected as well as the open-ended connections to the posterior parts of the brain, the brainstem and cerebellum. In addition to standard anatomical brain imaging, CSF analysis is required and very often, functional imaging with SPECT or PET brain scanning, at times magnetic resonance spectroscopy and less often brain biopsy. The challenge is usually in considering or entertaining these disorders in the first place as antibody testing, CSF analysis, cerebral catheter angiography, and at times brain biopsy as indicated are often diagnostic.

Many pathological processes may have as their earliest manifestations some degree of cognitive impairment, most commonly involving the frontal network systems with systemic and meningeal irritation manifestations either absent or appearing much later in the illness. Hence, frontal network syndromes may be the first sign of a potentially treatable disorder, which if missed is devastating. With CSF analysis of dementia currently a nonroutine investigation, it is possible to ascribe a slowly dementing state to the most common disorders such as AD or FTD and miss a treatable condition such as chronic cryptococcal meningitis. This has indeed been reported several times and may present a tip of the iceberg phenomenon [231–233]. Others seen much less frequently include Candida, Aspergillus, Blastomyces, Coccidioides, and Histoplasma (Table 12.11).

Vasculitides

In the largest series to date ($n = 101$) of cerebral vasculitis, cognitive impairment in general was present in over three quarters of the patients with

Table 12.11 Viral, bacterial, fungal, and parasitic brain infections with frontal subcortical circuit involvement

Viral
HIV encephalopathy
JC virus—progressive multifocal leucoencephalopathy (PML)
Herpes simplex encephalitis
West Nile virus
Tegmentothalamic syndrome (various)
Bacterial
Tuberculous meningitis
Neisseria meningitides
Hemophilus influenzae
Listeria monocytogenes
Whipple's disease
Spirochetal
Borreliosis (Lyme disease)
Neurolues
Fungal
Cryptococcal meningitis
Histoplasmosis
Coccidioides immitis
Blastomyces dermatitidis
Candida species
Prionopathies
Creutzfeldt–Jakob disease (CJD)
Variant Creutzfeldt–Jakob disease (V-CJD)
Kuru
Fatal familial insomnia (FFI)
Gerstmann–Sträussler–Scheinker syndrome (GSS)
Parasitic
Malaria
Bilharziasis
Cysticercosis
Toxoplasmosis
Amoebic meningitis (*Naegleria fowleri*)

altered cognition in 50 % and aphasia in 28 % of patients [234]. As with other subcortical and more diffuse brain processes, inattention, dysexecutive function, and working memory problems are the most common cognitive disturbances. Laboratory (ESR, CRP), cerebral angiography (often four vessel catheter cerebral angiography), and at times brain biopsy are required for diagnosis, but the most important clinical error is not to entertain the diagnosis in the first place in the appropriate clinical context (Table 12.12).

Table 12.12 Cerebral vasculitides and autoimmune disorders classification

Primary
Polyarteritis nodosa
Cogan's syndrome
Churg–Strauss
Temporal
Takayasu's
Granulomatous
Lymphomatoid
Wegener's
Kawasaki
Susac's
Hypersensitivity
Buergers
Acute posterior multifocal placoid pigment epitheliopathy
Kohlmeier Degos
Isolated angiitis of the CNS
Secondary to autoimmune and systemic diseases
Sarcoidosis
Rheumatoid arthritis
Systemic lupus erythematosis
Sjogrens
Behcet's
Scleroderma
Mixed connective tissue disease
Dermatomyositis
Ulcerative colitis
Celiac disease
Secondary—infectious related
Human immunodeficiency virus
Varicella Zoster
Herpes zoster
Cytomegalovirus
Mycotic
Lues
Borrelia burgdorferi
Tuberculosis
Cysticercosis
Bacterial meningitis
Secondary to neoplasia
Hodgkin's and non-Hodgkin's
Malignant histiocytosis
Hairy cell leukemia
Secondary to illicit drugs
Cocaine
Sympathomimetic agents

(continued)

Table 12.12 (continued)

Amphetamine
Conditions that may present with focal or diffuse arterial narrowing and in the differential of vasculitis
Radiotherapy
Vasospasm; acute hypertension, migraine, benign reversible cerebral angiopathy, or posterior reversible encephalopathy syndrome (PRES)
Lymphoma of the central nervous system
Intracranial dissection; traumatic, spontaneous, fibromuscular dysplasia
Intracranial atherosclerosis
Recanalizing embolus
Moya Moya disease and syndrome
Tumor encasement due to pituitary adenoma or meningioma
Sickle cell anemia
Neurofibromatosis [235, 236]

References

1. Karussis D, Leker RR, Abramsky O. Cognitive dysfunction following thalamic stroke: a study of 16 cases and review of the literature. J Neurol Sci. 2000;172(1):25–9.
2. Kumral E, Evyapan D, Balkir K. Acute caudate vascular lesions. Stroke. 1999;30(1):100–8.
3. Tullberg M, Fletcher E, DeCarli C, Mungas D, Reed BR, Harvey DJ, et al. White matter lesions impair frontal lobe function regardless of their location. Neurology. 2004;63(2):246–53.
4. Neau JP, Arroyo-Anllo E, Bonnaud V, Ingrand P, Gil R. Neuropsychological disturbances in cerebellar infarcts. Acta Neurol Scand. 2000;102(6):363–70.
5. Hoffmann M, Schmitt F. Cognitive impairment in isolated subtentorial stroke. Acta Neurol Scand. 2004;109(1):14–24.
6. Malm J, Kristensen B, Karlsson T. Cognitive impairment in young adults with infratentorial infarcts. Neurology. 1998;50:1418–22.
7. Garrard P, Bradshaw D, Jaeger HR, Thompson AJ, Losseff N, Playford D. Cognitive dysfunction after isolated brain stem insult. An underdiagnosed cause of long term morbidity. J Neurol Neurosurg Psychiatry. 2002;73:191–4.
8. Goldberg E. The executive brain. Frontal lobes and the civilized mind. London: Oxford University Press; 2001.
9. Chow TW, Cummings JL. Frontal subcortical circuits. In: Miller B, Cummings JL, editors. The human frontal lobes. 2nd ed. London: Guilford; 2009.
10. Lichter DG, Cummings JL. Frontal subcortical circuits in psychiatric and neurological disorders. New York: The Guilford Press; 2001.

11. Dobzhansky T. Nothing in biology makes sense except in the light of evolution. Am Biol Teach. 1973;35:125–9.

12. Dawkins R, Tale A's. The ancestor's tale. A pilgrimage to the dawn of evolution. New York, NY: Houghton Mifflin; 2004.

13. Zalc B, Goujet D, Colman DR. The origin of the myelination program in vertebrates. Curr Biol. 2008;18:R511–2.

14. Roots BI. The evolution of myelin. Adv Neural Sci. 1993;1:187–213.

15. Coppens Y. East side story: the origin of humankind. Scientific American. 1994

16. Schultz S, Nelson E, Dunbar RIM. Hominin cognitive evolution: identifying patterns and processes in the fossil and archaeological record. Phil Trans R Soc B. 2012;367:2130–40.

17. Previc FH. Dopamine and the origin of human intelligence. Brain Cogn. 1999;41:299–350.

18. Marean C. When the sea saved humanity. Scientific American. 2010.

19. DeLange F. The role of iodine in brain development. Proc Nutr Soc. 2000;59:75–9.

20. Wainwright PE. Dietary essential fatty acids and brain function: a developmental perspective on mechanisms. Proc Nutr Soc. 2002;61:61–9.

21. Klein RG. The human career: human biological and cultural origins. 3rd ed. Chicago, IL: University of Chicago Press; 2009.

22. Holloway RL. Human brain evolution: a search for units, models and synthesis. Can J Anthropol. 1983;3:215–32.

23. Carlson KJ, Stout D, Jashashvili T, de Ruiter DJ, Tafforeau P, Carlson K, et al. The endocast of MH1, Australopithecus sediba. Science. 2011;333: 1402–7.

24. Elston GN. Cortex, cognition and the cell: new insights into the pyramidal neuron and prefrontal function. Cereb Cortex. 2003;13:1124–38.

25. Passingham RE. The frontal cortex: does size matter? Nat Neurosci. 2002;5:190–2.

26. Semendeferi K, Damasio H, Frank R, Van Hoesen GW. The evolution of frontal lobes: a volumetric analysis based on three dimensional reconstructions of magnetic resonance scans of human and ape brains. J Hum Evol. 1997;32:375–88.

27. Semendeferi K, Lu A, Schenker N, Damasio H. Humans and apes share a large frontal cortex. Nat Neurosci. 2002;5:272–6.

28. Holloway RL. The human brain evolving. A personal perspective. In: Broadfield D, Yuan M, Schick K, Toth N, editors. The human brain evolving. Gosport, IN: Stone Age Publication. Stone Age Institute Press; 2010.

29. Passingham RE, Wise SP. The neurobiology of the prefrontal cortex. Oxford: Oxford University Press; 2012.

30. Hoffmann JN, Montag AG, Dominy JN. Meissner corpuscles and somatosensory acuity: the prehensile appendages of primates and elephants. Anat Rec A Discov Mol Cell Evol Biol. 2004;281:1138–47.

31. Nudo RJ, Masterton RB. Descending pathways to the spinal cord, IV: some factors related to the amount of cortex devoted to the corticopsinal tract. J Comp Neurol. 1990;296:584–97.

32. Kaas JH, Stepniewska I. Evolution of posterior parietal cortex and parietal-frontal networks for specific actions in primates. J Comp Neurol. 2015;524:595–608. doi:10.1002/cne.23838 [Epub ahead of print].

33. Mackey S, Petrides M. Quantitative demonstration of comparable cytoarchitectonic areas within the ventromedial and lateral orbital frontal cortex in human and macaque monkey brains. Eur J Neurosci. 2010;32:1940–50.

34. Semendeferi K, Armstrong E, Schleicher A, Zilles K, Van Hoesen GW. Prefrontal cortex in humans and apes: a comparative study of area 10. Am J Phys Anthropol. 2001;114:224–41.

35. Jacobs B, Schall M, Prather M, Kapler E, Driscoll L, Baca S, et al. Regional dendritic and spine variation in human cerebral cortex: a quantitative Golgi study. Cereb Cortex. 2001;11:558–71.

36. Ongur D, Ferry AT, Price JL. Architectonic subdivision of the human orbital and medial prefrontal cortex. J Comp Neurol. 2003;460:425–49.

37. Christoff K, Gabrieli JDE. The frontopolar cortex and human cognition: evidence for a rostrocaudal hierarchical organization within the human prefrontal cortex. Psychobiology. 2000;28:168–86.

38. Tsujimoto S, Genovesio A, Weiss SP. Frontal pole cortex: encoding ends at the end of the endbrain. Trends Cogn Sci. 2011;15:169–76.

39. Bookstein F, Schaefter K, Prossinger H, Seidler H, Fieder M, Stringer C, et al. Comparing frontal cranial profiles in archaic and modern homo by morphometric analysis. Anat Rec. 1999;257:217–24.

40. Smaers JB, Schleicher A, Zilles K, Vinicius L. Frontal white matter volume is associated with brain enlargement and higher structural connectivity in anthropoid primates. PLoS One. 2010;5:1–6. e9123.

41. Stephan H, Baron G, Frahm H. Comparative brain research in mammals, vol 1. Insectivora. New York, NY: Springer; 1991.

42. Zilles K. Evolution of the human brain and comparative cyto and receptor architecture. In: Dehaene S, Duhamel JR, Hauser MD, Rizzolatti G, editors. From money brain to human brain. Cambridge, MA: Fyssen Foundation, MIT; 2005.

43. Schoenemann PT, Sheehan MJ, Glotzer LD. Prefrontal white matter is disproportionately larger in humans than in other primates. Nat Neurosci. 2005;8:242–52.

44. Semendeferi K, Armstrong E, Schleicher A, Zilles K, Van Hoesen GW. Limbic frontal cortex in homi-

noids: a comparative study of area 13. Am J Phys Anthropol. 1998;106:129–55.

45. Semendeferi K, Armstrong E, Schleicher A, Zilles K, Van Hoesen GW. Prefrontal cortex in humans and apes: a comparative study of area 10. Am J Phys Anthrorpol. 2001;114:224–41.

46. Hof PR, Mufson EJ, Morrison JH. Human orbito-frontal cortex: cytoarchitecture and quantitative immuno-histochemical parcellation. J Comp Neurol. 1995;359:48–68.

47. Hof PR, Van Der Gucht E. Structure of the cerebral cortex of the humpback whale, Megaptera novae-angeliae. Anat Rec. 2007;290:1–31.

48. Schenker N, Desgouttes AM, Semendeferi K. Neural connectivity and cortical substrates of cognition in hominoids. J Hum Evol. 2005;49:547–69.

49. Schumann C, Amaral DG. Stereological estimation of the number of neurons in the human amygdaloid complex. J Comp Neurol. 2005;491:320–9.

50. Barton RA, Aggleton JP, Grenyer R. Evolutionary coherence of the mammalian amygdala. Proc Biol Sci. 2003;270:539–43.

51. Brothers L. The social brain: a project for integrating primate behavior and neurophysiology in a new domain. Concept Neurosci. 1990;1:27–51.

52. Bargar N, Stefanacci L, Semendeferi K. A compara-tive volumetric analysis of amygdaloid complex and basolateral division in the human and ape brain. Am J Phys Anthropol. 2007;134:392–403.

53. Passinghim RE, Stephan KE, Kotter R. The anatomi-cal basis of functional localization in the cortex. Nat Rev Neurosci. 2002;3:606–16.

54. Willamson PC, Allman JM. The human illnesses. Neuropsychiatric Disorders and the Nature of the Human Brain: Oxford University Press, Oxford; 2011.

55. Bortz II WM. Physical exercise as an evolutionary force. J Hum Evol. 1985;14:145–55.

56. Carrier DR. The energetic paradox of human running and hominid evolution. Curr Anthropol. 1984;25: 483–95.

57. Leonard WR, Robertson MS. Comparative primate energetics and hominid evolution. Am J Phys Anthropol. 1997;102:265–81.

58. Raghanti MA, Stimpson CD, Erwien JM, Hof PR, Sherwood CC. Cortical dopaminergic innervation of the frontal cortex: differences among humans, chim-panzees and macaque monkeys. Neuroscience. 2008;155:203–20.

59. Berger B, Gaspar P, Verney C. Dopaminergic inner-vation of the cerebral cortex: unexpected differences between rodents and primates. Trends Neurosci. 1991;14:21.

60. Lewis DA, Melchitzky DS, Sesack SR, Whitehead RE, Auh S, Sampson A. Dopamine transporter immunoreactivity in monkey cerebral cortex: regional, laminar and ultrastructural organization. J Comp Neurol. 2001;432:119–36.

61. Jakab RL, Goldman Rakic PS. Segregation of sero-tonin 5HT 2A and 5HT 3 receptors in inhibitory circuits in the primate cerebral cortex. J Comp Neurol. 2000;417:337–48.

62. Soubrie P. Reconciling the role of central serotonin neurons in human and animal behavior. Behav Brain Sci. 1986;9:319–64.

63. Sarter M, Parikh V. Choline transporters, cholinergic transmission and cognition. Nat Rev Neurosci. 2005;6:48–56.

64. Levin ED, Simon BB. Nicotinic acetylcholine involvement in cognitive function in animals. Psychopharmacology (Berl). 1998;138:217–30.

65. Subiaul F. Mosaic cognitive evolution: the case of imitation behavior. In: Broadfield D, Yuan M, Schick K, Toth N, editors. The human brain evolving. Gosport, IN: Stone Age Institute Press; 2010.

66. Brady ST, Siegel GJ, Albers RW, Price DL. Basic neurochemistry. Principles of molecular, cellular and medical neurobiology. 8th ed. Amsterdam: Academic; 2012.

67. Nestler EJ, Malenka R, Hyman S. Molecular neuro-pharmacology. A foundation for clinical neurosci-ence. New York, NY: McGraw Hill; 2009.

68. Mesulam MM. Principles of behavioural and cogni-tive neurology. London: Oxford University Press; 2000. p. 1–120.

69. Heilman KM, Valenstein E. Clinical neuropsychol-ogy. 5th ed. Oxford: Oxford University Press; 2012.

70. Coolidge FL, Wynn T. The rise of Homo sapiens. New York, NY: Wiley Blackwell; 2009.

71. Petrides M. Dissociable roles of the mid dorsolateral prefrontal cortex and anterior inferotemporal cortex in visual working memory. J Neurosci. 2000; 20:7496–503.

72. Champod AS, Petrides M. Dissociation within the frontoparietal network in verbal working memory: a parametric functional magnetic resonance imaging study. J Neurosci. 2010;30:3849–56.

73. Champod AS, Petrides M. Dissociable roles of the posterior parietal and the prefrontal cortex in manip-ulation and monitoring process. Proc Natl Acad Sci U S A. 2007;104:14837–42.

74. Medalla M, Barbas H. Diversit of laminar connec-tions linking periarcuate and lateral intraparietal areas dpends on cortical structure. Eur J Neurosci. 2006;23:161–79.

75. Petrides M. The mid-dorsolateral prefrontal-parietal network and the epoptic process. In: Stuss DT, Knight RT, editors. Principles of frontal lobe func-tion. Oxford: Oxford University Press; 2012.

76. Cummings JL. Frontal subcortical circuits and human behavior. Arch Neurol. 1993;50:873–80.

77. Catanin M, Jones DK, Ffytche DH. Perisylvian lan-guage networks of the human brain. Ann Neurol. 2005;57:8–16.

78. Mallet N, Micklem BR, Henny P, Brown MT, Williams C, Bolam JP, et al. Dichotomous organiza-tion of the external globus pallidus. Neuron. 2012; 74:1075–86.

79. Yeterian EH, Pandya DN. Striatal connections of the parietal association cortices in rhesus monkeys. J Comp Neurol. 1993;332:175–97.

80. Stuss DT, Floden D, Alexander MP, Levine B, Katz D. Stroop performance in focal lesion patients: dis-

sociation of process and frontal lobe lesion location. Neuropsychologica. 2001;39:771–86.

81. Fesenmeier JT, Kuzniecky R, Garcia JH. Akinetic mutism caused by bilateral anterior cereberal tuberculous obliterative arteritis. Neurology. 1990;40:1005–6.

82. Bogousslavsky J, Regli F. Anterior cerebral artery territory infarction in the Lausanne stroke registry. Arch Neurol. 1990;47:144–50.

83. Berthier ML, Kulisevsky J, Gironell A, Heras JA. Obsessive compulsive disorders associated with brain lesions: clinical phenomenology, cognitive function and anatomic correlates. Neurology. 1996;47:353–61.

84. Berthier ML, Starkstein SE, Robinson RG, Leiguarda R. Limbic lesions in a patient with recurrent mania. J Neuropsychiatr Clin Neurosci. 1990;2:235–6.

85. Seo JP, Kim OL, Kim SH, Chang MC, Kim MS, Son SM, et al. Neural injury of uncinate fasciculus in patients with diffuse axonal injury. NeuroRehabilitation. 2012;30:323–8.

86. Hirstein W, Ramachandran VS. Capgras syndrome: a novel probe for understanding the neural representation of the identity and familiarity of persons. Proc R Soc B. 1997;264:437–44.

87. Von Der Heide R, Skipper LM, Kobusicky E, Olsen IR. Dissecting the uncinate fasciculus: disorders, controversies and a hypothesis. Brain. 2013;136:1692–707.

88. Weaver AH. Reciprocal evolution of the cerebellum and neocortex in fossil humans. Proc Natl Acad Sci U S A. 2005;102:3576–80.

89. Ramnani N. The primate cortico-cerebellar system: anatomy and function. Nat Rev Neurosci. 2006;7:511–22.

90. Kamali A, Kramer LA, Frye RE, Butler IJ, Hasan KM. Diffusion tensor tractography of the human brain cortico-ponto-cerebellar pathways: a quantitative preliminary study. J Magn Reson Imaging. 2010;32:809–17.

91. Dehaene S. Reading in the brain. New York, NY: Penguin; 2009.

92. Tarkiainen A, Cornelissen PL, Salmelin R. Dynamics of visual feature analysis and object level processing in face versus letter-string perception. Brain. 2002;125:1125–36.

93. Nieder A, Millker EK. A parieto-frontal network for visual numerical information in the monkey. Proc Natl Acad Sci U S A. 2004;101:7457–62.

94. Tsang JM, Dougherty RF, Deutsch GK, Wandell BA, Ben-Shachar M. Frontoparietal white matter diffusion properties predict mental arithmetic skills in children. Proc Natl Acad Sci U S A. 2009;106:22546–51.

95. Wood JN, Grafman J. Human prefrontal cortex: processing and representational perspectives. Nat Rev Neurosci. 2003;4:139–47.

96. Foussias G, Remington G. Negative symptoms in schizophrenia: avolition and Occam's razor. Schizophr Bull. 2010;36:359–69.

97. Greene MF, Penn DL, Bentall R. Social cognition in schizophrenia: an NIMH workshop on definitions, assessment and research opportunities]. Schizophr Bull. 2008;34(6):1211–20.

98. Yung AR, McGorry PD. The prodromal phase of first episode psychosis: past and current conceptualizations. Schizophr Bull. 1996;22:353–70.

99. Hoffmann M. The panoply of field dependent behavior in 1436 stroke patients. The mirror neuron system uncoupled and the consequences of loss of personal autonomy. Neurocase. 2014;20(5):556–68.

100. Stuss DT. New approaches to prefrontal lobe testing. In: Miller B, Cummings JL, editors. The human frontal lobes. New York, NY: Guilford; 2009.

101. Stuss DT, Binns MA, Murphy KJ, Alexander MP. Dissociations within the anterior attentional system: effects of task complexity and irrelevant information on reaction time speech and accuracy. Neuropsychologica. 2002;16:500–13.

102. Stuss DT, Guberman A, Nelson R, Larochelle S. The neuropsychology of paramedian thalamic infarction. Brain Cogn. 1988;8:348–78.

103. Cicerone KD. Attention deficits and dual task demands after mild traumatic brain injury. Brain Inj. 1996;10:79–89.

104. Aharon-Peretz J, Tomer R. Traumatic brain injury. In: Miller B, Cummings JL, editors. The human frontal lobes. New York, NY: Guilford; 2009.

105. Bar-On R, Tranel D, Denburg NL, Bechara A. Exploring the neurological substrate of emotional and social intelligence. Brain. 2003;126:1790–800.

106. Shamay-Tsoory SG, Tomer R, Goldsher D, Berger BD, Aharon-Peretz J. Impairment in cognitive and affective empathy in patients with brain lesions: anatomical and cognitive correlates. J Clin Exp Neuropsychol. 2004;26:1113–27.

107. Pessoa L. On the relationship between emotion and cognition. Nat Rev Neurosci. 2008;9:148–58.

108. Hoffmann M, Benes Cases L, Hoffmann B, Chen R. The impact of stroke on emotional intelligence. BMC Neurol. 2010;10:103.

109. Denny-Brown D. The nature of apraxia. J Nerv Ment Dis. 1958;126:9–32.

110. Graff-Radford J, Rubin MN, Jones DT, Aksamit AJ, Ahlskog JE, Knopman DS, et al. The alien limb phenomenon. J Neurol. 2013;260(7):1880–8.

111. Scepkowski LA, Cronin-Golomb A. The alien hand: cases, categorizations, and anatomical correlates. Behav Cogn Neurosci Rev. 2003;2(4):261–77.

112. Carrazana E, Rey G, Rivas-Vazquez R, Tatum W. Ictal alien hand syndrome. Epilepsy Behav. 2001;2(1):61–4.

113. Kapur N. Paradoxical functional facilitation in brain behavior research. A critical review. Brain. 1996;119:1775–90.

114. Schott GD. Pictures as a neurological tool: lessons from enhanced and emergent artistry in brain disease. Brain. 2012;135:1947–63.

115. Miller B. The frontotemporal lobe dementias. New York, NY: Oxford University Press; 2013.

116. Christodoulou GN, Magariti M, Kontaxakis VP, Christodoulou NG. The delusion misidentification

syndromes: strange, fascinating and instructive. Curr Pyschiatry Rep. 2009;11:185–9.

117. Pell MD. Judging emotion and attitudes from prosody following brain damage. Prog Brain Res. 2006;156:303–17.

118. Pen`a-Casanova J, Roig-Rovira T, Bermudez A, Tolosa-Sarro E. Optic aphasia, optic apraxia and loss of dreaming. Brain Lang. 1985;26:63–71.

119. Treffert DA. Islands of genius. London: Jessica Kingsley Publishers; 2010.

120. Assessment: neuropsychological testing of adults. Considerations for neurologists. Report of the Therapeutics and Technology Assessment Subcommittee of the American Academy of Neurology, 1996.

121. Kramer JH, Reed BR, Mungas D, Weiner MW, Chui HC. Executive dysfunction in subcortical ischemic vascular disease. J Neurol Neurosurg Psychiatry. 2002;72:217–20.

122. Prins ND, van Dijk EJ, Heijer T, Vermeer SE, Jolles J, Koudstaal PJ, et al. Cerebral small-vessel disease and decline in information processing speed, executive function and memory. Brain. 2005;128:2034–41.

123. Nasreddine ZS, Phillips NA, Bedirian V, Charbonneau S, Whitehead V, Collin I, et al. The Montreal Cogntive Assessment (MoCA): a brief screening tool for mild cognitive impairment. J Am Geriatr Soc. 2005;53:695–9.

124. Dubois B, Slachevsky A, Litvan I, Pillon B. The FAB. A frontal assessment battery at the beside. Neurology. 2000;55:1621–6.

125. Royall DR, Mahurin RK, Gray KF. Bedside assessment of executive cognitive impairment: the executive interview. J Am Geriatr Soc. 1992;40:1221–6.

126. Hoffmann M, Schmitt F, Bromley E. Comprehensive cognitive neurological assessment in stroke. Acta Neurol Scand. 2009;119(3):162–71.

127. Hoffmann M, Schmitt F. Metacognition in stroke: bedside assessment and relation to location, size and stroke severity. Cogn Behav Neurol. 2006; 19(2):85–94.

128. Dwolatsky T, Whitehead V, Doniger GM, Simon ES, Schweiger A, Jaffe D, et al. Validity of a novel computerized cognitive battery for mild cognitive impairment. BMC Geriatr. 2003;3:4.

129. Robbins TW, James M, Owen AM, Sahakian BJ, McInnes L, Rabbitt P. Cambridge Neuropsychological Test Automated Battery (CANTAB): a factor analytic study of a large sample of normal elderly volunteers. Dementia. 1994;5:266–81.

130. Kiernan R, Mueller J, Langston JW, Van Dyke C. The neurobehavioral cognitive status examination. A brief but differentiated approach to cognitive assessment. Ann Intern Med. 1987;107:481–5.

131. Gualtieri CT, Johnson LG. Reliability and validity of a computerized neurocognitive test battery, CNS vital signs. Arch Clin Neuropsychol. 2006;21:623–43.

132. Delis DC, Kaplan E, Kramer JH. Delis–Kaplan executive function system. San Antonio, TX: The Psychological Corporation. A Harcourt Assessment Company; 2001.

133. Wechsler D. Wechsler adult intelligence scale. 4th ed. San Antonio, TX: The Psychological Corporation. Harcourt Brace and Company; 2008.

134. Grace J, Malloy PF. Frontal systems behavior scale. Lutz, FL: PAR Neuropsychological Assessment Resources Inc; 2001.

135. Roth RM, Isquith PK, Gioia GA. BRIEF-A behavior rating inventory of executive function—adult version. Lutz, FL: Neuropsychological Assessment Resources Inc (PAR); 2005.

136. Kertesz A, Davidson W, Fox H. Frontal behavioural inventory: diagnostic criteria for frontal lobe dementia. Can J Neurol Sci. 1997;24:29–36.

137. Reynolds CR. Comprehensive trail making test. Austin, TX: Pro-ed; 2002.

138. Gladsjo JA, Walden Miller W, Heaton RK. Norms for letter and category fluency: demographic corrections for age, education and ethnicity. Lutz, FL: Psychological Assessment Resources Inc; 1999.

139. Heaton RK. Wisconsin card sorting test computer version 4. Lutz, FL: PAR Psychological Assessment Resources; 2003.

140. Culbertson WC, Zillmer EA. Tower of London. Toronto, ON: Multi Health Systems Inc; 2001.

141. Bar-On R. The bar-on emotional quotient inventory (EQ-i): technical manual. Toronto, ON: Multi-Health Systems; 1997.

142. Mayer JD, Salovey P, Caruso GR. MSCEIT. Toronto, ON: Multi Health Systems Inc; 2002.

143. Trenerry MR, Crosson B, DeBoe J, Leber WR. Stroop neuropsychological screening test. Lutz, FL: Psychological Assessment Resources (PAR); 1989.

144. Bechara A. Iowa gambling test. Lutz, FL: Psychological Assessment Resources Inc; 2007.

145. Rutter M, Le Couteur A. ADI-R. Los Angeles, WA: Western Psychological Services; 2005.

146. Baron-Cohen S, Wheelwright S, Skinner R, Martin J, Clubley E. The autism spectrum quotient (AQ): evidence from Asperger syndrome/high-functioning autism, males and females, scientists and mathematicians. J Autism Dev Disord. 2001;31:5–17.

147. Brown TE. Brown attention deficit disorder scales. Hamden, CT: The Psychological Corporation. A Harcourt Assessment Company; 1996.

148. Radloff L. The CES-D scale: a self-report depression scale for research in the general population. Appl Psychol Meas. 1977;1(3):385–401.

149. Beck AT, Steer RA, Brown GK. Beck depression inventory II. San Antonio, TX: Psychological Corporation; 1996.

150. Reynolds WM, Kobak KA. Hamilton depression inventory. Lutz, FL: Psychological Assessment Resources Inc; 1995.

151. Stone VE, Baron-Cohen S, Knight RT. Frontal lobe contribution to theory of mind. J Cogn Neurosci. 1998;10:640–56.

152. Baron-Cohen S, Joliffe T, Mortimore C, Robertson M. Another advanced test of theory of mind: evidence from very high functioning adults with autism or Asperger syndrome. J Child Psychol Psychiatry. 1997;38:813–22.

153. Manly T, Hawkins K, Evans J, Woldt K, Robertson IH. Rehabilitation of executive function: facilitation of effective goal management on complex tasks using periodic auditory alerts. Neuropsychologia. 2002;40:271–81.

154. Knight C, Alderman N, Burgess PW. Development of a simplified version of the multiple errrands test for use in hospital settings. Neuropsychol Rehab. 2002;12:231–55.

155. Windmann S, Wehrmann M, Calabrese P, Guntuerken O. Role of the prefrontal cortex in attentional control over bistable vision. J Cogn Neurosci. 2006;18:456–71.

156. Torrance EP. Influence of dyadic interaction on creative functioning. Psychol Rep. 1970;26:391–4.

157. Trenerry MR, Cross B, De Boe J, Leber WR. Visual Search and Attention Test (VSAT). Lutz, FL: Psychological Assessment Resources Inc; 1990.

158. Holmes Bernstein J, Waber DP. Developmental scoring system for the Rey Osterrieth complex figure. Lutz, FL: Psychological Assessment Resources Inc; 1996.

159. Goldberg E, Podelle K, Bilder R, Jaeger J. The executive control battery. Melbourne, VIC: Psych Press; 1999.

160. Kertesz A. The western aphasia battery. Hamden, CT: The Psychological Corporation. A Harcourt Assessment Company; 1982.

161. Goodglass H, Kaplan E, Barresi B. Boston diagnostic aphasia test. 3rd ed. Philadelphia, PA: Lippincott Williams Wilkins; 2001.

162. Doty R. Sensonics Inc. Haddon Heights, NJ. 2006.

163. VanSwearingham M, Paschal KA, Bonino P, Yang J-F. The modified gait abnormality rating scale for recognizing the risk of recurrent falls in community dwelling elderly adults. Phys Ther. 1996;76: 994–1002.

164. Jong's D. The neurologic examination. Philadelphia, PA: Lippincott Williams and Wilkins; 2005.

165. Mesulam M-M. Large scale neurocognitive networks and distributed processing for attention, language and memory. Ann Neurol. 1990;28:597–613.

166. Winter B, Bert B, Fink H, Dirnagl U, Endres M. Dysexecutive syndrome after mild cerebral ischemia? Stroke. 2004;35:191–5.

167. Gillman PK. A review of serotonin toxicity data: implications for the mechanisms of antidepressant action. Biol Psychiatry. 2006;59:1046–51.

168. Gnanadesignan N, Espinoza RT, Smith RL. The serotonin syndrome. N Engl J Med. 2005;352:2454–6.

169. Whyte IM. Serotonin toxicity/syndrome. Philadelphia: Medical Toxicology. Williams and Wilkins; 2004.

170. Strawn JR, Keck PE, Caroff SN. Neuroleptic malignant syndrome. Am J Psychiatry. 2007;164:870–6.

171. Litman R, Rosenberg H. Malignant hyperthermia: update on susceptibility testing. JAMA. 2005;293: 2918–24.

172. Ochs KL, Zell-Kanter M, Mycyk MB. Toxikon consortium. Hot, blind and mad: avoidable geriatric anticholinergic delirium. Am J Emerg Med. 2012;30:514.e1–3.

173. Blackman JA, Patrick PD, Buck ML, Rust RS. Paroxysmal autonomic instability with dystonia after brain injury. Arch Neurol. 2004;61:321–8.

174. Kelly BJ, Boeve BF, Josephs KA. Young onset dementia: demographic and etiological characteristics of 235 patients. Arch Neurol. 2008;65:1502–8.

175. Seeley WW, Menon V, Schatzberg AF, Keller J, Glover GH, Kenna H, et al. Dissociable intrinsic connectivity networks for salience processing and executive control. J Neurosci. 2007;27:2349–56.

176. Fuster JM. Neuroimaging. In: Fuster JM, editor. The prefrontal cortex. 4th ed. New York, NY: Elesevier; 2009. p. 285–332.

177. Johnson JK, Diehl J, Mendez MF, Neuhaus J, Shapira JS, Forman M, et al. Frontotemporal lobar degeneration: demographic characteristics of 353 patients. Arch Neurol. 2005;62:925–30.

178. Prayson RA, Goldblum JR. Neuropathology. Amsterdam: Elsevier; 2005.

179. Brunnstrom H, Gustafson L, Passant U, Englund E. Prevalence of dementia subtypes: a 30-year retrospective survey of neuropathological reports. Arch Gerontol Geriatr. 2009;49:146–9.

180. Mercy L, Hodges JR, Dawson K, Barker RA, Brayne C. Incidence of early onset dementia in Cambridgshire, United Kingdom. Neurology. 2008;71:1496–9.

181. Rosso SM, Donker Kaat L, Baks T, Joosse M, de Koning I, Pijnenburg Y, et al. Frontotemporal dementia in the Netherlands: patient characteristics and prevalence estimates from a population based study. Brain. 2003;126:2016–22.

182. Miller BL, Cummings JL, Villanueva-Meyer J, Boone K, Mehringer CM, Lesser IM, et al. Frontal lobe degeneration; clinical neuropsychological and SPECT characteristics. Neurology. 1991;41: 1374–82.

183. Neary D, Snowden JS, Gustafson L, Passant U, Stuss D, Black S, et al. Frontotemporal lobar degeneration: a consensus on clinical diagnostic criteria. Neurology. 1998;51:1546–54.

184. Rascovsky K, Hodges JR, Knopman D, Mendez MF, Kramer JH, Neuhaus J, et al. Sensitivity of revised diagnostic criteria for the behavioral variant of frontotemproal dementia. Brain. 2011;134:2456–77.

185. Coste CP, Sadaghiani S, Friston KJ, Kleinschmidt A. Ongoing brain activity fluctuations directly account for intertrial and indirectly for intersubject variability in Stroop task performance. Cereb Cortex. 2011;21:2612–9.

186. Hatanpaa KJ, Blass DM, Pletnikova O, Crain BJ, Bigio EH, Hedreen JC, et al. Most cases of dementia with hippocampal sclerosis may represent fronto-

temporal lobe dementia. Neurology. 2004;63: 538–42.

187. Neurann M, Tolnay M, Mackenzie IR. The molecular basis of frontotemporal dementia. Expert Rev Mol Med. 2009;11:23.

188. Seeley WW. Selective functional, regional and neuronal vulnerability in frontotemporal dementia. Curr Opin Neurol. 2008;21:701–7.

189. Lomen-Hoerth C. Characterization of amyotrophic lateral sclerosis and frontotemporal lobe dementia. Dement Geriatr Cogn Disord. 2004;17:337–41.

190. Foulds P, McAuley E, Gibbons L, Davidson Y, Pickering-Brown SM, Neary D, et al. TDP 43 protein in plasma may index TDP 43 brain pathology in Alzheimer's disease and frontotemporal lobe degeneration. Acta Neuropathol. 2008;116:141–6.

191. Arai T, Mackenzie IR, Hasegawa M, Nonoka T, Niizato K, Tsuchiya K, et al. Phosphorylated TDP 43 in Alzheimer's disease and dementia with Lewy bodies. Acta Neuropathol. 2009;117:125–36.

192. Spillantini MG, Yoshida H, Rizzini C, Lantos PL, Khan N, Rossor MN, et al. A novel tau mutation (N296N) in familial dementia with swollen achromatic neurons and corticobasal inclusion bodies. Ann Neurol. 2000;48:939–43.

193. Kim EJ, Sidhu M, Gaus SE, Huang EJ, Hof PR, Miller BL, et al. Selective frontoinsular von Economo neuron and fork cell loss in early behavioral variant of frontotemporal dementia. Cereb Cortex. 2012;22: 251–9.

194. Rohrer JD, Guerreiro R, Vandrovcova J, Uphill J, Reiman D, Beck J, et al. The heritability and genetics of frontotemporal lobe degeneration. Neurology. 2009;73:1451–6.

195. Hu WT, Wang Z, Lee VM, Trojanowski JQ, Detre JA, Grossman M. Distinct cerebral perfusion patterns in FTLD and AD. Neurology. 2010;75:881–8.

196. Brian H, Van Swieten JC, Leight S, Massimo L, Wood E, Forman M, et al. CSF biomarkers in frontotemporal lobar degeneration with known pathology. Neurology. 2008;70:1827–35.

197. Huey ED, Putnam KT, Grafman J. A systematic review of neurotransmitter deficits and treatments in frontotemporal dementia. Neurology. 2006;66:17–22.

198. Yenner GG, Rosen HJ, Papatriantafyllou L. Frontotemporal lobe degeneration. Continuum. 2010;16:191–211.

199. Mungas D, Jagust WJ, Reed BR, Kramer JH, Weiner MW, Schuff N, et al. MRI predictors of cognition in subcortical ischemic vascular disease and AD. Neurology. 2000;57:2229–35.

200. Looi JC, Sachdev PS. Differentiation of vascular dementia from Alzheimers on neuropsychological tests. Neurology. 1999;53:670–8.

201. Gold G, Giannakopoulos P, Montes-Paixao Junior C, Herrmann FR, Mulligan R, Michel JP, et al. Sensitivity and specificity of newly proposed clinical criteria for possible vascular dementia. Neurology. 1997;49:690–4.

202. Ingles JL, Wentzel C, Fish JD, Rockwood K. Neuropsychological predictors of incident dementia in patients with vascular cognitive impairment, without dementia. Stroke. 2002;33:1999–2002.

203. Varma AR, Laitt R, Lloyd JJ, Carson KJ, Snowden JS, Neary D, et al. Diagnostic value of high signal abnormalities on T2 weighted MRI in the differentiation of Alzheimer's, frontotemporal ad vascular dementias. Acta Neurol Scand. 2002;105:355–64.

204. Josephs KA. Frontotemporal dementia and related disorders: deciphering the enigma. Ann Neurol. 2008;64:4–14.

205. Hauw JJ. The neuropathology of vascular and mixed dementia and vascular cognitive impairment. In: Aminoff MJ, Boller F, Swaab DF, editors. Handbook of clinical neurology dementias, vol 89, series 3. New York, NY: Elsevier; 2003.

206. Delano-Wood L, Abeles N, Sacco JM, Wieranga CE, Horne NR, Bozoki A. Regional white matter pathology in mild cognitive impairment. Differential influence of lesion type on neuropsychological functioning. Stroke. 2008;39:794–9.

207. Yatsuya H, Folsom AR, Wong TY, Klein R, Klein BE, Sharrett AR. Retinal microvascular abnormalities and risk of lacunar stroke. Stroke. 2010;41: 1349–55.

208. Libon D, Price CC, Giovannetti T, Swenson R, Bettcher BM, Heilman KM, et al. Linking MRI hyperintensities with patterns of neuropsychological impairments. Stroke. 2008;39:806–13.

209. Viswanathan A, Rocca WA, Tzourio C. The vascular – dementia continuum. Neurology. 2009;72:368–74.

210. Deschaintre Y, Richard F, Leys D, Pasquier F. Treatment of vascular risk factors is associated with a slower decline in Alzheimer disease. Neurology. 2009;73:674–80.

211. Becker JT, Huff RJ, Nebes RD, Holland A, Boller F. Neuropsychological function in Alzheimer's disease: pattern of impairment and rates of progression. Arch Neurol. 1988;45:263–8.

212. Nebes RD, Brady CB. Focused and divided attention in Alzheimer's disease. Cortex. 1989;25: 305–15.

213. Baddeley AD, Della Sala S, Spinnler H. The two component hypothesis of memory deficit in Alzheimer's disease. J Clin Exp Neuropsychol. 1991;13:372–80.

214. Collete F, Vand der Linden M, Delrue G, Salmon E. Frontal hypometabolism does not explain inhibitory dysfunction in Alzheimer disease. Alzheimer Dis Assoc Disord. 2002;16:228–38.

215. Chiaravalloti ND, DeLuca J. Cognitive impairment in multiple sclerosis. Lancet Neurol. 2008;7: 1139–51.

216. Roca M, Torralva T, Meli F, Fiol M, Calcagno ML, Carpintiero S, et al. Cognitive deficits in multiple sclerosis correlate with changes in fronto-subcortical tracts. Mult Scler. 2008;14:364–9.

217. Junque C, Pujol J, Vendrell P, Bruna O, Jodar M, Ribas JC, et al. Leukoaraiosis on magnetic resonance imaging and speed of mental processing. Arch Neurol. 1990;47:151–6.

218. Len TK, Neary JP. Cerebrovascular pathophysiology following mild traumatic brain injury. Clin Physiol Funct Imaging. 2011;31:85–93.

219. Johnson B, Zhang K, Gay M, Horovitz S, Hallett M, Sebastianelli W, et al. Alteration of brain default network in subacute phase of injury in concussed individuals: resting-state fMRI study. Neuroimage. 2012;59:511–8.

220. Skuja S, Groma V, Smane L. Alocholism and cellular variability in different brain regions. Ultrastruct Pathol. 2012;36:40–7.

221. Sabeti J. Ethanol exposure in early adolescence inhibits intrinsic neuronal plasticity via sigma 1 receptor activation in hippocampal CA1 neurons. Alcohol Clin Exp Res. 2011;35:885–904.

222. Kazui H. Cognitive impairment in patients with idiopathic normal pressure hydrocephalus. Brain Nerve. 2008;60:225–31.

223. Gleichgerrcht E, Cervio A, Salvat J, Rodríguez Loffredo A, Vita L, Roca M, et al. Executive function improvement in normal pressure hydrocephalus following shunt surgery. Behav Neurol. 2009;21:181–5.

224. Tarnaris A, Kitchen ND, Watkins LD. Noninvasive biomarkers in normal pressure hydrocephalus: evidence for the role of neuroimaging. J Neurosurg. 2009;110:837–51.

225. Graus F, Saiz A, Lai M, Bruna J, López F, Sabater L, et al. Neuronal surface antigen antibodies in limbic encephalitis: clinical-immunologic associations. Neurology. 2008;71:930–6.

226. Gizza CC, Hovda DA. Neurometabolic cascade of concussion. J Athl Train. 2001;36:228–35.

227. Barkhoudarian G, Hovda DA, Giza CC. The molecular pathophysiology of concussive brain injury. Clin Sports Med. 2011;30:33–48.

228. McKeon A, Lennon VA, Pittock SJ. Immunotherapy responsive dementias and encephalopathies. Contin Life Long Learn Neurol. 2010;16:80–101.

229. Flanagan EP, McKeon A, Lennon VA, Boeve BF, Trenerry MR, Tan KM, et al. Autoimmune dementia: clinical course and predictors of immunotherapy response. Mayo Clin Proc. 2010;85(10):881–97.

230. Vernino S, Geschwind M, Boeve B. Autoimmune encephalopathies. Neurologist. 2007;13:140–7.

231. Rafael H. Secondary Alzheimer started by cryptococcal meningitis. J Alzheimer Dis. 2005;7:99–100.

232. Ala TA, Doss RC, Sullivan CJ. Reversible dementia: a case of cryptococcal meningitis masquerading as Alzheimer's disease. J Alzheimers Dis. 2004;6(5):503–8.

233. Hoffmann M, Muniz J, Carroll E, De Villasante JM. Cryptococcal meningitis masquerading as Alzheimer's disease: complete neurological and cognitive recovery with treatment. J Alzheimers Dis. 2009;16:517–20.

234. Salvarani C, Brown CR, Calamia KT, Christianson TJH, Weigand SD, Miller DV, et al. Primary central nervous system vasculitis: analysis of 101 patients. Ann Neurol. 2007;62:442–51.

235. Hajj-Ali RA, Singhal AB, Benseler S, Molloy E, Calabrese LH. Primary angiitis of the CNS. Lancet Neurol. 2011;10:561–72.

236. Moore PM, Calabrese LH. Neurologic manifestations of systemic vasculitides. Semin Neurol. 1994;14:300–6.

Neuroimaging and Treatments Perspectives

13

Anatomical and Functional Imaging

Brain imaging in the context of both pathology and normal functioning is increasingly moving toward imaging of complex networks such as resting state networks (RSN) or intrinsic connectivity network (ICN) imaging and positron emission tomography (PET) brain imaging. Although largely the domain of functional imaging, contributions from certain anatomical imaging techniques such as diffusion tensor imaging (DTI) also image brain networks at the macroscopic level. Traumatic brain injury and depression are two examples where standard magnetic resonance imaging (MRI) is usually normal but ICN has for the first time shown objective abnormalities.

The first step with neuroimaging is to exclude emergency neurological conditions such as various mass lesions, cerebral infarction, brain bleeds, and HSV-1 encephalitis meningitis for example, using anatomical imaging with computerized tomography (CT) brain scanning or multimodality MR imaging. Thereafter the pursuit of underlying etiological processes is attended, often requiring and complemented by functional imaging usually such as DTI, f-MRI, [18]Fluorodeoxyglucose position emission tomography ([18]FDG-PET) brain, single photon emission computed tomography (SPECT), brain PET Pittsburgh Compound B (PIB), and PET receptor

(Dopa) imaging (Tables 13.1, 13.2, 13.3, and 13.4). [18]FDG-PET brain in particular has been an important tool in the early diagnosis of mild cognitive impairment (MCI) and in differentiating types of dementia, with frontotemporal disorders (FTD) and Alzheimer's disease (AD) [1]. Functional imaging is increasingly able to detect pathology, long before the clinical state emerges, with PET brain imaging being the most accurate diagnostic method for most common dementia categories [2]. PET brain scan patterns reliably differentiate the major dementia subtypes including the AD variants, posterior cortical atrophy syndrome (Benson syndrome) [3], FTD (frontotemporal hypometabolism), and AD (temporal, parietal, posterior cingulate hypometabolism) being relatively easily identified. There are also overlap syndromes such as AD and cognitive vascular disorders (CVD), the frontal variant of Alzheimer's and bv-FTD which cannot be differentiated easily clinically [4]. Other conditions that present predominantly as a frontal network systems (FNS) syndrome include autoimmune dementias, toxic dementias, HIV dementia, and the prefrontal atrophy secondary to chronic stimulation of the pain matrix (chronic pain syndrome) [5–8]. Positron emission tomography (PET) [9–11] and functional magnetic resonance imaging (fMRI) [12] implicate the anterior cingulate cortex (ACC) and posterior cingulate cortex (PCC) having key roles in processing of pain perception [13, 14].

© Springer International Publishing Switzerland 2016
M. Hoffmann, *Cognitive, Conative and Behavioral Neurology*,
DOI 10.1007/978-3-319-33181-2_13

Table 13.1 Anatomical and functional imaging categories and examples of major disease entities [16–19]. Adapted from [20]

(a) Magnetic resonance multimodality sequences for specific pathologies

MRI (routine series)

T1/T2, fluid attenuated inversion recovery (FLAIR), gradient echo (GRE), magnetic resonance angiography (MRA) to detect degree of concomitant vascular disease, atrophy patterns, and other secondary pathologies

MRI-DTI

Diffusion Tensor Imaging (DTI) has become the imaging modality of choice to objectively quantify the anatomical pathology which predominantly affects the fiber tracts that occurs with traumatic brain injury and multiple sclerosis, often in the context of normal standard MRI scans [21, 22]

MRI Quantitative atrophy estimation

Different patterns of the major dementia syndromes [23]

MRI: perfusion

Perfusion as a reflection of hypometabolism, similar to SPECT (perfusion) and PET (metabolism) patterns of abnormality

MR spectroscopy

Biochemical analysis of N-acetylaspartate (NAA), choline, lactate particularly useful in brain tumor diagnosis

(b) SPECT

Hypoperfusion (occurs in vascular compromise or due to hypometabolism)

Hyperperfusion (focal hyperperfusion may occur with ictal foci)

(c) PET brain

Differing hypometabolic patterns may be discerned in the principal dementia syndromes. FDG PET increases diagnostic accuracy beyond that derived from clinical evaluation [24, 25]

MR perfusion scanning gives similar information to PET brain scanning and being based on MRI techniques, lack of radiation may give this modality preference in the near future [15].

Neurotransmitter and Neurotransmitter Receptor PET

In AD for example, cholinergic (nicotinic receptors) and dopaminergic systems measurements have revealed increased ^{11}C nicotinic binding sites associated with cognitive improvement after

Table 13.2 PET brain patterns in dementias (Fig. 13.1)

Dementia subtype	[18]FDG PET hypometabolism pattern
Alzheimer	Relatively symmetric parietotemporal, medial temporal, posterior cingulate, frontal association cortex to lesser degree
AD variant (PCAS)	Occipital hypometabolism predominates
FTD behavioral variant	Frontal and anterior temporal hypometabolism
PDD	Temporoparietal, may be similar to AD
DLBD	Occipital and temporal hypometabolism
CVD	Cortical and subcortical, singular or multifocal, correlating with structural imaging abnormality
CBD	Global reduction in metabolism as well as asymmetric prefrontal, premotor, sensorimotor superior temporal, parietal hypometabolism with thalamic hypometabolism contralateral to limb apraxia
Huntington's	Caudate nucleus hypometabolism, frontal association cortex to a lesser degree
PSP	Caudate nucleus, putamen, thalamus, pons, superior and anterior frontal cortex

PCAS posterior cortical atrophy syndrome, *PSP* progressive supranuclear palsy

Table 13.3 Intrinsic connectivity network patterns in dementias [20, 28, 29]

Dementia subtype	Intrinsic connectivity pattern
Alzheimer	Default mode network shows reduced connectivity
FTD behavioral variant	Salience network shows reduced connectivity
Parkinson's	BN-thalamocortical loops show increased connectivity
DLBD	Uncertain at present but may show ascending brainstem projection system
CBG	Uncertain

FTD frontotemporal lobe disorder, *DLBD* diffuse Lewy body disease, *BN* basal nuclei (basal ganglia), *CBG* cortico-basal-ganglionic disorder

rivastigmine for 3 months [26]. In Parkinson's using ^{11}C methyl-4-piperidyl acetate (MP4A), dopaminergic system imaging with ^{18}F fluoro-dopa (FDOPA) showed decreased uptake in the striatum [27].

Table 13.4 Cellular and molecular responses to brain injury (modified from Cramer) [35]

Cellular
Increased angiogenesis
Increased synaptogenesis
Increased dendritic branching and spine density
Increased neuronal sprouting
Receptor
GABA downregulation
Increased *N*-methyl-ᴅ-aspartate receptor binding
Molecular
Increased growth factors
Increased cell cycle proteins
Increased growth associate proteins
Increased inflammatory markers
Hyperexcitability with long-term potentiation facilitation
Future treatment strategies have been proposed for the stroke model but these may be equally applicable to other brain pathologies

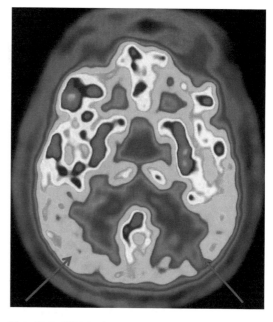

Fig. 13.1 PET brain metabolic imaging of occipital hypometabolism with progressive Lewy body disease (*arrows*)

Resting State Network (RSN), or Intrinsic Connectivity Networks (ICN) Imaging

The Default Mode Network (DMN) for example reflects the basal or default mode activity of the brain (without activation procedures). Regions metabolically active include the posterior cingulate, the precuneus, lateral parietal, lateral temporal, and medial frontal areas. Hence the DMN is active during rest and becomes less active during task engagement. DMN connectivity disruption has been documented in AD, FTD, epilepsy, autism, schizophrenia, and depression. Interestingly the distribution of the DMN impairment is similar to the fibrillar amyloid deposition seen with AD (amyloid PET scanning). DMN disruption was accurate in identifying major depression with a 94 % correct classification with the amygdala, anterior cingulate cortex, parahippocampal gyrus, and hippocampus exhibiting the highest discriminative power in classifying major depression (Fig. 13.2). Using functional connectivity MRI of the DMN and other RSN such the salience (for FTD) as well attentional networks allows RSN patterns to differentiate AD and FTD [17–19].

Intrinsic State Connectivity Networks (Fig. 13.3) [19]

Default Mode
Salience Network
Attentional network
Visual network
Auditory network

Quantitative Electroencephalography (Q-EEG) and Magnetoencephalography (MEG)

Mostly experimental, both modalities have good temporal rather than spatial resolution. Q-EEG allows pattern analysis in Alzheimer's disease with reduced alpha and beta connectivity in frontoparietal and frontotemporal regions. In Parkinson's, increased connectivity of alpha and beta frequencies is evident both locally and globally, whereas in Diffuse Lewy Body Dementia a reduced connectivity in alpha range locally and globally has been documented. MEG improves spatial resolution to

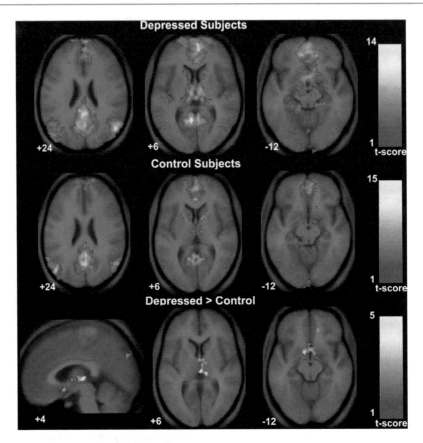

Fig. 13.2 Resting state network abnormalities in depression. The networks with the most discriminating functional connections were located within the default mode network (DMN), affective network, visual cortical areas, and the cerebellum. The figure shows increased functional connectivity in the DMN network in depressed versus control patients. Figure with permission: Greicius MD, Flores BH, Menon V et al. Resting-State Functional Connectivity in Major Depression: Abnormally Increased Contributions from Subgenual Cingulate Cortex and Thalamus. Biol Psychiatry 2007;62:429–437

~5 mm, with less artifact interference, but disadvantages are the requirement of shielded rooms to record the extremely weak magnetic fields measured in femtoteslas (10^{-15} T) [30, 31].

An Approach to the Evaluation of Cognitive and Behavioral Impairment in Patients

Some brain pathological processes may take the form of transient neurochemical aberrations, synaptopathies or network-opathies. These are typi-cally associated with normal standard MRI multimodality imaging. Functional scanning with ICA, DTI, or the various PET scan modalities are more likely to yield objective and quantifiable changes that can be monitored and tracked for improvement or worsening. Receptor imaging, beta amyloid imaging, and CSF assays form part of this group of imaging and investigative approaches. As pathology tends towards macroscopic appreciation, such as stroke, brain tumor, brain hemorrhage, demyelinating diseases, hydrocephalus, and atrophy, standard MRI sequences are more appropriate (Fig. 13.4).

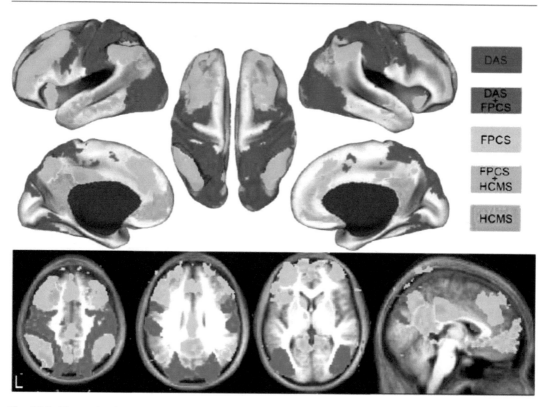

Fig. 13.3 The major intrinsic connectivity networks: Dorsal attention system (DAS; *blue*), frontoparietal control salience (FPCS; *light green*), default mode network (DMN; *orange*) systems, overlap of salience and DMN (*dark green*) and DAS and salience (*red*). Figure with permission: Vincent JL, Kahn I, Snyder AC et al. Evidence for a Frontoparietal Control System Revealed by Intrinsic Functional Connectivity. J Neurophysiol. 2008;100(6):3328–3342

Fig. 13.4
Neuroimaging approach in cognitive and behavioral disorders

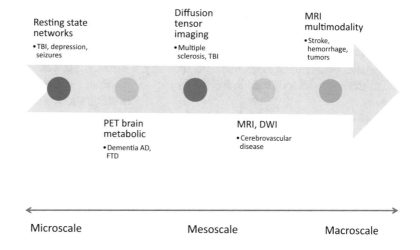

Treatment and Management Perspectives

There are many different patho-mechanisms associated with brain injury or lesions and an understanding of these may lead to avenues of improved brain function after injury. In the endeavor of promoting improvement after brain injury, consideration needs to be given to:

1. Augmenting and supporting mechanisms of spontaneous recovery
2. Avoiding interventions (particularly medications) that may worsen the condition
3. Pharmacotherapy—mainly the ascending monoaminergic systems
4. Behavioral therapies—using the top down influence of the prefrontal cortex
5. Overcoming inhibitory influences post-injury (mirror visual feedback type therapy)

Pharmacotherapy is mainly concerned with the neuromodulatory systems, which are mainly concerned with adjusting signal to noise ratios and so influence processing [32]. Neuromodulation may be associated with augmenting, diminishing, or prolonging signaling in neuronal networks. There is also a top down regulatory control over the ascending modulatory systems from the prefrontal cortex (PFC) to the brainstem neuronal cells groups of NE, DA, 5HT, and Ach for example [33, 34].

Information gleaned mostly from animal models has revealed the cellular and molecular responses to brain injury. Currently known processes that are involved in spontaneous recovery in the stroke model for example include:

Basic Science Evidence from Animal and Human Studies of Treatment Effects on Core Frontal Functions

An important principal revealed by basic science animal models has been the realization that neurotransmitter systems function in phasic as well as tonic modes [37]. This applies to the modulatory

ascending systems and correlate with the concept of a bell-shaped curve or the Yerkes-Dodson inverted U-shaped response seen in animals and humans. This psychological concept relates to the task performance, with the horizontal axis representing level of arousal and vertical axis representing a person's particular performance with the peak or top of the bell being the site of optimal performance [38]. This implies that the level of monoaminergic function optimal for one particular task may not be so for another task and may be either sub- or supraoptimal [39]. For example, different levels of norepinephrine (NE) may affect different NE receptors, with moderate levels of NE release affecting high affinity alpha 2 A receptors and even higher NE levels as encountered during stress, for example, involve alpha 1 adrenoreceptors and Beta adrenoreceptors [40]. These receptors have opposing functions in the PFC, the former improving and the latter impairing PFC function. The same applies to D1 and D2 receptors where different levels of presynaptic dopamine levels may either improve components of cognition or impair others [41]. Both NE and DA are also considered to have complementary actions affecting cognition function in the PFC as has been reported with respect to spatial WM function for example [42]. The specific mechanisms of the monoamines of regulation of working memory have implicated the HCN channel (hyperpolarization activated cyclic nucleotide gate cation channels). These are localized on the heads and necks of dendritic spines near incoming synapses in the superficial layers of monkey PFC. These layers form the cortico-cortical networks [43]. The other functions of the monoamines on the PFC include excitatory and neuroplasticity effects [44, 45]. The NE component has been associated with sustained attention when in its phasic mode and distractibility in its tonic mode in nonhuman primates performing a go-no-go visual attentional task. In addition, single unit recordings of the locus coeruleus, that were associated with optimal performance on a gono-go visual target detection paradigm in rhesus monkeys, was correlated with phasic firing of NE cells [46].

The Contribution of Genetics to Potential Treatment of Neuropsychiatric States

These include alterations in gene encoding molecules associated with glutamate signaling, cortical development, and the ascending mono-aminergic systems.

DISC1 (Disrupted in Schizophrenia 1): Major Susceptibility Gene for Mental Illness Including Schizophrenia, Bipolar Disease, and Depression

DISC1 interacts with phosphodiesterase 4B (PDE4B) and impaired DISC1 function likely leads to overactivation of cAMP-HCN signaling and weakening of the PFC network connections [47].

RGS4 (Regulator of G-Protein Signaling 4)

One of the regulatory proteins acting as GTP-ase activating proteins that drive G alpha subunits into inactive GDP form, decreasing their activity. RGS4 inhibits Gq and Gi signaling. Reduced RGS4 leads to an excess of PKC signaling and impaired PFC cognitive function [48].

DGKH (Diacyl Glycerol Kinase Isoenzyme)

One of the lipid kinases catalyzing conversion of diacylglycerol (DAG) to phosphatidic acid with an overall reduction in DAG which is a cofactor in the activation of PKC. Loss of DGKH leads to an increase in PKC signaling and mutations are linked to bipolar disease [49] and mania [50]. Of note is that the treatments with valproate, lithium, and tamoxifen inhibit PKC signaling [51].

Clinical Studies from Case Reports, Case Series of Animal and Human Data

DLPFC Syndrome and Dopaminergic and Noradrenergic Therapy

Using executive function tasks such as Word fluency and Trails tests, these have been associated with a positive response to clonidine in schizophrenia and Korsakoff's patients [52, 53]. Medications augmenting DA and NE systems have been shown to improve executive function in Tourette's syndrome and attention deficit hyperactivity disorder. A number of medications have been used including tricyclic antidepressants, guanfacine, clonidine, and deprenyl [54, 55]. In another example, idazoxan has been correlated with improved executive function in a frontotemporal lobe degeneration patient [56].

The Anterior Cingulate Syndrome of Apathy and Akinetic Mutism

Based on animal and human data, dopamine agonists such as apomorphine and bromocriptine appear effective in the treatment of the condition akinetic mutism spectrum of syndromes for example. On the other hand, presynaptic dopaminergic agents (methylphenidate, carbidopa/levodopa) seem ineffectual [57–59]. Midbrain infarction with damage to the dopaminergic neurons also causes akinetic mutism and is responsive to DA agonists [60]. In patients where there is damage to the anterior cingulate gyrus with DA receptor damage, however, it has been speculated that DA agonists may fail. Hence loss of dopaminergic input from cortical structures such as the anterior cingulate gyrus as opposed to the striatum may be a factor in determining the type of DA treatment. However other clinical studies in patients with apathy (as a component of various psychiatric conditions, stroke, Wilson's disease,

HIV dementia), have revealed benefit from an array of dopaminergic medications, including bromocriptine, amantadine, methylphenidate, bupropion, and selegiline [61–65].

Medial Orbitofrontal Syndrome, Disinhibition, and Behavioral Abnormalities: Serotonergic Agents, Serenics (Anti-aggressive Therapy Agents), and Some Antiepileptic Drugs May Be Beneficial

Behavioral disinhibition correlates with a central serotonergic deficiency [66] and serotonergic treatment has been reported beneficial in treatment of aggression. One hypothesis regarding aggression is that it may be due to a downregulation of 5-HT2 receptors in the striatum and nucleus accumbens where they occur in abundance [67, 68]. Serotonergics such as fluoxetine and clomipramine may also be useful in disinhibited, impulsive, and aggressive behavior that may be present despite normal or near normal neuropsychological test results [69]. Serenics (5HT 1A agonists) that bind to postsynaptic 5HT 1A receptors have been successful in treating aggression in animals. Propranolol, pindolol, and buspirone are examples [70]. In behavioral syndromes that may include mania and noradrenaline overactivity, adrenergic therapy may be beneficial in certain syndromes associated with bilateral inferior orbitofrontal contusions and respond to clonidine [71]. Other agents that have been useful include carbamazepine, sodium valproate, propranolol, clonidine, and lithium [72].

Lateral Orbitofrontal Syndrome and Obsessive Compulsive Disorder (OCD)

Aided recently by improved neuroimaging, namely functional neuroimaging studies, these have delineated increased activity, either metabolism or blood flow related, in the orbitofrontal cortex, the head of caudate nucleus, and anterior cingulate gyrus [73]. In general, serotonin reuptake inhibitors as well as clomipramine have been the most advocated pharmacotherapeutic approaches and right caudate head glucose metabolism (PET brain scan) was reduced with successful fluoxetine therapy for obsessive compulsive disorder (OCD) [74]. The interaction in cerebral DA and 5 HT may account for the improvement seen in some OCD with neuroleptics [75]. In cases where hypometabolism (PET brain scan) occurs in the anterior cingulate region and right OFC, this too has been correlated with an improved response to clomipramine therapy [76]. In refractory cases sumatriptan (5HT 1D agonist) has improved both depression and OCD [77]. Both cognitive behavioral therapy and SSRI therapy have been shown effective in OCD treatment and their combination potentiated [78].

Randomized Controlled Trials: Human Studies

Amantadine and Severe TBI

In a landmark international study, randomized, double-blinded, placebo-controlled trial of inpatient rehabilitation patients ($n = 184$) with minimally conscious or vegetative state were given amantadine 100 mg bid and increased to 200 mg bid by week 4. Outcome determined by the disability rating scale (DRS) and recovery was faster in the amantadine group as recorded by 0.24 units difference in the DRS per week over the period week 4 to week 16, in the DRS ($p = 0.007$). With discontinuation of amantadine there was loss of function. The beneficial effects of amantadine were attributed to presynaptic release facilitation and postsynaptic reuptake blockade, thereby augmenting dopaminergic transmission in the mesolimbic, nigrostriatal, and mesocortical circuitry that subserve attention, conation, and arousal [79].

Methylphenidate and Moderate to Severe Traumatic Brain Injury (TBI)

In one of the few randomized controlled trials, the core frontal component of attention was found to be significantly improved, the surrogate of speed of information processing in 40 participants with moderate to severe TBI receiving methylphenidate at a dose of 0.3 mg/kg twice daily [80].

Trazodone and Frontotemporal Lobe Disorder (FTLD)

In a meta-analysis, FTLD patients were presumed to have predominant serotonergic deficit as well as dopaminergic deficit with little evidence for Ach and NE related impairment. A double blinded, placebo controlled, cross over trial of trazodone with 300 mg daily revealed a significant improvement using the neuropsychiatric inventory score. Trazodone is a selective serotonin reuptake inhibitor (SSRI), with agonist function of 5HT1A, 5HT1C, 5HT2 receptors. The active metabolite is a direct serotonin receptor agonist, as well as an adrenergic (alpha 1, alpha 2) and histamine (H1) blocking agent. The effects were noted in the domains of behavior rather than cognition [81, 82].

Serotonergic Therapy and Stroke (Motor Deficit)

In the FLAME (Fluoxetine for motor recovery after acute ischemic stroke) trial patients with ischemic stroke and moderate to severe motor deficit, the early use of fluoxetine in combination with physiotherapy enhanced motor recovery at 3 months evaluation. The mechanism of action is suggested to be a modulation of spontaneous brain plasticity by drugs attributed to brain-derived neurotrophic factor [83].

Neuropsychiatric Component Treatment

Partly preempted by the advent of the new DSM-V criteria, there is increased emphasis toward a neurobiological models of disease, a renewed neuroimaging focus and using dimensional scales as opposed to categorical diagnoses only (DSM-IV R) on traditional neuropsychiatric conditions [84]. The neuropsychiatric disorders include a diverse collection of syndromes affecting behavior, emotion, executive function, and other core frontal network functions that primarily affect emotion, executive function, higher functions and their circuitry. With a distinct paucity of biomarkers for these syndromes, similar to the approach in neurology, it seems essential to integrate basic neuroscience, neurogenetics, epigenetics, and neuroradiology in order to establish a foundation for diagnostic based on pathophysiology and presumed etiology. The current psychiatric classifications (DSM IV) have had the effect of dichotomizing disease when they are better configured as dimensional traits that often overlap with normality. This latter premise is in keeping with the polygenic mode of inheritance.

Many patients, diagnosed according to the DSM IV or DSM V, receive multiple diagnoses that are termed comorbidities, probably reflective of a diagnostic artifact due to symptom splitting and lumping. The true underlying neurobiological process may, however, be due to a single pathophysiological entity. Furthermore, current psychiatric medications do not respect DSM-IV/V defined boundaries and disorders. Both antipsychotic and antidepressant agents for example are used to treat many different psychiatric disorders [85].

Thinking in terms of symptom clusters and the core components of the frontal subcortical network circuitry may help construct a more neurobiological and pathophysiological relevant approach. This does however combine neurological, cerebrovascular, psychiatric, neuropsychological, general medical, and neuroradiological information and consequently the disciplines to advantage (Tables 13.5 and 13.6).

Table 13.5 Future treatment strategies proposed for the stroke model (modified from Cramer) [36]

1. Small molecules (monoaminergic systems, antibodies against axonal growth inhibitor Nogo-A)
2. Growth factors (fibroblast growth factor, brain-derived neurotrophic factor, hematopoietic growth factor, granulocyte colony stimulating factor)
3. Cell-based therapies (endothelial progenitor cells, intracerebral transplantation of cultured neuronal cells, intravenous mesenchymal stromal cells)
4. Electromagnetic stimulation
5. Device-based therapies
6. Task orientated and repetitive training-based interventions

Table 13.6 Proposed approach of neurological/neuro-psychiatric disorders

1. Use the list of symptoms and signs to form an overall generic categorical diagnostic syndromes such as abulic/apathetic, disinhibited/dysexecutive, depression, and obsessive compulsive disorders. This is a clinical assessment that may be aided by inventories, scales, or diagnostic manuals such as DSM V
2. Component analysis in terms of the core frontal functions embedded in the five currently appreciated frontal subcortical behavioral circuits
3. Establish the cerebrovascular component and its specific treatment
4. Establish whether medical conditions (hypothyroidism, low B12, Vit D, Folate) are contributing to the cognitive impairment
5. Establish contribution of impaired sleep (sleep apnea, dyssomnia)
6. Establish contribution of centrally acting drugs and discontinue, reduce dosage, or change to another less conflicting drug if possible
7. Use known information about neurotransmitter deficiencies in these syndromes and/or FSCs and target with specific pharmacological and behavioral treatment approaches
8. Counsel wrt the five principal components of brain health

Depression

Serotonergic agents, electroconvulsive therapy, transcranial magnetic stimulation physical exercise, and cognitive behavioral therapy have all been shown to benefit major depressive disorder [86–88]. An important study using PET brain scanning before and after treatment with cognitive behavioral therapy (CBT) and a comparison group with the serotonergic agent paroxetine revealed changes in brain metabolism. The CBT group had increased activity in the hippocampal and dorsal cingulum and decreases in the frontal cortex whereas the paroxetine group had increased metabolism in the prefrontal cortex and decreased activity in the subgenual cingulate and hippocampus. These findings have been interpreted to suggest that CBT has a top down effect on the medial frontal and cingulate cortex and the pharmacological group work in a bottom up manner [89, 90]. Interpersonal psychotherapy (ITP) compared to venlafaxine similarly showed activation of the right posterior cingulate and right basal ganglia and of the right posterior temporal lobe and right basal ganglia in the venlafaxine group, this time assessed by SPECT scans [91]. Similarly the psychosurgical treatment, namely anterior cingulotomy, reserved for severe treatment resistant depression revealed a decrease of metabolism measured by PET brain scanning in the left subgenual PFC and left posterior thalamus, from the preoperative values [92].

Attention Deficit Hyperactivity Disorder

Stimulants such as methylphenidate increase NE and to a lesser extent DA in the PFC, while producing lesser effects in the subcortical regions. Atomoxetine also increases NE and DA, with less effect on striatal DA, and may have a beneficial effect on impulsivity as well. Atomoxetine is an important new treatment option for adults with ADHD and is particularly so for those who are at risk for substance abuse. Atomoxetine is effective and well tolerated and the first ADHD treatment approved specifically for adult use, administered as a single daily dose and is not a controlled substance [93].

Emotional Component Treatment in Relation to Neurological or Psychiatric Conditions

Involuntary emotional expression disorders (IEED) is a more frequently diagnosed condition especially post-stroke, traumatic brain injury, multiple sclerosis, and neurodegenerative diseases. Recently, pharmacological treatment has been successfully demonstrated with the efficacy of the dextromethorphan-quinidine combination (Nuedexta) [94, 95]. There are also behavioral programs that endeavor to improve one's emotional responses. Although in its infancy, one example of a regimen to improve one's emotional style, comprising, outlook, self-awareness, attention, resilience, social intuition, and sensitivity to context, has been detailed by Davidson and Begley [96–98].

Task Orientated and Repetitive Training Based Interventions

Currently there is support for aphasia therapy, attentional training, rehabilitation of unilateral spatial neglect, and compensatory strategies for apraxia. The following is a summary, using information from randomized controlled trials, case series and single case reports to help classify recommendations (A, B, and C) for several different forms of cognitive rehabilitation modalities.

Randomized Controlled Trials

Attention post-TBI—attentional training—improvement
Apraxia—apraxia training—improvement

Systematic Reviews

Aphasia—intensive treatment. Improvement based on systemic reviews
Neglect post-stroke—visual scanning and visuospatial motor training

Attention disorders post-stroke—attention task improvement
Memory post-stroke—errorless learning (electronic aids)—improvement [99]

Constraint Therapy

This rationale of this mode of therapy proposes that in some patients there is a kind of learned nonuse of the paretic or paralyzed hand or arm after stroke or TBI. Physical therapy is applied to this limb while the unaffected limb is deliberately restrained. A phase III trial of constraint therapy of 2 weeks duration, resulted in significant gains that endured for approximately 2 years [100]. This was subsequently analyzed further, comparing early (3–9 months) and later (15–21 months) initiation of CIMT after stroke. The outcome was notable for both groups achieving similar level of significant arm motor function 24 months after enrollment [101].

Devices

Transcranial Magnetic Stimulation (TMS)

Depending on the pulse frequency, either hypofunction or improved function may ensue due to the inhibitory or excitatory effects on cortical function. This may have application in those instances where inhibitory cortical circuits are operative and if diminished function may return. A randomized trial of treatment resistant depression with TMS has established this as a therapeutic component [102].

Mirror Visual Feedback Therapy (MVF)

In controlled case series studies, MVF has been effective in treating post-stroke paresis (arm or leg), phantom limb pain, complex regional pain syndrome, and anxiety. There is also evidence that the modality can modulate pain and reverse objective signs such as inflammation and paralysis [103, 104]. Both arm and leg paresis have been studied post-stroke. In patients with arm paresis, that were treated with MVF (*n* = 17, controls *n* = 19) significant improvement scoers were reported for hand FIM and arm Brunnstrom scores at baseline, 4 weeks and at 6 months [105]. In a study of 40 patients with leg weakness post-stroke, compared to best rehabilitation therapy and placebo controlled with opaque glass, a statistically significant improvement was documented. Proposed postulated mechanisms included visuomotor tract restoration leading to an "unlearning of the learned paralysis" component after stroke [106]. This mechanism may also be attributed to a function of the mirror neuron system that involves interactions between the motor, vision, and proprioception modalities. Limb weakness after stroke may be related to both fiber tract damage and so-called learned paralysis whereby neurons and their fiber tracts are inhibited and that this can be unlearned using a mirror [107].

Mirror Neuron Therapy and Rehabilitation

Also called action observation treatment, this therapy is based on the premise that circuits are activated by observation, similar to those that perform the movement. Initial results from observational studies as an add-on therapy appear promising [108].

References

1. Foster NL, Heidebrink JL, Clark CM, Jagust WJ, Arnold SE, Barbas NR, et al. FDG PET improves accuracy in distinguishing frontotemporal dementia and Alzheimer's disease. Brain. 2007;130:2616–35.
2. Berti V, Pupi A, Mosconi L. PET/CT in diagnosis of dementia. Ann N Y Acad Sci. 2011;1288:81–92.
3. Migliaccio R, Agosta F, Rascovsky K, Karydas A, Bonasera S, Rabinovici GD, et al. Clinical syndromes associated with posterior atrophy: early age at onset AD spectrum. Neurology. 2009;73:1571–8.
4. Stern Y. Cognitive reserve. Alzheimer Dis Assoc Disord. 2006;20:112–7.
5. Robinson ME, Craggs JG, Price DD, Perlstein WM, Staud R. Gray matter volumes of pain-related brain areas are decreased in fibromyalgia syndrome. J Pain. 2011;12(4):436–43.
6. Obermann M, Nebel K, Schumann C, Holle D, Gizewski ER, Maschke M, et al. Gray matter changes related to chronic posttraumatic headache. Neurology. 2009;73(12):978–83.
7. Geha PY, Baliki MN, Harden RN, Bauer WR, Parrish TB, Apkarian AV. The brain in chronic CRPS pain: abnormal gray-white matter interactions in emotional and autonomic regions. Neuron. 2008;60(4):570–81.
8. Apkarian AV, Sosa Y, Sonty S, Levy RM, Harden RN, Parrish TB, et al. Chronic back pain is associated with decreased prefrontal and thalamic gray matter density. J Neurosci. 2004;24(46):10410–5.
9. Rainville P, Duncan GH, Price DD, Carrier B, Bushnell MC. Pain affect encoded in human anterior cingulate but not somatosensory cortex. Science. 1997;277:968–71.
10. Coghill RC, Sang CN, Maisog JM, Iadarola MJ. Pain intensity processing within the human brain: a bilateral, distributed mechanism. J Neurophysiol. 1999;82:1934–43.
11. Casey KL. Concepts of pain mechanisms: the contribution of functional imaging of the human brain. Prog Brain Res. 2000;129:277–87.
12. Kwan CL, Crawley AP, Mikulis DJ, Davis KD. An fMRI study of the anterior cingulate cortex and surrounding medial wall activations evoked by noxious cutaneous heat and cold stimuli. Pain. 2000;85: 359–74.
13. Sawamoto N, Honda M, Okada T, Hanakawa T, Kanda M, Fukuyama H, et al. Expectation of pain enhances responses to nonpainful somatosensory stimulation in the anterior cingulate cortex and parietal operculum/posterior insula: an event-related functional magnetic resonance imaging study. J Neurosci. 2000;20:7438–45.
14. Nielsen FA, Balslev D, Hansen LK. Mining the posterior cingulate: segregation between memory and pain components. Neuroimage. 2005;27(3):520–32.
15. Hu WT, Wang Z, Lee VM, Trojanowski JQ, Detre JA, Grossman M. Distinct cerebral perfusion patterns in FTLD and AD. Neurology. 2010;75:881–8.
16. Small SA, Schobel SA, Buxton RB, Witter MR, Barnes CA. A pathophysiological framework of hippocampal dysfunction in ageing and disease. Nat Rev Neurosci. 2011;12:585–601.
17. Petrella JR, Sheldon FC, Prince SE, Calhoun VD, Doraiswamy PM. Default mode network connectivity in stable vs progressive mild cognitive impairment. Neurology. 2011;76:511–7.
18. Zeng L, Shen H, Liu L, Wang L, Li B, Fang P, et al. Identifying major depression using whole brain functional connectivity: a multivariate pattern analysis. Brain. 2012;135:1498–507.
19. Seeley WW, Menon V, Schatzberg AF, Keller J, Glover GH, Kenna H, et al. Dissociable intrinsic

connectivity networks for salience processing and executive control. J Neurosci. 2007;27:2349–56.

20. Standley K, Brock C, Hoffmann M. Advances in functional neuroimaging in dementias and potential pitfalls. Neurol Int. 2012;4(1), e7. doi:10.4081/ni.2012.e7.

21. Hayes JP, Miller DR, Lafleche G, Salat DH, Verfaellie M. The nature of white matter abnormalities in blast-related mild traumatic brain injury. Neuroimage Clin. 2015;8:148–56.

22. Koenig KA, Sakaie KE, Lowe MJ, Lin J, Stone L, Bermel RA, et al. The relationship between cognitive function and high-resolution diffusion tensor MRI of the cingulum bundle in multiple sclerosis. Mult Scler. 2015;pii:1352458515576983.

23. Ossenkoppele R, Cohn-Sheehy BI, La Joie R, Vogel JW, Möller C, Lehmann M, et al. Atrophy patterns in early clinical stages across distinct phenotypes of Alzheimer's disease. Hum Brain Mapp. 2015; 36(11):4421–37.

24. Barthel H, Schroeter ML, Hoffmann KT, Sabri O. PET/MR in dementia and other neurodegenerative diseases. Semin Nucl Med. 2015;45(3):224–33. doi:10.1053/j.semnuclmed.2014.12.003. Review.

25. Morbelli S, Garibotto V, Van De Giessen E, Arbizu J, Chételat G, Drezgza A, et al. A cochrane review on brain [18F]FDG PET in dementia: limitations and future perspectives. European Association of Nuclear Medicine. Eur J Nucl Med Mol Imaging. 2015;42(10):1487–91.

26. Kadir A, Darreh-Shori T, Almkvist O, Wall A, Långström B, Nordberg A. Changes in brain 11C nicotine binding sites in patients with mild Alzheimer's disease following rivastigmine treatment as assessed by PET. Psychopharmacology (Berl). 2007;191:1005–14.

27. Hilker R, Thomas AV, Klein JC, Weisenbach S, Kalbe E, Burghaus L, et al. Dementia in Parkinson disease: functional imaging of cholinergic and dopaminergic pathways. Neurology. 2005;65:1716–22.

28. Zhou J, Greicius MD, Gennatas ED, Growdon ME, Jang JY, Rabinovici GD, et al. Divergent network connectivity changes in behavioural variant frontotemporal dementia and Alzheimer's disease. Brain. 2010;133:1352–67.

29. La Joie R, Landeau B, Perrotin A, Bejanin A, Egret S, Pélerin A, et al. Intrinsic connectivity identifies the hippocampus as a main crossroad between Alzheimer's and semantic dementia-targeted networks. Neuron. 2014;81(6):1417–28.

30. Liang Y, Chen Y, Li H, Zhao T, Sun X, Shu N, et al. Disrupted functional connectivity related to differential degeneration of the cingulum bundle in mild cognitive impairment patients. Curr Alzheimer Res. 2015;12(3):255–65.

31. Garn H, Waser M, Deistler M, Benke T, Dal-Bianco P, Ransmayr G, et al. Quantitative EEG markers relate to Alzheimer's disease severity in the Prospective Dementia Registry Austria (PRODEM). Clin Neurophysiol. 2015;126(3):505–13.

32. Sarter M, Hasselmo ME, Bruno JP, Givens B. Unraveling the attentional functions of cortical cholinergic inputs; interactions between signal driven and cognitive modulation of signal detection. Brain Res Rev. 2005;48:98–111.

33. Sara SJ, Herve-Minvielle A. Inhibitory influences of frontal cortex on locus coeruleus. Proc Natl Acad Sci U S A. 1995;92:6032–6.

34. Amat J, Baratta MV, Paul E, Bland ST, Watkins LR, Maier SF. Medical prefrontal cortex determines how stressor controllability affects behavior and dorsal raphe nucleus. Nat Neurosci. 2005;8:365–71.

35. Cramer SC. Repairing the human brain after stroke: 1. Mechanisms of spontaneous recovery. Ann Neurol. 2008;63:272–87.

36. Cramer SC. Repairing the human brain after stroke. II Restorative therapies. Ann Neurol. 2008;63:549–60.

37. Robbins TW, Arnsten AFT. The neuropsychopharmacology of fronto-executive function: monoaminergic modulation. Annu Rev Neurosci. 2009;32:267–87. doi:10.1146/annurev.neuro.051508.135535.

38. Yerkes RM, Dodson JD. The relation of strength of stimulus to rapidity of habit formation. J Comp Neurol Psychol. 1908;18:459–82.

39. Arnsten AFT. Through the looking glass: differential noradrenergic modulation of the prefrontal cortical function. Neural Plast. 2000;7:133–46.

40. Floresco SB, Magyar O, Ghods-Sharifi S, Vexelman C, Tse MT. Multiple dopamine receptor subtypes in the medial prefrontal cortex of the rat regulate set shifting. Neuropsychopharmacology. 2006;31:297–309.

41. Floresco SB, Magyar O. Mesocortical dopamine modulation of executive functions: beyond working memory. Psychopharmacology (Berl). 2006;188: 5670585.

42. Li B-M, Mei Z-T. Delayed response deficit induced by local injection of the alpha 2 adrenergic antagonist yohimbine into the dorsolateral prefrontal cortex in young adult monkeys. Behav Neural Biol. 1994;62:134–9.

43. Wang M, Ramos B, Paspalas C, Shu Y, Simen A, Duque A, et al. Alpha 2A adrenoreceptor stimulation strengthens working memory networks by inhibiting cAMP-HCN channel signaling in the prefrontal cortex. Cell. 2007;129:397–410.

44. Seamans JK, Robbins TW. Dopamine modulation of the prefrontal cortex and cognition function. In: Neve K, editor. Dopamine receptors. Totowa, NJ: Humana; 2009.

45. Seamns JK, Durstewitz D, Christie BR, Stevens CF, Sejnowski TJ. Dopamine D1/D5 receptor modulation of excitatory synaptic inputs to layer V prefrontal cortical neurons. Proc Natl Acad Sci U S A. 2001;98:301–6.

46. Aston-Jones G, Cohen JD. An integrative theory of locus coeruleus-norepinephrine function: adaptive gain and optimal performance. Annu Rev Neurosci. 2005;28:403–50.

47. Millar JK, Mackie S, Clapcote SJ, Murdoch H, Pickard BS, Christie S, et al. Disrupted in schizo-

phrenia 1 and phosphodiesterase 4B: towards an understanding of psychiatric illness. J Physiol. 2007;584:401–5.

48. Mirnics K, Middleton FA, Stanwood GD, Lewis DA, Levitt P. Disease specific changes in regulator of G-protein signaling 4 (RGS4) expression in schizophrenia. Mol Psychiatry. 2001;6:293–301.

49. Baum AE, Ajula N, Cabanero M, Cardon I, Corona W, Klemens B, et al. A genome wide association study implicates diacylglycerol kinase eta (DGKH) and several other genes in the etiology of bipolar disorder. Mol Psychiatry. 2008;13:197–207.

50. Manji HK, Lenox RH. Protein kinase C signaling in the brain: molecular transduction of mood stabilization in the treatment of manic depressive illness. Biol Psychiatry. 1999;46:1328–51.

51. Yildiz A, Guleryuz S, Ankerst DP, Ongur D, Renshaw PF. Protein kinase C inhibition in the treatment of mania: a double blind, placebo controlled trial of tamoxifen. Arch Gen Psychiatry. 2008;65:255–63.

52. Fields RB, Van Kammen DP, Peters JL, Rosen K, Van Kammen WB, Nugent A, et al. Clonidine improves memory function in schizophrenia independently from change in psychosis. Schizophr Res. 1988;1:417–23.

53. Mair RG, McEntee WJ. Cognitive enhancement in Korsakoff's psychosis by clonidine: A comparison with L-Dopa and ephedrine. Psychopharmacology. 1986;88:374–80.

54. Arnsten AFT, Goldman Rakic PS. Alpha 2 adrenergic mechanism in prefrontal cortex associated with cognitive decline in aged non human primates. Science. 1985;230:1273–6.

55. Arnsten AFT, Steere JC, Hunt RD. The contribution of alpha 2 noradrenergic mechanisms to prefrontal cortical function: potential significance for attention deficit hyperactivity disorder. Arch Gen Psychiatry. 1995;53:448–55.

56. Sahakian BJ, Coull JJ, Hodges JR. Selective enhancement of executive function by idazoxan in a patient with dementia of the frontal lobe type. J Neurol Neurosurg Psychiatry. 1994;57:120–1.

57. Ljungberg T, Ståhle L, Ungerstedt U. Effects of repeated administration of low doses of apomorphine in three behavioural models in the rat. J Neural Transm Park Dis Dement Sect. 1989;1:165–75.

58. Ljunberg T, Ungerstedt U. Reinstatement of eating by dopamine agonists in aphagic dopamine denervated rats. Physiol Behav. 1976;16:277–83.

59. Ross ED, Stewart RM. Akinetic mutism from hypothalamic damage: successful treatment with dopamine agonists. Neurology. 1981;31:1435–9.

60. Alexander MP. Reversal of chronic akinetic mutism after mesencephalic injury with dopaminergic agents. Neurology. 1995;45:330.

61. Marin RS, Fogel BS, Hawkins J, Duffy J, Krupp B, Tolosa E, et al. Apathy: a treatable syndrome. J Neuropsychiatry Clin Neurosci. 1995;7:23–30.

62. Wantanabe MD, Martin EM, DeLeon OA, Gaviria M, Pavel DG, Trepashko DW. Successful methylphenidate treatment of apathy after subcortical infarcts. J Neuropsychiatry Clin Neurosci. 1995; 7:502–4.

63. Barrett K. Treating organic abulia with bromocriptine and lisuride: four studies. J Neurol Neurosurg Psychiatry. 1991;54:7180721.

64. Holmes VF, Fernandes F, Levy JK. Psychostimulant therapy in AIDS related complex patients. J Clin Psychiatry. 1989;50:5–8.

65. Parks RW, Crockett DJ, Manji HK, Ammann W. Assessment of bromocriptine intervention for the treatment of frontal lobe syndrome: a case study. J Neuropsychiatry Clin Neurosci. 1992;4:109–11.

66. Brown GL, Linnoila MI. CSF serotonin metabolite (5-HIAA) studies in depression, impulsivity, and violence. J Clin Psychiatry. 1990;51:31–41.

67. Coccaro EF, Siever LJ, Klar HM, Maurer G, Cochrane K, Cooper TB, et al. Serotonergic studies in patients with affective and personality disorders. Correlates with suicidal and impulsive aggressive behavior. Arch Gen Psychiatry. 1989;46(7):587–99.

68. Coccaro EF. Central serotonin and impulsive aggression. Br J Psychiatry. 1989;8:52–62.

69. Hollander E, Wong CM. Body dysmorphic disorder, pathological gambling, and sexual compulsions. J Clin Psychiatry. 1995;56:7–12.

70. Olivier B, Mos J. Serenics and aggression. Stress Med. 1986;2:197–209.

71. Bakchine S, Lacomblez L, Benoit N, Parisot D, Chain F, Lhermitte F. Manic-like state after bilateral orbitofrontal and right temporoparietal injury: efficacy of clonidine. Neurology. 1989;39:777–81.

72. Tariot PN, Schneider LS, Cummings J, Thomas RG, Raman R, Jakimovich LJ, et al. Chronic divalproex sodium to attenuate agitation and clinical progression of Alzheimer disease. Alzheimer's Disease Cooperative Study Group. Arch Gen Psychiatry. 2011;68:853–61.

73. Baxter LR, Clark EC, Iqbal M, Ackerman RF. Cortical subcortical system in the mediation of obsessive compulsive disorder. In: Lichter EG, Cummings JL, editors. Frontal subcortical circuits in psychiatric and neurological disorders. New York, NY: Guilford; 2001.

74. Baxter LR, Schwartz JM, Bergman KS, Szuba MP, Guze BH, Mazziotta JC, et al. Caudate glucose metabolic rate changes with both drug and behavior therapy for obsessive compulsive disorder. Arch Gen Psychiatry. 1992;49:681–9.

75. Wong DF, Brasić JR, Singer HS, Schretlen DJ, Kuwabara H, Zhou Y, et al. Mechanisms of dopaminergic and serotonergic neurotransmission in Tourette syndrome: clues from an in vivo neurochemistry study with PET. Neuropsychopharmacology. 2008; 33:1239–51.

76. Brody AL, Saxena S, Schwartz JM, Stoessel PW, Maidment K, Phelps ME, et al. FDG-PET predic-

tors of response to behavioral therapy and pharmacotherapy in obsessive compulsive disorder. Psychiatry Res. 1998;84:1–6.

77. Stern L, Zohar J, Cohen R, Sasson Y. Treatment of severe, drug resistant obsessive compulsive disorder with the 5HT1D agonist sumatriptan. Eur Neuropsychopharmacol. 1998;8:325–8.

78. O'Connor K, Todorov C, Robillard S, Borgeat F, Brault M. Cognitive-behaviour therapy and medication in the treatment of obsessive-compulsive disorder: a controlled study. Can J Psychiatry. 1999;44:64–71.

79. Giacino JT, Whyte J, Bagiella E, Kalmar K, Childs N, Khademi A, et al. Placebo-controlled trial of amantadine for severe traumatic brain injury. N Engl J Med. 2012;366:819–26.

80. Willmott C, Ponsford J. Efficacy of methylphenidate in the rehabilitation of attention following traumatic brain injury: a randomised, crossover, double blind, placebo controlled inpatient trial. J Neurol Neurosurg Psychiatry. 2009;80:552–7.

81. Huey E, Putnam K, Grafman J. A systematic review of neurotransmitter deficits and treatments in frontotemporal dementia. Neurology. 2006;66:17–22.

82. Lebert F, Stekke W, Hasenbroek C, Paquir F. Frontotemporal dementia: a randomized controlled trial with trazodone. Dement Geriatr Cogn Disord. 2004;17:355–9.

83. Chollet F, Tardy J, Albucher JF, Thalamas C, Berard E, Lamy C, et al. Fluoxetine for motor recovery after acute ischaemic stroke (FLAME): a randomised placebo-controlled trial. Lancet Neurol. 2011;10:123–30.

84. Hyman S. DSM IV and V and integration of neuroscience. Nat Rev Neurosci. 2007;8:725–32.

85. Carlat DJ. Unhinged the trouble with psychiatry – a doctor's revelations about a profession in crisis. New York, NY: Free Press; 2010.

86. Rimer J, Dwan K, Lawlor DA, Greig CA, McMurdo M, Morley W, et al. Exercise for depression. Cochrane Database Syst Rev. 2012;7:CD004366.

87. Lee JC, Blumberger DM, Fitzgerald P, Daskalakis Z, Levinson A. The role of transcranial magnetic stimulation in treatment-resistant depression: a review. Curr Pharm Des. 2012;18(36):5846–52.

88. Farahani A, Correll CU. Are antipsychotics or antidepressants needed for psychotic depression? A systematic review and meta-analysis of trials comparing antidepressant or antipsychotic monotherapy with combination treatment. J Clin Psychiatry. 2012;73(4):486–96.

89. Apostolova I, Block S, Buchert R, Osen B, Conradi M, Tabrizian S, et al. Effects of behavioral therapy or pharmacotherapy on brain glucose metabolism in subjects with obsessive-compulsive disorder as assessed by brain FDG PET. Psychiatry Res. 2010;184:105–16.

90. Butler AC, Chapman JE, Forman EM, Beck AT. The empirical status of cognitive behavioural therapy: a

review of meta- analyses. Clin Psychol Rev. 2006;26:17–31.

91. Martin SD, Martin E, Rai SS, Richardson MA, Royall R. Brain blood flow changes in depressed patients treated with interpersonal psychotherapy or venlafaxine hydrochloride: preliminary findings. Arch Gen Psychiatry. 2001;58:641–8.

92. Dougherty DD, Weiss AP, Cosgrove GR, Alpert NM, Cassem EH, Nierenberg AA, et al. Cerebral metabolic correlates as potential predictors of response to anterior cingulotomy for treatment of major depression. J Neurosurg. 2003;99:1010–7.

93. Chamberlain SR, Del-Campo N, Dowson J, Muller U, Clark L, Robbins TW, et al. Atomoxetine improved response inhibition in adults with attention deficit hyperactivity disorder. Biol Psychiatry. 2007;62:977–84.

94. Panitch HS, Thisted RA, Smith RA, Wynn DR, Wymer JP, Achiron A, et al. Pseudobulbar affect in multiple sclerosis study group. Ann Neurol. 2006;59:780–7.

95. Miller A, Pratt H, Schiffer RB. Pseudobulbar affect: the spectrum of clinical presentations, etiologies and treatments. Expert Rev Neurother. 2011;11:1077–88.

96. Davidson RJ, Begley S. The emotional life of your brain. New York, NY: Hudson Street Press; 2012.

97. Fava GA, Tomba E. Increasing psychological well-being and resilience by psychotherapeutic methods. J Pers. 2009;77:1903–34.

98. Hoelzel BK, Ott U, Gard T, Hempel H, Weygandt M, Morgen K, et al. Investigation of mindfulness meditation practitioners with voxel based morphometry. Soc Cogn Affect Neurosci. 2008;3:55–61.

99. Cappa SF, Benke T, Clarke S, Rossi B, Stemmer B, van Heugten CM. EFNS guidelines on cognitive rehabilitation: report of an EFNS task force. Eur Neurol. 2005;12:665–80.

100. Wolf SL, Winstein CJ, Miller JP, Taub E, Uswatte G, Morris D, et al. Effect of constraint induced movement therapy on upper extremity function 3 to 9 months after stroke: the EXCITE randomized clinical trial. JAMA. 2006;296:2095–104.

101. Wolf SL, Thompson PA, Winstein CJ, Miller JP, Blanton SR, Nichols-Larsen DS, et al. The EXCITE stroke trial: comparing early and delayed constraint induced movement therapy. Stroke. 2010;41:2309–15.

102. George MS, Lisanby SH, Avery D, McDonald WM, Durkalski V, Pavlicova M, et al. Daily left prefrontal transcranial magnetic stimulation therapy for major depressive disorder: a sham controlled randomized trial. Arch Gen Psychiatry. 2010;67:507–16.

103. Ramachandran VS, Altschuler EL. The use of visual feedback, in particular mirror visual feedback, in restoring brain function. Brain. 2009;132:1693–710.

104. McCabe CS, Haigh RC, Ring EF, Halligan PW, Wall PD, Blake DR. A controlled pilot study of the utility of mirror visual feedback in the treatment of com-

plex regional pain syndrome (type 1). Rheumatology (Oxford). 2003;42:97–101.

105. Yavuzer G, Selles R, Sezer N, Sütbeyaz S, Bussmann JB, Köseoğlu F, et al. Mirror therapy improves hand function in subacute stroke: a randomized controlled trial. Arch Phys Med Rehabil. 2008;89:393–8.

106. Sütbeyaz S, Yavuzer G, Sezer N, Koseoglu BF. Mirror therapy enhances lower-extremity motor recovery and motor functioning after stroke: a randomized controlled trial. Arch Phys Med Rehabil. 2007;88:555–9.

107. Franceschini M, Agosti M, Cantagallo A, Sale P, Mancuso M, Buccino G. Mirror neurons: action observation treatment as a tool in stroke rehabilitation. Eur J Phys Rehabil Med. 2010;46:517–23.

108. Sale P, Franceschini M. Action observation and mirror neuron network: a tool for motor stroke rehabilitation. Eur J Phys Rehabil Med. 2012;48:313–8.

Index

A

Abulia, 261, 268, 274, 279, 282, 283
Acalculias, 228
    asymbolic, 230
    study and rehabilitation, 231
    treatment, 231
Acetylcholine (Ach), 44, 258–259
Achromatopsia, 63, 64
Action observation treatment, 308
Adrenaline, 44
Aesthetics
    laws of, 236–237
Agmatine, 42, 45–46
Agnosias (apperceptive and associative), 60, 61
Agraphias
    bedside testing, 235
    central (aphasic), 233
    classification, 233
    dystypia, 235
    hypergraphia, 235
    left hemisphere language, 233
    micrographia, 235
    movement disorder associated dysgraphia, 234
    neuropsychological testing, 235
    paretic, 234
    pure/isolated, 233, 234
Akinetic Mutism
    anterior cingulate syndrome, 303–304
Akinetopsias, 67
Alexias
    central, 231
    description, 231
    dyslexia, 233
    frontal, 232
    hemialexia, 232
    hyperlexia, 232–233
    intact elementary visual function, 231
    spatial (neglect), 232
    subtype classification, 231–233
    with agraphia, 232
    without agraphia, 231, 232
Alexithymia, 268
Algodiaphoria, 152

Alien Hand and Anarchic Hand Syndrome, 270
Allesthesia, 171
Allochiria, 171
Allokinesia, 171
Alternating attention, 269
Alzheimer Disease (AD), 6, 75, 146, 282–283
Amantadine, 304
Amazonian group *vs.* French control group, 227, 229
Amgydala, 257
AMH working memory capacity, 22
Amino acids, 42
AMPA receptor, 43
Amusia, 90
    bedside testing, 242
    clinical classification, 240
    music therapy, 243
    musical variable disorders, 240–242
    neuropsychological musical testing, 242
    testing musical competence, 242
Amygdala, Occipito-Temporal Cortex, 69
Amyloid angiopathy, 148
Angular gyrus syndrome, 149, 152, 209
Animal and human studies
    anterior cingulate syndrome, 303–304
    DLPFC Syndrome, 303
    dopaminergic therapy, 303
    noradrenergic therapy, 303
    RCT, 304–306
Anomias
    callosal, 206
    category-specific, 206
    definition, 205
    disconnection, 206
    modality-specific, 206
    semantic, 205–206
    semantic anomia, 206
    word production, 205, 206
    word selection, 205, 206
Anorexia syndrome, 69
Anosodiaphoria, 170
Anosognosia
    Bisiach scale, 170
    hemiplegia, 169

© Springer International Publishing Switzerland 2016
M. Hoffmann, *Cognitive, Conative and Behavioral Neurology*,
DOI 10.1007/978-3-319-33181-2

Anterior cingulate cortex (ACC), 257, 297
 akinetic mutism, 303–304
 apathy, 303–304
Anterior or Frontal Variant Alien Hand Syndromes, 270
Anterograde amnesia, 110–111
Anti-aggressive therapy agents, 304
Anticholinergic toxidrome, 47
Antiepileptic drugs, 304
Anton's syndrome, 61, 62, 76, 77
AP1 transcription factors, 41
Apathy
 anterior cingulate syndrome, 303–304
Aphasia
 frontal dynamic aphasia, 211
 ictal aphasia syndromes, 211
Aphasias, 212–215
 Broca's aphasia, 199
 Broca's area, 198
 Broca-Wernicke brain-language model, 198
 conduction, 199–201
 global, 199, 201–203
 lesion studies in neurology, 198
 neurobiological process, 198
 perisylvian, 199
 progressive, 211
 subcortical, 208
 subtype classification, 196–199
 syndromes, 197, 198, 211–212
 TCM, 199, 203
 TCS, 199, 203–204
 testing
  bedside, 212
  metric, 212
 TMA, 199, 204
 treatment
  CIMT, 215
  device-based therapies, 214
  gestural therapy, 214–215
  intensive speech-language therapy, 213
  MIT, 213–214
  MNS in Rehabilitation, 215
  nonvascular syndromes, 213
  pharmacotherapy, 213
  Right Hemisphere Engagement, 215
  Situational Therapy, Wernicke's Aphasia, 215
  vascular aphasic syndromes, 213
 Wernicke's aphasia, 199, 200
Aphemia
 core linguistic features, 206
 elementary neurological findings, 207
 infarct locations, 207
 neuroradiology, 207
 prognosis, 207
Aphonagnosia/phonagnosia, 153
Apraxia of Speech, 206–207
Apraxia technology, 28–29
Apraxias, 139–141
 axial, 136
 bifaces, 132
 buccofacial apraxia, 136–137
 CC, 135

classification, 134
conceptual, 136
conduction (gesture imitation), 135
definition, 131
dissociation, 136
ideational, 137
ideomotor prosodic, 137
IMA, 134, 135
limb kinetic apraxia, 135
melokinetic dyspraxia, 135
ocular, 137
Oldowan techonology, 131
pantomime, 136
pathophysioly, 132–134
sympathetic, 135
tactile motor dissociation, 136
testing
 buccofacial, 140
 comprehensive, 141
 ideational, 140–141
 ideomotor limb intransitive, 139
 ideomotor limb transitive, 140
 melokinetic, 140
 short tests, 140–141
treatment/management, 141–142
verbal motor dissociation, 136
visuomotor, 136
visuomotor integration, 131, 132
Aprosodias
 bedside test, 174
 dysprosody, 173, 174
 hyperprosodia, 173
 left hemisphere, 173
 mandarin tonal language, 172
 right hemisphere, 173
Arbi and Rizzolatti model
 language evolution, 189
Arcuate fasciculus (AF) circuitry, 188
Arithmetic operations
 activation study analysis, 228
 clinical presentations, 228
 dyscalculia, acalculias, 228
 skills measured in children, 228
Arithmetics
 and language, 226–227
Art and brain
 ages, 235
 brushwork techniques, 236
 Color and happiness: Color–emotion synesthesia, 237, 239
 communication, 235
 culinary arts, 235
 figural and emotional primitives, 235
 Franco-Cantabrian cave art, 235, 236
 FTD, 237, 238
 Judith and Holofernes" painted, 236
 Lascaux painting, 235
 laws of aesthetics, 236–237
 neuroesthetics, 236, 237
 'The Art Instinct", 235
 vestigial fitness indicator, 235

visual art, 235, 237
Ascending Neurotransmitter, 265
Asomatognosia, 171
Associative agnosia, 61, 62
Astereopsis, 62–63
Astrocytes, 46
Astroglial networks, 39–41
Asymbolia, 152
Asymbolic acalculia, 230
Athymhormia spectrum disorders, 279
Attention deficit hyperactivity disorder (ADHD), 306
Auditory agnosia, 87, 90, 208
Auditory hallucination, 87, 90
Auditory paracusias, 90
Australopithecus sediba, 249
Autoimmune dementia, 286
Autoimmune disorders, 285–288
Autoimmune encephalitis, 112–113
Autoscopic hallucinations, 67
Autotopagnosia, 171
Axial apraxia, 136
Axonal disconnection, 286

B
Balint's syndrome, 26, 138, 149
    and simultanagnosia, 65, 66
Basal ganglia anatomy, 263
Basic science animal models, 302
Basilar syndrome, 76
Bear-Fedio Inventory, 88, 89
Bedside testing
    agraphias, 235
    amusia, 242
    selective anarithmetria, 230–231
    semiquantitative test battery, 212
    speech and language assessment, 212
Behavioral syndromes, 304
Benson's syndrome, 67, 74
Benton bedside memory test, 116
Benton's Battery of Arithmetic tests, 231
Biface technology, 28–29
Bipedalism, 187
Blindsight, 61, 62
    extensive occipital networks, 55
    extensive visual networks, 55
    perception of image, 55
    type 1, 55
    type 2, 55
    unconscious processing, 55
    Visual radiations, 55, 56
Bonobos chimpanzees, 188
Boring Billion, 38
Boston Diagnostic Aphasia Examination (Version 3) (BADE version 3), 212
Boston Naming Test (BNT Version 2), 212
Brain evolution, Primate, 251
Brain injury
    treatment and management, 302
Brain natriuretic peptide, 43
Brain Reorganization and Mosaic Systems, 257

Brain scan, white matter hyperintensities on, 281
Breathing control, 193–194
Broca's area, 192
Broca's dysphasia (expressive aphasia)
    associated cognitive neurological signs, 199
    associated elementary neurological signs, 199
    core linguistic features, 199
    neuroradiology and lesion location, 199
    prognosis, 199
Broca-Wernicke brain-language model, 198
Brodmann areas (BA), 188
Buccal hemineglect, 176
Buccofacial apraxia, 136–137

C
Calcium Ion Channel, 46
California verbal learning test version II (CVLT-II), 117
Call Fleming syndrome, 73
Callosal anomia, 206
Camouflage, 59
Cannabinoids, 43, 45
Capgras and Fregoli's syndromes, 69
Capgras syndrome, 176
Carbon monoxide, 76
Catecholamines, 42
Category-specific anomia, 206
CC. *See* Corpus callosal (CC)
Central (aphasic) agraphia, 233
Central alexias, 231
Central Deep alexia (paralexia), 231
Cerebellar cognitive affective syndromes, 28
Cerebellum evolution, 266
Cerebral leukodystrophies
    Parieto-Occipital Disease, 75–76
Cerebral vasculitides, 287, 288
Cerebrovascular/cognitive vascular disorders, 278–282
    neurobiology of, 280–282
Ceroid neuronal/lipoid proteinosis, 114
Channelopathies, 46
Charcot Willbrand syndrome, 69
Charles Bonnet syndrome (CBS), 70
Chloride, 46
Chloride Channels, 47
Cholinergic Toxidrome, 47
Chronic alterations in neurotransmission, 286
Chronic inflammatory disorders, 287
Claustrum, 150–151
Clinical brain sciences, 1
Clinical system approach, 107–108
Cognition disease
    dopamine, 39
Cognitive behavioral therapy (CBT), 29, 306
Cognitive missing link, 7
Cognitive neuroscience, 1, 5, 7
Cognitive rehabilitation modalities
    constraint therapy, 307
    RCT, 307
    systematic reviews, 307
Cognitive vascular disorders (CVD), 273
Cognitively lean system, 93

Complex apraxia forms, 137
Computerized Screening Tests, 271–272
Conation, 247, 268
Conceptual apraxia, 136
Conduction aphasia, 199
    and arcuate fasciculus imaging, 201, 202
    associated cognitive neurological signs, 201
    associated elementary neurological signs, 201
    core linguistic features, 200
    neuroradiology and lesion location, 201, 202
    prognosis, 201
Conduction apraxia. *See* Gesture imitation apraxia
Connectom
    diaschisis, 23, 24
    hub failure hypothesis, 23–25
    neocortical hyperscaling, 23
Constraint therapy, 307
Constraint-induced therapy for aphasia (CIMT), 215
Core frontal functions
    treatment effects, 302
Corpus callosal (CC), 135
Cortical areas, 62
    Agnosias (apperceptive and associative), 60, 61
    Anton's syndrome, 61, 62
    astereopsis, 62–63
    blindsight, 61, 62
    cortical blindness, 61, 62
    gnosanopsia and Riddoch syndrome, 62
    gnosopsia and agnosopsia, 62
    heminanopias, scotomas visual hallucinations, 60
    in associative agnosia, 61, 62
    inverse Anton's syndrome, 61, 62
    perceptual categorization deficit, 61
    scieropia, 62
    stereopsis, 62–63
Cortical blindness, 61, 62
Cortical deafness, 90
Cortical neural circuitry
    AF, 188
    Arbi and Rizzolatti model, 189
    BA, 188
    Broca's region, 188, 190
    fiber connectivity, 188
    gestural control circuits "collateralized" speech
        control area, 190, 191
    gestural protolanguage, 189
    hominid communication, 189
    human language evolution, 188, 189
    language evolution theories, 189, 190
    language gene, 189
    lexical protolanguage theory, 192
    mirror neuron system, 189–190
    musical protolanguage (prosodic protolanguage),
        191–192
    synesthetic boot strapping, 192–193
    type of communication, 189
    vocal grooming, 190–191
Cortical sensory syndromes, 148
Cortical subcortical circuits, 187

Cortico-cortical networks, 302
Cosmochemistry, 35
Cotard's syndrome, 177
Culinary arts, 235
Cultural circuits, 231–235
    agraphias (*see* Agraphias)
    alexias (*see* Alexias)
    art and brain, 235–237
    asymbolic acalculia, 230
    basic arithmetic operations, 228
    brain regions, 221, 222
    concept of subitizing, 227
    Dehaene's recycling, 221
    developmental dyscalculia, 228
    human letter box area, 221, 223
    magnetoencephalography, 221, 223
    music and brain, 237–243
    mysticism, 243–244
    neuronal alphabet and protoletters, 221–224
    neurophysiology, 224–227
    religion, 243–244
    selective anarithmetria, 230–231
    spirituality, 243–244
    transcoding/syntactic processing impairment, 229
    visual arts, 221, 224
    visual detectors, 221, 223
    working memory capacity, 221, 222
Cyclic AMP response element-binding (CREB), 41
Cytoarchitectonic, 55

D
De Clerambault's syndrome, 176
Default Mode Network (DMN), 299
Delayed palinopsia, 67
Delusional Misidentification Syndromes, 90
Dentate gyrus (DG), 102
Depression, 306
Descent of the larynx, 193, 194
DG. *See* Dentate gyrus (DG)
DGKH (Diacyl Glycerol Kinase Isoenzyme), 303
DHA. *See* Docosahexaenoic acid (DHA)
Diaschisis syndromes, 23, 29
Diffuse Lewy body dementia (DLDB), 75
Diffusible gases, 40
Direct Current Transcranial Stimulation (d-TCS), 214
DISC1 (Disrupted in Schizophrenia 1), 303
Disconnection anomia, 206
Disinhibition, 261, 262, 268, 272, 274, 275, 278,
        282–284, 304
Disinhibitory Control Syndromes
    Alien Hand and Anarchic Hand Syndrome, 270
    Anterior or Frontal Variant Alien Hand Syndromes,
        270
    Corpus Callosal Type, 270
    Ictal Alien Hand, 270
    Posterior Parietal Type, 270
Dissociation apraxia, 136
Divided attention, 269

DLPFC Syndrome, 303
Docosahexanoic acid (DHA), 17–19
Dopamine (DA), 44, 258
   cognition disease, 39
   in humans disease, 39
   neuropsychiatric disease, 39
Dopaminergic therapy, 303
Dorsal stream syndromes, 65, 66
   Akinetopsias and Zeitraffer phenomenon, 67
   autoscopic hallucinations, 67
   delayed palinopsia, 67
   entomopias, 67
   illusory visual spread, 67
   Palinopsia, 67
   Perky effect, 68
   polyopias, 67
   Simultanagnosia
      Balint's syndrome, 65
      Dorsal and ventral, 65, 66
   visual alloesthesia, 67
Dorsolateral prefrontal cortex (DLPFC), 39, 254
DSM IV, 305
DSM IV/V, 305
DSM-IV R, 305
DSM-V criteria, 305
Dysarthria, 210
   cerebellar abnormalities, 210
   dysphonia, 210
   Echolalia, 211
   Foreign Accent Syndrome, 211
   Frontal Dynamic Aphasia, 211
   Ictal Aphasia Syndromes, 211
   Palilalia, 211
   progressive Aphasias, 211
   stuttering, 211
   subcortical/basal ganglia, 210
Dyscalculia, 228
   developmental, 228
   treatment, 231
Dysexecutive, 261, 271, 273, 282–284, 287
Dyslexia, 233
Dysphasia
   dysarthria, 210
   dysphonia, 210
   mutism, 210
Dysphonia, 210
Dysprosody, 173, 174
Dystypia, 235

E
Earth's history, 35
East African Rift Valley, 15, 16
Echographia, 175–176
Eclampsia, 74
EI. *See* Emotional intelligence (EI)
Eidetic imagery, 68
Ekbom Syndrome, 176
Electrochemical signaling process, 40

Emotion disorders
   and Linking vision, 91, 92
   brain region, 91, 92
   clinical, 95
   cortical and subcortical regions, 91
   evolutionary insights, 91–95
   frontopolar cortex subregion activation patterns, 93
   IEED, 95, 181
   metacognitive control system, 93
   meta-cognitive multi-agent cognitive control system, 93, 94
   meta-cognitive system, 94
   occipitotemporal pathway, 91, 93, 178
   process, 177, 178
Emotional intelligence (EI), 95, 179, 180
Entomopias, 67
Epigenetics, 40–41
Episodic memory (EM)
   basolateral limbic circuit (amygdaloid), 108, 109
   hippocampus anatomy, 99, 100
   neuro-anatomical components, 100, 108
   neurobiological components, 111
   papez circuit (hippocampal), 108
   pathological process, 111
   testing, 116
Errorless naming therapy (ENT), 215
Eukaryotic cells, 38
Evolutionary psychiatry, 1
Executive Interview bedside test (EXIT), 271
Exner's area, 233, 234
Experimental psychology, 1, 4
Expressive aphasia, 199
External speech (exophasia), 194

F
FDB. *See* Field-dependent behavior (FDB)
Feline and primate visual cortex, 53
Field-dependent behavior (FDB), 26
FNS. *See* Frontal network syndromes (FNS)
Focused attention, 269
Foreign Accent Syndrome, 211
FPC. *See* Function of the frontopolar (FPC)
Franco-Cantabrian cave art, 235, 236
Fregoli's syndrome, 176
Frontal alexia, 232
Frontal Assessment Battery (FAB), 271
Frontal cortical hubs, 270
Frontal dynamic aphasia
   Luria, 211
Frontal lobe episodic memory, 111
Frontal lobe evolution, 13, 14
Frontal lobes, 259
   and alcohol, 283, 284
Frontal network syndromes (FNS), 247, 267, 269, 274–276, 282, 285
   common cause, 279
   common clinical disorders, 276
Frontal subcortical circuit involvement infections, 287

Frontal Subcortical Networks (FSC), 262–264
Frontal syndromes, 247
Frontal system syndromes, 248–255, 273, 278
    evolutionary aspects and clinical syndromes, 247–257
        brain volume, 248–249
        frontal lobe size, 249
        prefrontal cortex, 249–255
    investigations, 278
        clinical cognitive/behavioral, 278
        laboratory, 278
        neuroimaging, 278
        treatment, 278
Frontoparietal Working Memory (WM) Group, 261–262
Frontopolar, 249, 250, 253–255
Frontopolar cortex, 254
Frontopontocerebellar circuit, 265
Frontotemporal disorders (FTD) phenotypes, 277
Frontotemporal lobe degeneration, 274
Frontotemporal lobe dementia (FTLD), 4, 6, 237, 238, 271, 305
Function of the frontopolar (FPC), 14
Functional MRI (f-MRI), 297

G
G proteins
    activate/inactivate certain proteins, 40
    classification, 40
    neurological conditions, 41
    neurological, cognitive and neuropsychiatric disorders, 40–41
GABA, 43–44
Gait apraxia, 137–138
Gait ignition apraxia (GIA), 137, 138
Gases (gasotransmitters), 42
Gasotransmitters
    agmatine, 45–46
    cannabinoids, 45
    H₂S, 45
    NO, 45
Gastrins, 42
Gelada baboon, 188
Geochemistry, 35
Gerstmann's syndrome, 209, 210
Gerstmann–Sträussler–Scheinker syndrome (GSS), 287
Geschwind–Gastaut syndrome (GGS), 26, 27, 85, 87, 88
Gestural protolanguage, 189
Gestural therapy, 214–215
Gesture imitation apraxia, 135
GGS. See Geschwind–Gastaut syndrome (GGS)
GIA. See Gait ignition apraxia (GIA)
Glial cells, 46
Global aphasia
    associated cognitive neurological signs, 202
    associated elementary neurological signs, 202
    core linguistic features, 202
    neuroradiology, 202
    prognosis, 202–203
Global cooling consequent, 253

Glutamate, 43
Gnosanopsia
    and Riddoch syndrome, 62
    and agnosopsia, 62
Gondwana, 38
Gourmand syndrome, 175
Graded Difficulty Arithmetic Test (GDA), 231
Grammaticalization, 192
Granular cell layer (IV), 252
Graphomania, 175–176
Green Earth
    time period, 38
Grouping, 58–59
Gyral white matter, 83

H
Hadean period, 36, 37
Hadean-Archean Earth, 36
Hazen's theory, 36
HCN channel, 302
Hemialexia, 232
Heminanopias, scotomas visual hallucinations, 60
Herpes simplex encephalitis, 112–113
Heschl's gyrus, 84, 90, 207
Higher cortical functions (HCF), 257
Histamine, 42, 45
Histone acetyltranferase (HAT) inhibitor, 41
Hodological effects, 23
Homo ergaster, 189
Homo heidelbergensis, 13, 16
Hopkins verbal learning test, 117
5-HT2 receptors, 304
Human evolution, 16, 18, 20, 29, 30
Humans disease
    dopamine, 39
Hydrogen Sulfide (H₂S), 45
Hyperactivity/neurotransmitter deficiency syndromes
    anticholinergic toxidrome, 47
    cholinergic deficiency syndromes, 48
    cholinergic toxidrome, 47
    dopamine deficiency syndromes, 47
    malignant hyperthermia, 47
    neuroleptic malignant syndrome, 47
    norepinephrine deficiency syndromes, 48
    serotonin deficiency syndromes (depression), 47
    serotonin syndrome/toxidrome, 47
Hyperempathy, 175
Hypergraphia, 175–176, 235
Hyperlexia, 232–233
Hyperprosodia, 173
Hyperthymesia, 124, 125

I
ICA. See Internal carotid artery (ICA)
Ictal Alien Hand, 270
Ictal Aphasia Syndromes, 211
Ideational apraxia, 137

Ideograms, 224
Ideomotor apraxia (IMA), 134–135
Ideomotor prosodic apraxia, 137
IEED. *See* Involuntary emotional expression disorder (IEED)
Inferior parietal lobe function test, 154
Initiation, 269
Inner speech (endophasia), 194
Insulins, 42
Intensive speech-language therapy, 213
Intercortical connectivity, 30
Internal carotid artery (ICA), 121
Interpersonal psychotherapy (ITP), 306
Intracortical connectivity, 30
Intraparietal sulcus (IPS), 145, 224
Intraparietal sulcus areas (ventral posterior component), 188
Intrinsic Connectivity Networks (ICN), 299
Intrinsic State Connectivity Networks, 299, 301
Inverse Anton's syndrome, 61, 62
Involuntary emotional expression disorder (IEED), 90, 95, 181, 307
Ion channels
    calcium, 46
    cardiac conducting tissues, 46
    channelopathies, 46
    chloride, 46, 47
    heart beat, 46
    ligand-gated, 46
    nerve impulses, 46
    potassium, 46, 47
    sodium, 46
    voltage-gated, 46
IPS. *See* Intraparietal sulcus (IPS)
Isolation of the speech area, 204

J
Johns Hopkins Acalculia Battery, 231
Judith and Holofernes" painted, 236
Junque leukoaraiosis grading, 284

K
Kakopsia, kalopsia, 69
Kluver–Bucy syndrome (KBS), 69, 85, 87

L
Language, 187–193
    activities, 195
    breathing control, 193–194
    concept of lemma, 195
    definitions of language components, 195
    descent of the larynx, 193, 194
    development, cortical neural circuitry (*see* Cortical neural circuitry)
    dominance and left handedness, 195–196
    DWI, 195, 197

external speech (exophasia), 194
functional imaging of language networks, 195, 196
and human speech
    archeological fossil, 187
    bipedalism, 187
    comparative primate studies, 187
    cortical subcortical circuits, 187
    gelada baboon, 188
    Hmmmm theory, 188
    music therapy, 187
    origins, 188
    protophone, 188
    rhythm, 187
    sensorimotor control, 187
ideas, 187
inner speech (endophasia), 194
neurobiology, 195
perfusion imaging mismatch, 195, 197
pharynx development, 193
shorter trachea, 193
vs. speech, 194–195
syndromes, 205, 208–209
Language gene, 189
Larynx
    descent, 193, 194
Lascaux painting, 235
Lateral geniculate nucleus (LGN), 52
Lateral Orbitofrontal Syndrome, 304
Left handedness
    and language dominance, 195–196
Left hemisphere, language, 209, 210
    -related syndromes
        Angular Gyrus Syndrome, 209
        Gerstmann's syndrome, 209, 210
Leukoaraiosis, 138
Lexical protolanguage theory, 192
Limb kinetic apraxia, 135
Limbic Encephalitis, 285–287
Linguistics, 17
Lipid neurotransmitters, 40
Logopenic progressive aphasia (LPA), 275
Long-term potentiation (LTP), 109
Luria
    Frontal Dynamic Aphasia, 211
Lycanthropy, 177

M
Macrosaccades, 54
Macroscale level, 35
Magnetoencephalography (MEG), 299–300
Magnetoencephalography study, 221
Malignant hyperthermia, 47
Mantle transition zone, 36
Marchiafava–Bignami syndrome, 285
MEA. *See* Medial entorhinal areas (MEA)
Medial entorhinal areas (MEA), 109
Medial orbitofrontal syndrome, 304
Medial prefrontal cortex (MPFC), 13, 253

Meissner's corpuscles, 250
Melatonin, 42
Melodic intonation therapy (MIT), 29, 213–214
Memory syndromes
    episodic memory rehabilitation, 125
    neurochemistry, 109–110
    neurological and psychiatric conditions, 114–115
    neurophysiology, 109–110
    physical exercise, 126
    prosopagnosia, 90
    sleep disruption, 126
    social impairment, 90
    TBI, 91
    testing, 115–116
    Williams Syndrome, 91
Mercury poisoning, 76
Meridional overturning circulation (AMOC), 12
Mesoscale level, 35
Metacognitive test, 271
Metamemory, 111
Metamorphopsias or dysmetropsia, 65
Methylphenidate, 303–306
Metric testing, 71–72, 272–273
    ADHD, 272
    autism, 272
    BADE version 3, 212
    Behavioral Neurological Tests, 272
    BNT Version 2, 212
    clinical syndrome orientated, questionnaire based,
        272
    creativity tests, 272
    emotional intelligence tests, 272
    global tests, 272
    WAB-Revised, 212
    working memory/executive function tests, 272
Micro RNAs, 41
Microelectrode recordings, 226
Micrographia, 235
Microsaccades, 54
Microscopic level, 35
Mirror neuron circuitry, 84
Mirror neuron system (MNS), 7, 26
Mirror Neuron Therapy and Rehabilitation, 215, 308
Mirror Visual Feedback Therapy (MVF), 307–308
Misoplegia, 170
MIT. See Melodic intonation therapy (MIT)
Mithen's hmmmmm model, 191
Mitochondrial cytopathies, 11
Mitochondrial DNA (mtDNA), 17, 18
MNS. See Mirror neuron system (MNS)
Modality-specific anomia, 206
Modulatory ascending systems, 302
Monoaminergic function, 302
Monoamines, 42, 302
Montreal Cognitive Assessment (MOCA), 271
Mosaic Cognitive Evolution, 259
Movement disorder associated dysgraphia, 234
MPFC. See Medial prefrontal cortex (MPFC)
Multiple sclerosis (MS), 283

Music therapy, 187, 243
Musical protolanguage (prosodic protolanguage), 191–192
Musicality and brain, 239–243
    ACC, 237
    emotions and Plutchik, 238
    human auditory processing, 237, 240
    and language, 237, 239
    musicolanguage, 237
    neural circuitry, 237
    neurobiology, 237
    neurophysiology, 237
    pathology
        Amusia Classification, 240–243
        neuropathological processes, 239, 242
        right hemisphere traumatic brain injury, 239, 243
    pedalism and rhythm, 237
    Plutchik's cone model of emotion, 238, 241
    principal circuits, 237, 240
    universal sense, 237
Musicolanguage, 237
Mutism, 210
Mycoplasma pneumonia encephalitis, 113
Mysticism, 243–244
Mythograms, 224

N
Neglect syndromes
    allesthesia, 171
    allochiria, 171
    allokinesia, 171
    anosodiaphoria, 170
    anosognosia, 169–170
    asomatognosia, 171
    attentional arousal hypothesis, 168
    autotopagnosia, 171
    aynchiria, 171
    clinical cases, 167
    definition, 163
    hemispatial, 165–166
    management, 169
    misoplegia, 170
    motor, 164
    pathobiology, 167
    rehabilitation, 170
    representational, 166–167
    rubber hand illusion, 171–172
    sensory, 164
    somatoparaphrenia, 170
    spatial, 164–165
    supernumerary phantom limb, 171
    treatment, 169
    Troxler fading effect, 168
    xenomelia, 171
Neuroanatomy and Neurophysiological Functional
        Systems, 259–260
Neurobiological approach, 107–108
Neurochemical tracts, 259
Neurochemistry, 39

brain functions, 35
and classification system, 35
Earth evolution, 36, 37
elements, 35, 36
eukaryotic cells, 38
evolutionary origin, 35
formation of RNA strands, 37
global climate cycle, 38
Hadean period, 36, 37
Hazen's theory, 36
intracellular messenger cascades, 39, 40
macroscale level, 35
mantle transition zone, 36
mesoscale level, 35
microscopic level, 35
mineral-rich fluids, 38
minerals, 35, 36
neurochemicals, 35, 36
neurotransmitters (*see* Neurotransmitters)
organic molecules, 35, 36
primordial soup theory, 38
prokaryotes, 38
time period of Green Earth, 38
vertical tectonics, 36
white Earth/snowball Earth, 38
zircon crystals, 36
Neuroesthetics, 236, 237
Neurohypophyseal hormones, 43
Neuroimaging, 297–299
anatomical and functional
ACC, 297
categories and examples, 297, 298
cellular and molecular responses to brain injury, 297, 299
complex networks, 297
detect pathology, 297
DTI, 297
emergency neurological conditions, 297
[18]FDG-PET brain, 297
f-MRI, 297
FNS syndrome, 297
Intrinsic connectivity network patterns in dementias, 297, 298
MR perfusion scanning, 298
overlap syndromes, 297
PET brain, 297
PET brain patterns in dementias, 297, 298
PET receptor (Dopa) imaging, 297
PIB, 297
posterior cingulate cortex (PCC), 297
SPECT, 297
animal models, 302
cognitive and behavioral disorders, 300–302
ICN, 299
Intrinsic State Connectivity Networks, 299, 301
MEG, 299–300
neurotransmitter and neurotransmitter receptor PET, 298
Q-EEG, 299–300

RSN, 299, 300
Neuroleptic Malignant Syndrome, 47
Neurological/neuropsychiatric disorders, 305, 306
Neurological/psychiatric conditions
emotional component treatment, 307
Neurology
behavioral, 25
cognitive, 25
Neuromodulation, 302
Neuromodulators (NM), 257
Neuronal alphabet and protoletters
NRH, 222
numerosity and calculation, 224, 226
reading annexes, 222
shapes, 221, 222, 224
writing evolution, 224, 225
Neuronal recycling hypothesis (NRH), 222
Neuropathologic subtypes, 281
Neuropeptides, 42
Neuropharmacological agents, 268
Neurophysiological Frontal Functional Systems, 260–261
Neurophysiology, cultural circuits
Amazonian group *vs.* French control group, 227, 229
arithmetic and language, 226–227
microelectrode recordings, 226
numerosity in brain, 226, 227
Neurophysiology, visual processing
areas V1–V8, 55, 56
basic, intermediate and higher, 55–57
brain activation, 57, 58
change blindness, 57–58
cytoarchitectonic, 55
grouping, 58–59
higher level processing, 57
illusions, 56, 57
inattention blindness, 57–58
peak shift effect, 59
Rubin Vase and Necker Cube, 57, 58
supernormal stimulus, 59
V1 (BA 17), 55
ventral and dorsal streams, 55
Neuropil, architectural changes, 255–257
Amgydala, 257
frontal lobes, 255–257
temporal lobe, 257
Neuropsychiatric Component Treatment, 305–306
Neuropsychiatric disease
dopamine, 39
Neuropsychiatric disorders, 305
Neuropsychiatric states
DGKH, 303
DISC1, 303
RGS4, 303
Neuropsychological testing
Agraphias, 235
selective anarithmetia, 231
Neurotoxicological Syndromes, 76
Neurotransmitter and neurotransmitter receptor PET, 298

Neurotransmitter systems, 257–259
  acetylcholine, 258–259
  dopamine, 258
  mosaic cognitive evolution, 259
  serotonin, 258
Neurotransmitters, 46
  Ach, 44
  adrenaline, 44
  Agmatine, 42
  amino acids, 42
  AP1 transcription factors, 41
  brain natriuretic peptide, 43
  cannabinoids (lipids), 43
  Catecholamines, 42
  CREB, 41
  DA, 44
  deficiency syndromes/hyperactivity, 47–48
  dopamine, 39
  evolution, 39
  G proteins, 40–41
  GABA, 43–44
  gases (gasotransmitters), 42
  gasotransmitters, 45–46
  gastrins, 42
  gene expression, 41
  glutamate, 43
  heart, 43
  histamine, 42, 45
  hypothalamic releasing hormones, 43
  insulins, 42
  intercellular messaging, 40
  intracellular amplification, 39
  ion channel (see Ion channels)
  living organisms, 39
  melatonin, 42
  micro RNAs, 41
  monoamines, 42
  neurohypophyseal hormones, 43
  neuropeptides, 42
  noradrenaline, 44
  octopamine, 42
  orexins, 45
  peptides, 42
  pituitary peptides, 43
  principal cellular signal transduction, 40, 41
  purine and pyrimidine, 42
  rapid effects and slower effects, 39
  second messenger, 39, 40
  secretins, 42
  serotonin, 44
  signal transduction, 39
  slower transmission, 40
  somatostatins, 42
  tachykinins, 42
  tryptamine, 42
  tyramine, 42
  ultrafast transmission, 40
Nicotinic and muscarinic receptors, 44
Nitric Oxide (NO), 45

NMDA receptors, 43
Noradrenaline, 44
Noradrenergic Therapy, 303
Norepinephrine (NE), 302
Nuedexta, 307
Number Processing and Calculation (NPC) Battery, 231
Numerosity
  and calculation, 224, 226

O
Obsessive Compulsive Disorder (OCD), 304
Occipital seizures, 76
Occipito-frontal fasciculus
  anoneirognosis, 70
  bedside, 71
  Capgras and Fregoli's syndromes, 69
  CBS, 70
  Metric Tests, 71–72
  peduncular hallucinations, 71
  topographical or hodological pathophysiological
    features, 69
Occipito-parietal (When) pathway, 68–69
Occipito-parietal pathway, 68
Occipito-Temporal Cortex
  amygdala, 69
  Kakopsia, kalopsia, 69
  Kluver-Bucy syndrome, 69
  PTSD, 69
Occipito-temporal fasciculus, 91
Occipito-temporal pathway, 91, 93
Octopamine, 42
Ocular apraxia, 137, 138
Oldowan techonology, 131
Orbitofrontal, 249, 253, 257, 269, 271, 274, 283
Orexins, 45
Othello syndrome, 176

P
Paleoneurological, 1
Palinopsia, 67
Palmitoylethanolamide (PEA), 45
Pantomime agnosia, 136
Parahippocampal (PHC), 109
Paralexia, 231
Parasitosis. See Ekbom Syndrome
Paretic agraphia, 234
Parietal expansion, 20, 22
Parietal lobe syndromes, 146–150
  Alzheimer's disease, 146
  claustrum, 151
  episodic pain, 150
  fMRI and PET brain scanning, 153
  IPS, 145, 146
  knapping techniques, 145
  neuro-anatomical and neurophysiological aspects
    association somatosensory cortex, 146
    frontoparietal networks, 146, 147

IPS, 146
multimodal association area, 146
primary somatosensory cortex, 146
precuneus, 151
right/left parietal lesion
attention and consciousness hubs, 149
cortical sensory function and deficits, 150
cortical sensory impairment, 148
TOP junction functions, 149
visuospatial dysfunctions, 149
visuospatial function, 150
Parieto-occipital disease
cerebral leukodystrophies, 75–76
Parinaud's syndrome, 113
Parkinson's disease (PD), 47, 75
Peak shift effect, 59
Peduncular hallucinations, 71
Peptides, 42
Perceptual categorization deficit, 61
Perisylvian aphasias
Broca's dysphasia (expressive aphasia), 199
Perky effect, 68
PET brain, 297–299, 304, 306
Petrochemistry, 35
PFC, 52
Phantom boarder, 177
Pharmacotherapy, 213, 302
Pharynx development, 193
PHC. See Parahippocampal (PHC)
Pick's disease, 281
Pictograms, 224, 225
Pitres law, 195
Pituitary peptides, 43
Planotopokinesia, 175
Planum polare, 84
Plate tectonics, 11, 13
Plutchik's cone model of emotion, 238, 241
Polyopias, 67
Polyunsaturated fatty acids (PUFAs), 19
Posterior cingulate cortex (PCC), 297
Posterior cortical atrophy, 230
Posterior Cortical Atrophy Syndrome (PCAS), 74
Posterior reversible encephalopathy syndrome (PRES), 66, 73–74
Post-traumatic stress disorder (PTSD), 69, 74–75
Potassium, 46
Potassium channels, 47
Precuneus
and claustrum, 150–151
posterior parietal lobule, 146
Prefrontal cortex (PFC), 249, 250, 252–254, 262, 263, 265–267, 269, 271, 302
and White Matter, 255
Prefrontal function
current theories of, 267–270
Prefrontal lobe
size, 250
Presynaptic autoreceptors, 44
Presynaptic dopamine levels, 302

Primordial soup theory, 38
Procedural memory
basal ganglia, 123
cerebellum, 123
functional neuroanatomy, 123
Homo erectus, 106
metric tests, 124
neurophysiology, 123
pathophysiology, 123–124
tests, 124
Progressive Aphasias, 211
Prokaryotes, 38
Prosodic protolanguage, 191–192
Prosody. See Aprosodias
Prosopagnosia, 63, 64, 69, 71, 74, 76
Protophone, 188
PUFA. See Polyunsaturated fatty acids (PUFA)
Pulvinar, 52
Pure word deafness (PWD)
associated cognitive neurological signs, 207
associated elementary neurological signs, 207
and auditory agnosia, 208
core linguistic components, 207
lesion location, 207, 208
neuroradiology and lesion location, 207
prognosis, 207
Pure Word Mutism, 206–207
Pure/isolated agraphia, 233, 234
Purine and Pyrimidine, 42

Q
Quantitative Electroencephalography (Q-EEG), 299–300

R
Randomized controlled trials (RCT)
cognitive rehabilitation modalities, 307
FTLD, 305
Neuropsychiatric Component Treatment, 305–306
Serotonergic Therapy and Stroke (Motor Deficit), 305
TBI, 304, 305
Trazodone, 305
RBANS. See Repeatable battery for the assessment of neuropsychological status (RBANS)
Red Earth, 38
Religion, 243–244
Repeatable battery for the assessment of neuropsychological status (RBANS), 117
Representational neglect, 165, 166
Resting state network (RSN), 299, 300
Retrograde amnesia, 110
Rey auditory verbal learning test, 117
RGS4 (Regulator of G-Protein Signaling 4), 303
Rhythmic auditory stimulation therapy, 237
Ribot's law, 195
Riddoch syndrome, 73
and gnosanopsia, 62
Right hemisphere engagement, 215

Right hemisphere syndromes
   ACC, 159, 160
   acetylcholine, 163
   acute confusional states, 163
   altered mental status, 163
   apraxia, 174–175
   attentional systems, 157, 161
   buccal hemineglect, 176
   dominant *vs.* non-dominant hemisphere, 157
   emotion, 161
   Gourmand syndrome, 175
   hyperempathy, 175
   MPFC, 159
   neurological and neuropsychiatric conditions, 161–163
   nosagnosia Overestimation, 176
   planotopokinesia, 175
Rubber hand illusion, 171–172

S
Sahelanthropus tchadensis, 15
Sapient paradox
   archeological evidence, 16–17
   cerebellum enlargement, 19
   DHA, 17, 19
   genetic evidence, 17
   linguistics, 17
   parietal lobe expansion, 20
   SBH, 20
   synaptic bandwidth, 17, 19
Sauropsids, 51
Savant syndromes, 124, 125
SBH. *See* Social brain hypothesis (SBH)
Scieropia, 62
Second messenger systems, 39, 40
Secretins, 42
Selective anarithmetria
   acalculia and dyscalculia treatments, 231
   bedside testing, 230–231
   bilateral IPS representation of numerosity, 230
   left hemisphere lesions, 230
   neuropsychological batteries, 230
   neuropsychological testing, 231
   parietal damage, 230
   pathophysiology, 230
   posterior cortical atrophy, 230
   rTMS, 230
   selective calculation impairment/dissociation of
      impairments, 230
   testing calculation at bedside, 230
Selective attention, 269
Selective serotonin reuptake inhibitor (SSRI), 305
Semantic anomia, 205–206
Semantic memory
   Alzheimer's pathology, 122
   aphasic anomia, 122
   DLPFC, 121
   testing, 122–123
Sensorimotor control, 187
Serenics (Anti-aggressive Therapy Agents), 304

Serotonergic agents, 304
Serotonergic therapy, 305
Serotonin, 44, 258
Serotonin Syndrome/Toxidrome, 47
Shorter trachea, 193
Short-term working memory capacity (STWMC), 103
Simultanagnosia, 26, 149
   Balint's syndrome, 65
   CTPT, 74
   dorsal and ventral, 65, 66
   posterior watershed lesions, 77
   and prosopagnosia, 71
Situational therapy, Wernicke's Aphasia, 215
SMG. *See* Supramarginal gyrus (SMG)
Social brain hypothesis (SBH), 20
Sodium Channels, 46
Somatoparaphrenia, 170
Somatostatins, 42
Spatial (Neglect) alexia, 232
Spirituality, 221, 235, 243–244
Stereopsis, 62–63
STIMA test, 140
Stroke (Motor Deficit), 305
Stroke model, 305, 306
STWMC. *See* Short- term working memory capacity
      (STWMC)
Subcortical aphasia
   core linguistic features, 208
   neuroradiology and lesion locations, 208, 209
   prognosis, 208
Subitizing, 227
Superior colliculus, 52–53, 66
Superior parietal lobe function test, 154
Superior parietal lobule, 146
Supernormal stimulus, 59
Supernumerary phantom limb, 171
Supramarginal gyrus (SMG), 146, 152
Supramarginal gyrus component, 188
Sustained attention, 269
Sympathetic apraxia, 135
Synapsida, 51
Synapsids, 51
Synaptopathies, 285–287
Synchiria, 171
Syndromes
   brain and cognitive, 3
Synesthesia, 151–152
Synkinesis, 193

T
Tachykinins, 42
Tactile motor dissociation, 136
Task- apecific apraxia
   apraxic, 139
   constructional, 138–139
   dressing, 138
   eyelid opening, 138
   gait, 137–138
   ocular, 138

Temporal lobe, 257
Temporal lobe syndromes
    amusia, 90
    amygdaloid complex, 83
    anterior, 84
    Disorders of Time Perception, 90
    emotion disorders, 91–95
    Frontal Network Syndromes, 90
    GGS, 88
    gyral white matter, 83
    infero-occpitotemporal activation, 84, 86
    intracerebral hemorrhage, 87
    lateral and medial, 85
    Memory Syndromes, 90–91
    metaphysical preoccupation, 88
    mirror neuron circuitry, 84
    neuroanatomy, 84–85
    neuropathological process, 87–88
    neurophysiology, 84–85
    posterior, 84
    right/left and bilateral, 85–87
    sensory visual and auditory, 84
    social brain circuitry hubs, 84, 86
    social brain hypothesis, 83–84
    Social Circuitry, 83–84
    UF, 88–90
Temporo-occipito-parietal (TOP), 149, 151
Tetrahydrocannabinol (THC), 45
TGA. See Transient global amnesia (TGA)
Thalamic dazzle syndrome, 64
Thrombolytic therapy, 5
TOP. See Temporo-occipito-parietal (TOP)
Tourette's syndrome, 303
Transcoding/syntactic processing impairment, 229
Transcortical Aphasic Syndromes, 203
Transcortical mixed aphasia (TMA), 199
    approximate lesion sites, 204
    associated cognitive signs, 204
    associated elementary neurological signs, 204
    Broca's, conduction and Wernicke aphasia, 204
    core linguistic features, 204
    neuroradiology and lesion locations, 204
    Perisylvian areas, 204
    prognosis, 204
Transcortical motor (TCM) aphasia
    associated cognitive neurological signs, 203
    associated elementary neurological findings, 203
    core linguistic features, 203
    neuroradiology and lesion location, 203
    prognosis, 203
    suprasylvian aphasia centers, 203
Transcortical sensory (TCS) aphasia
    associated cognitive neurological signs, 203
    associated elementary neurological signs, 204
    core linguistic features, 203
    neuroradiology and typical lesion location, 204
    prognosis, 204
    WAB, 203
Transcranial Magnetic Stimulation (TMS), 214, 307
Transient global amnesia (TGA), 113, 114

Traumatic brain injury (TBI), 91, 283, 284, 286
    amantadine and severe, 304
    methylphenidate and moderate, 305
Trazodone, 305
Troxler fading effect, 168
Tryptamine, 42
Tyramine, 42

U
Ultrafast transmission, 40
Uncinate fasciculus (UF)
    anterior temporal lobes, 88
    auditory agnosia, 90
    auditory paracusias, 90
    cortical deafness, 90
    Delusional Misidentification Syndromes, 90
    lesions, 27–28
    neurophysiological functions, 90
    OFC, 88
    Social-Emotional Processing Impairment, 90
    Uncinate Fits, 90
    White matter tract expansion, 89
Uncinate Fasciculus (UF) Group, 264–265
Unconscious brain processing, vision
    blindness, 54
    blindsight, 54–55
    central function, 54
    change blindness, 54
    inattention blindness, 54
    macrosaccades, 54
    microsaccades, 54
Universal sense, 237
Urbach–Wiethe disease, 114

V
Van Heugten Test, 141
Vascular aphasic syndromes, 197, 213
Vasculitides, 287–288
ventral premotor cortex (VPMC), 250
Ventral radiations, 26
Ventral stream disorders
    Color, 63–64
    environmental agnosia, 64
    hallucinations/illusions, 63
    hypofunction, 63
    metamorphopsias or dysmetropsia, 65
    prosopagnosia, 64
    synesthesia, 65
    thalamic dazzle syndrome, 64
    visual text hallucinations, 64
    visuospatial agnosia, 65
Ventrolateral prefrontal cortex (VLPFC), 14, 254
Verbal motor dissociation, 136
Vertical tectonics, 36
Vertiginous syndromes, 85
Visual agnosia, 60, 63, 65, 66, 69, 71, 73, 74, 76
Visual alloesthesia, 67
Visual and verbal memory, 116

Visual arts, 221, 224, 235–237, 240
Visual cortex
    and contiguous optic radiations, 61
    feline and primate, 53
    and memory, 68
    and radiations, 72–78
    and retina, 60
Visual disorders
    classification, 60
    cortical areas, 60–63
    symptoms and syndrome, 59–60
Visual hallucinations, 60, 64, 68–71, 73, 75
Visual processing, 52–59, 69–72
    Amygdala, Occipito-Temporal Cortex, 69
    and prefrontal cortex evolution, 52
    Basilar syndrome, 76
    Benson's Syndrome, 74
    brain evolution, 52
    Cerebral Leukodystrophies with Parieto-Occipital
        Disease, 75–76
    Cerebrovascular Watershed Lesions, 76–78
    conceptual sensory, 52
    cortex, 51
    DLDB, 75
    dorsal stream syndromes, 65–68
    clampsia, 74
    encephalization, 51, 52
    human sensory system, 51
    neocortex, 51
    neurophysiology (see Neurophysiology, visual
        processing)
    Neurotoxicological Syndromes, 76
    occipital seizures, 76
    occipito-frontal fasciculus (see Occipito-Frontal
        Fasciculus)
    occipito-parietal (When) pathway, 68–69
    occipito-parietal pathway, 68
    PCAS, 74
    PRES, 73–74
    PTSD, 74–75
    Riddoch syndrome, 73
    Sauropsids, 51
    sensory faculties, 51
    symptoms and syndrome, 59–60
    Synapsida, 51
    Synapsids, 51
    treatment, 72
    two-tier system, brain
        association (extrastriate) cortex, V2 (BA 18), 53
        colliculus, 52
        cortical and subcortical pathways, 52, 53
        Feline and primate visual cortex, 53
        freezing, 53
        LGN, 52
        MT/V5, 53
        primary (striate) visual cortex, V1 (BA 17), 53
        prosimians, 54
        pulvinar, 52
        retinal cones function, 53

        sensory–environment interface, 52
        superior colliculus, 53
        unconscious system, 52
    unconscious brain processing, 54–55
    ventral stream disorders, 63–65
Visual text hallucinations, 64
Visual-auditory mapping, 193
Visuomotor apraxia, 136
Visuospatial agnosia, 65
Visuospatial dysfunction, 158, 167, 172
Visuospatial function, 172
VLPFC. See Ventrolateral prefrontal cortex (VLPFC)
Vocal grooming, 190–191
VOSP test, 71

W
Wallenberg's syndrome, 2
Watershed lesions, 76–78
Watershed-type lesions, 203
Wechsler memory scale IV, 117
Wernicke's aphasia
    associated cognitive neurological signs, 200
    associated elementary neurological deficits, 200
    core linguistic features, 200
    neuroradiology and lesion location, 200
    prognosis, 200
    situational therapy, 215
Western Aphasia Battery (WAB), 203
Western Aphasia Battery (WAB-Revised), 212
White Earth/snowball Earth, 38
Williams Syndrome, 91
Wisconsin Card Sorting Test, 274
Word production anomia, 205, 206
Word selection anomia, 205, 206
Working memory (WM)
    bedside testing, 121
    capacity, 221
    cardinal function, 103
    computerized testing, 121
    definition, 102
    foraging systems, 104
    frontoparietal subcortical circuitry, 103, 118
    fronto-parieto-occipitotemporal circuitry, 118
    ICA, 121
    leukoaraiosis, 120
    neuropsychological tests, 121
    petrides, 103
    STWMC, 103
    traps and snares, 104

X
Xenomelia, 171

Z
Zeitraffer phenomenon, 67
Zircon crystals, 36

Printed in the United States
By Bookmasters